Stetson University
3 4369 00512408 5

W9-BVP-253

Dedicated to my daughter Nadya

and to my friends

Yuri and Liza, Moshe and Inna,
Mark and Lena, Alex and Genya

Vladimir Rovenski

Geometry of Curves and Surfaces with MAPLE

Birkhäuser
Boston • Basel • Berlin

Vladimir Rovenski
Department of Mathematics
University of Haifa
and Technion
Haifa, Israel

Library of Congress Cataloging-in-Publication Data

Rovenskii, Vladimir Y., 1953-
Geometry of curves and surfaces with MAPLE / Rovenski, V. Yu.
p. cm.
Includes bibliographical references and index.
ISBN 0-8176-4074-6 (acid-free paper) – ISBN 3-7643-4074-6 (alk. paper)
1. Geometry–Data processing. 2. Maple (computer file) I. Title.

QA448.D38.R68 1999
516.3'52'02855369–dc21 99-012491
CIP

AMS Subject Classifications: 51-01, 51M15, 52B, 53

Printed on acid-free paper.

Birkhäuser 

ISBN 0-8176-4074-6 SPIN 10746771
ISBN 3-7643-4074-6

Typeset by the author.
Cover design by Jeff Cosloy, Newton, MA.
Printed and bound by Hamilton Printing, Rensselaer, NY.
Printed in the United States of America.

9 8 7 6 5 4 3 2 1

Preface

We hope that students and practitioners of mathematics, engineering, computer science and scientific visualization will enjoy this excursion into the beautiful world of curves, surfaces and polyhedra using the computer algebra system (CAS) `MAPLE`.

The work is divided into several parts. Readers are first introduced to elementary methods of modeling and visualizing curves and surfaces in space. More complicated concepts and problems in differential geometry are presented afterwards.

All the `MAPLE` code segments may be easily seen and downloaded from the author's web page:

`http://math2.haifa.ac.il/ROVENSKI/rovenski-compgeom.html.`

The above may be linked back to the Birkhäuser site as well:

`http://math2.haifa.ac.il/ROVENSKI/rovenski/Birkhauser.html`

or from http://www.birkhauser.com/cgi-win/ISBN/0-8176-4074-6, which provides additional resources and updates.

Acknowledgments. We would like to thank Wolak R. (Mathematical Institute, Krakow), Borisenko A.A., Yampol'sky A.L., Masaltsev L.A., Kourinny G.Ch., and Semenihina E.V. (University of Kharkov), Blanc D. (University of Haifa) for their helpful corrections and suggestions for improvements concerning the entire manuscript. Finally, I would like to warmly thank Ann Kostant for human support and help in the publishing process.

Vladimir Rovenski
Haifa, June 1999

Contents

MAPLE V: A Quick Reference

The Computer Algebra System (CAS) MAPLE V Releases 4, 5 and/or 6 (below we write simply MAPLE) consists of a kernel written in **C**, a library written in the language of MAPLE, and an interface. The kernel executes most of the basic operations. The library contains commands (procedures) that work in interpretation mode. Built-in programming language support for writing our own procedures allows us to build on the basic capabilities of MAPLE. Although MAPLE can run on various platforms, we assume throughout the book that the reader uses this package with Microsoft Windows 95 or later. We work in MAPLE in session mode: after the prompt symbol > the user enters expressions (commands, procedures, terms)

```
> sum(i, i=1..1999);    # example of a command
```

that are interpreted in MAPLE to produce and display results. (We use the symbol # to denote comments at the ends of command lines).

A good way to become acquainted with MAPLE is through a general excursion (12 tours) touching on the capabilities of the package. (Open the file **user-menu.mws** for Release 4). We recommend that the reader take this trip. Also the demonstration files accompanying this text may be used. Detailed help with individual commands is available (appearing in a special window) via a menu or directly from a command line. To obtain help, you may either place the cursor (|) above a word and press **Ctrl-F1** or type ? and the word on a command line, then press **Enter**. For instance, to find out about the linalg package, one executes the following expression:

```
> ?linalg  # example of a help request
```

The simplest objects in MAPLE are numbers, constants and names. Some important constants are the following: Pi ($\pi \approx 3.14$), E ($e \approx 2.7$), I ($i = \sqrt{-1}$), gamma (Euler's constant ≈ 0.577), true, false.

There are many kinds of variables in MAPLE: float, integer, array, string, vector, function, exprseq, series, set and others.

We form expressions using arithmetic operations +, -, *, /, >, <, =, <=, >=, !, ^ and operations with sets intersect, minus, union, etc. Expressions and variables usually play the role of parameters of commands, that are separated by ; or : (in the latter case we do not see results on a display). We use := to define values of variables. (Note that capital and small letters are case sensitive in MAPLE).

A standard command has the form

```
> command(par 1, par 2, ... par n);
```

Here command is a name of a command and par 1, par 2, ... par n are its parameters. There is a long list of elementary functions. To use them, write their names with parameters:

```
> sin(Pi/3); evalf(%);  # calling function sin(x)
```

We transform mathematical expressions (in a symbolic manner) using the operations factor, expand, simplify, subs, and so on.

```
> f:=x^3+a^3: g:=factor(f); subs({x=2,a=1}, f);
```

$$g := (x + a)\,(x^2 - ax + a^2) \qquad 28$$

Fractions are normalized using the command normal; the commands numer and denom return the numerator and denumerator of a fraction:

```
> f:=(x^2-y^4)/(x+y^2); normal(f); numer(f); denom(f);
```

$$f := \frac{x^2 - y^4}{x + y^2} \qquad x - y^2 \qquad x^2 - y^4 \qquad x + y^2$$

We often perform various operations on polynomials, and apply other operations taken from analysis and linear algebra, which can be understood using help, but we assume the reader has little background in these areas of mathematics.

Graphics commands include plot, display, plot3d, textplot, animate, and so on. The simplest of all graphics commands plot has two main configurations:

```
> plot(expression, range of variable, options);
> plot([expression_1,expression_2,range of variable],options);
```

These are a *nonparametric plot* (graph) and a *parametric plot* (plane curve), respectively. For instance, if the variable x varies between two values a and b (where $a < b$), then we write simply x=a..b.

Very little familiarity with MAPLE is assumed, and so we provide three tables for quick MAPLE reference. These tables can be extended to meet individual requirements.

Symbols and Abbreviations

Symbol	Description	Example
:=	assignment	f:=x^2/y^3;
:	terminate command; hide result	int(x^2, x):
;	terminate command; display result	int(x^2, x);
..	specify a range or interval	plot(t*exp(t),t=0..3);
{ }	set delimiter	{x, y, z};
[ ]	list delimiter	[p[1], q[2]];
%	refers to previous result (double quote " for Release 4)	Int(exp(x^2),x=0..1): value(%);
` `	string delimiter (back quote)	TITLE:=`cycloid`
´ ´ (also ? uneval)	delayed evaluation (single quote) *Note*: back quote (`) is different	x:=´x´; not needed in normal usage
->	mapping (procedure) definition	f:=(x,y) -> sin(x)*y^2; f(Pi/4, 0);

Mathematical Operations, Functions, Constants

<table>
<tr><th>Symbol</th><th>Description</th><th>Example</th></tr>
<tr><td>+ - * / ^</td><td>arithmetic operations</td><td>x^(-3)*3+y/Pi;</td></tr>
<tr><td>sin, cos, tan
cot, sec, csc</td><td>trigonometric
functions</td><td>sin(x-Pi/3)-cot(x^2);</td></tr>
<tr><td>arcsin, arccos
arctan, arccot
arcsec, arccsc</td><td>inverse
trigonometric
functions</td><td>arctan(x*t);</td></tr>
<tr><td>exp</td><td>exponential function</td><td>exp(2*t); exp(1) $= e \approx 2.7$</td></tr>
<tr><td>ln</td><td>natural logarithm</td><td>ln(x*y/4);</td></tr>
<tr><td>log10</td><td>logarithm with base 10</td><td>log10(10000);</td></tr>
<tr><td>abs</td><td>absolute value</td><td>abs((-2)^7);</td></tr>
<tr><td>sqrt</td><td>square root</td><td>sqrt(2);</td></tr>
<tr><td>!</td><td>factorial</td><td>k!;</td></tr>
<tr><td>=, <>, <,</td><td>equations</td><td>diff(y(x),x)+x*y(x)=F(x);</td></tr>
<tr><td><=, >, >=</td><td>and inequalities</td><td>exp(Pi) > Pi^exp(1);</td></tr>
<tr><td>Pi, I</td><td>π, i (math. constants)</td><td>exp(Pi*I);</td></tr>
<tr><td>infinity</td><td>infinity ∞</td><td>int(x^(-2),x=1..infinity);</td></tr>
<tr><td>.(|| in Maple 6)</td><td>concatenation</td><td>p.i (p ||i)</td></tr>
</table>

Commands

Command	Description	Example
restart	clear all MAPLE definitions	restart:
with	load package	with(plots): with(linalg):
help (?)	display MAPLE on-line help	?plottools;
example	provide examples for functions	example(plot);
limit	calculate a limit	limit(sin(x)/x, x=0);
diff	compute the derivative	diff(a*x*exp(b*x), x);
int	(in) definite integration	int(cos(x), x=0..Pi);
value	evaluate an inert expression (used with Limit, Diff, Int)	G:= Int(cos(x), x=0..Pi); value(G);
Limit	inert (uneval.) form of limit	Limit(sin(x)/x, x=0);
Diff	inert (uneval.) form of diff	Diff(a*x*exp(b*x), x);
Int	inert (uneval.) form of int	Int(cos(x), x=0..Pi);
plot	create a 2-dimensional plot of functions, points, polygon	plot(t,t=0..2,title=`T`); plot({x/2, 2*x}, x=0..Pi);
plot3d	create a 3-dimensional plot of functions (surface-graph)	plot3d(x^2-y^2,x=-1..1, y=-1..sqrt(1-x^2));
display	display a list of plot structures	a:=plot([[0,0],[3,1],[6,0]]): b:=plot(cos(x),x=0..2*Pi): plots[display]([a,b]);
solve	solve equations	solve(x^4-5*x^2+6*x=2, x);
fsolve	solve (floating-point arithmetic)	fsolve(t+t*exp(-t)=1, t);
subs	substitute values	subs(x=1, cos(cos(x)));
simplify	apply simplification rules	simplify(exp(ln(exp(c))));
factor	factor a polynomial	factor((x^3-y^3)/(x^4-y^4));
expand	inverse to factor	expand((a+b)^5);
convert	convert to different form	convert(x/(x^2-1),parfrac,x);
collect	collect coefficients (of like powers)	collect((x+a)^3*(x+b)^2,x);
rhs (lhs)	right-hand (left-hand) side	rhs(y=a*x^2+b*x+c);
numer	extract the numerator	numer((x+a)^3/(x+b)^2);
denom	extract the denominator	denom((x+a)^3/(x+b)^2);
evalf	evaluate (floating-point arithmetic)	evalf(exp(Pi));
evalc	evaluate complex-valued term	evalc(exp(a+I*b));
evalb	evaluate a Boolean term	evalb(exp(Pi)>Pi^exp(1));
evalm	evaluate a matrix (vector)	evalm(linalg[dotprod](a,b));
assign	perform assignments (often used for solve or dsolve)	s:=solve({x+y=1, 2*x+y=3}, {x, y}); assign(s);
seq	create a sequence	[seq([0, i^2], i=-3..3)];
assume	additional properties	assume(a>0);
about	check on objects	about(a);

Part I

Functions and Graphs with MAPLE

1
Graphs of Tabular and Continuous Functions

Section 1.1 is a short introductory excursion into the basic rules of using MAPLE for symbolic transformations and two-dimensional plots. In Sections 1.2 and 1.3, we plot graphs of some elementary and special functions. In Section 1.3, we play with transformations and obtain animated graphs (or "movies") using the command `animate`. In Section 1.4, MAPLE helps us to investigate functions using derivatives.

In Chapter 1 the reader will become acquainted with the commands

`evalf, unapply, seq, map, array, simplify, convert, fsolve, subs, sum, max, min, op, nops, Digits, limit, diff, taylor;`

`plot, polygonplot, textplot, animate, animatecurve, arrow, display, with(<name of library>);`

`cos, sin, exp, Ei, Si, Ci, surd, arcsin, tan, ithprime, erf, GAMMA, J, EllipticE, EllipticF, EllipticK;`

`assume, for i from n to m do ... od, if A then B else C.`

1.1 Basic Two-Dimensional Plots

Let us play with some basic MAPLE commands.

Let the function $y = f(x)$ be given by the table of values (x_i, y_i), where $i = 1, 2, \ldots, n$. For example, we measure the morning air temperature (Celsius) on 10 days in May.

Date :	12	13	14	15	16	17	18	19	20	21
Temp. :	15	17	17.5	19	20	19.5	18	17	17	19

```
> DateTemp := [[12,15], [13,17], [14,17.5], [15,19],
  [16,20], [17,19.5], [18,18], [19,17], [20,17], [21,19]];
```

We plot the polygon through these points using the command `plot`, labeling the coordinate axes.

```
> Line:=plot(DateTemp, labels=[T, D]): %;  # Fig. 1.1.
```

We plot the points separately, using the options `style=point, symbol=circle`.

```
> Points:=plot(DateTemp, style=point, symbol=circle): %;
```

Sometimes it is important to form ordered lists with the x- and y-coordinates of the given points.

```
> Days:=[seq(i+11, i=1..10)];
  Temp:=[seq(op(2, DateTemp[i]), i=1..nops(Days))];
```

Using the command `map`, we can transform the temperature from *Celsius* to *Fahrenheit* by the formula $F = F(C)$. First we fix the number of significant digits to 3 using the system variable `Digits`.

```
> F:=C->evalf(9/5*C+32): Digits:=3: map(F,Temp);
```

$$[59., 62.6, 63.5, 66.2, 68., 67.1, 64.4, 62.6, 62.6, 64.4]$$

Now we plot the polygon `Line` by a different method, i.e., using the lists `Days, Temp`.

```
> plot([seq([Days[i], Temp[i]], i=1..10)]);
```

Also we can plot the diagram of vertical segments, Fig. 1.2.

```
> p:=i->plot([[Days[i],0], [Days[i],Temp[i]]]):
> plots[display]([seq(p(i), i=1..10), Points]);
```

Exercise 1.1.1 Plot the polygon passing through 20 points that lie on the parabola $y = x^2$: $\{x_i = -2 + \frac{4}{19}i,\ 0 \le i \le 19\}$, Fig. 1.3.

Sometimes one uses the command `unapply` (*f*ormulas, variables), which transforms the *formulas* into functions of *variables*:

```
> f:=x^2:g:=unapply(f, x); g(2);
```

Definition 1.1.1 The *graph of a function* $y = f(x)$ $(x \in I)$ is a set in the plane $\mathbb{R}^2$: $G_f = \{(x, y) \in \mathbb{R}^2 : y = f(x),\ x \in I\}$.

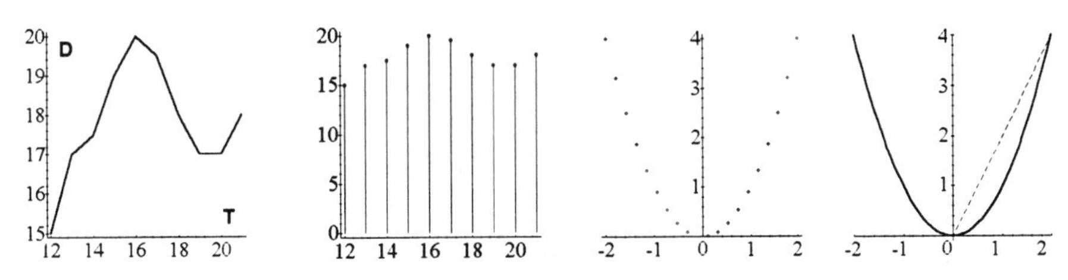

Figs. 1.1–1.4. Temperature during 10 days. Points on a parabola. Graphs of the function x^2 and its derivative

The graph of $f(x)$ of the class C^1 (i.e., $f'(x)$ exists and is continuous) is a *smooth curve* (a tangent line $y = f(x_0)+f'(x_0)(x-x_0)$ continuously depends on points x_0).

Graph can be plotted using the command `plot(expr,range,options)`.

```
> plot(x^2, x=-2..2, scaling=constrained);    # parabola
```

By assumption `x=-10..10`. By clicking the mouse at any point of the graph we can see its coordinates in the upper left corner boxes. The option `scaling=constrained` (or clicking the *button* [1 : 1]) is sometimes necessary, for example, to see the right angles of a rectangle.

```
> plot([[0,0],[16,12],[13,16],[-3,4],[0,0]],
  scaling=constrained);  # rectangle
```

Sometimes the same option actively prevents us from seeing details of the graph; for example,

```
> plot(sin(x), x=0..10*Pi, numpoints=999);
```

The command `plot(`[$expr_1, \ldots, expr_n$]`,range,color=`[$c_1, \ldots, c_n$]`,...)` allows us to plot several graphs on one figure, for example, graphs of $f(x)$ and its derivative $f'(x)$:

```
> f:=x^2: plot([f, diff(f, x)], x=-2..2, y=-.1..4, color=
  [blue,red], thickness=[2,1], linestyle=[1,2]); # Fig. 1.4
```

One can place text over a figure (and add names to graphs) using the command `textplot` (and the option `title`), and then `display` from the library `plots`. For example, see Fig. 1.5.

```
> with(plots):
> graph:=plot([-x/2,-x,-2*x,x/2,x,2*x],x=-1..1,y=-1..1):
> text:=textplot({[1,.5,`y=x/2`],[1,1,`y=x`],[.4,1,`y=
  2x`],[-.4,1,`y=-2x`], [-1,1,`y=-x`], [-1,.5,`y=-x/2`]},
```

```
  align=ABOVE):
> plots[display]([graph, text], font=[TIMES, ROMAN, 18]);
```

Exercise 1.1.2

1. Solve graphically and analytically the equation $e^{-x} = \sin(x)$.

```
> plot([exp(-x), sin(x)], x=0..Pi);
```

Hint. The intersection of the two graphs consists of two points, see Fig. 1.6, which correspond to the two roots $x_1 \approx 0.59$, $x_2 \approx 3.1$ of the equation $\exp(-x) = \sin(x)$. One can find them numerically by clicking the mouse. The largest root, 3.10, is returned by the command

```
> fsolve(exp(-x)-sin(x)=0, x);
```

For obtaining the second root, 0.589, we restrict the search by using the option `0..1`.

```
> fsolve(exp(-x)-sin(x)=0, x, 0..1);
```

2. From a piece of pasteboard 4 in. × 5 in. construct a rectangular box without a roof having maximal volume.

Hint. Let x be the height of the box. Obviously, $x \in [0, 2]$. We calculate the volume by the formula $V(x) = x(4-3x)(5-2x)$ and plot the graph of this polynomial.

```
> V:=x -> x*(4-2*x)*(5-2*x): plot(V(x), x=0..2);
```

By clicking the mouse we find the highest point of the graph, $\approx (0.74, 6.56)$. For an exact solution we solve the equation that sets the derivative of the polynomial equal to zero.

```
> Vx:=simplify(diff(V(x),x)); fsolve(Vx,x); subs(x=%[1],V(x));
```

$$V_x := 20 - 36x + 12x^2$$

$$0.736 \qquad 2.26 \qquad 6.57$$

3. Find the maximal surface area of a cylinder with radius R if its volume V is equal to 1.

Hint. Show that $V(R) = \frac{2}{R} + 2\pi R^2$, Fig. 1.7.

4. Plot the graph of the rational function $y = \frac{x^6 - x^2 + 1}{x^5 - 6x^4 + 12x^3 - 12x^2 + 11x - 6}$ and decompose it into partial fractions.

```
> f:=x-> (x^6-x^2+1)/ (x^5-6*x^4+12*x^3-12*x^2+11*x-6);
> plot(f(x), x=0.5..4, y=-200..200, discont=true); # Fig. 1.8
> convert(f(x), parfrac, x);
```

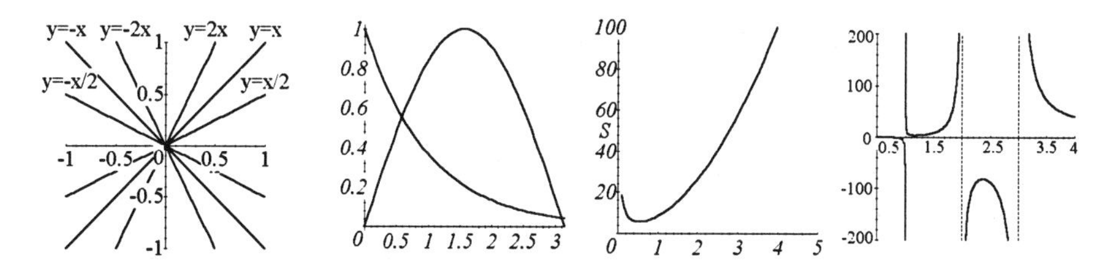

Figs. 1.5–1.8. Placing text. Solving equations. Rational function

$$x+6+\frac{1}{4}\frac{1}{x-1}-\frac{61}{5}\frac{1}{x-2}+\frac{721}{20}\frac{1}{x-3}-\frac{1}{10}\frac{x}{x^2+1}$$

Remark 1.1.1 The option `discont=true` cancels joining parts of the graph with its vertical asymptotes (we also use it for plotting graphs of piecewise continuous functions, see Chapter 2). The options `scaling=constrained`, `discont=true`, `linestyle=n`, `axes=boxed`, `numpoints=n`, `thickness=n`, `color=N` (`N` – name of a color `red`, `green`, `blue`, etc.) or `COLOR(RGB, R, G, B)`, and also `COLOR(HUE, H)`, etc., can be entered globally using the command `setoptions` *(options through commas)*

Let us consider some further examples. (Compare the results of the following commands and state your conclusions).

```
> plot(tan, -Pi..Pi);
> plots[display](%, view=[-4..4,-10..10], xtickmarks=
   [-3.14=`-Pi`, -1.57=`-Pi/2`, 1.57=`Pi/2`, 3.14=`Pi`]);
```

Plot the square diagram "colors".

```
> P:=seq(seq(polygonplot([[i,j],[i+1,j],[i+1,j+1],[i,j+1]],
  color=COLOR(RGB, 0.1*i, 0.1*j, 0)), i=0..10), j=0..10):
> plots[display]({P},scaling=constrained,labels=[`R`,`G`],
   title=`RGB color (R,G,0)`, tickmarks=[[seq(i+0.5=`.`.i,
   i=0..9),10.5=`1`],[seq(j+0.5=`.`.j,j=0..9),10.5=`1`]]);
```

Plot the diagram "width of line" (analogously for "type of line").

```
> for i from 0 to 15 do
  thick[i]:=plot(i, x=0..1, thickness=i) od:
> plots[display](convert(thick, set), axes=boxed,
  tickmarks=[0,15],title=`thickness option: [0,1,..,15]`);
```

Note that the option `filled=true` paints in the domain between the graph and the axis OX. Paint the domain between two graphs using the following program.

```
> sine:=plot(sin, 0..4*Pi, color=black, thickness=3):
```

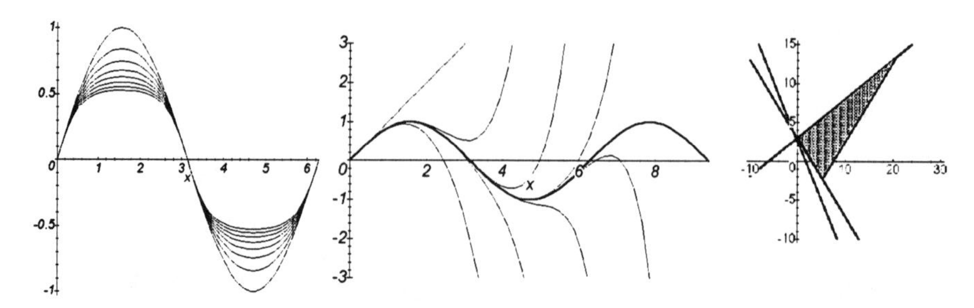

Figs. 1.9–1.11. Iterations of $y = \sin(x)$ and Taylor polynomials. Inequalities

```
> s:=plot(sin, 0..4*Pi, filled=true, color=red):
> cosine:=plot(cos, 0..4*Pi, color=black, thickness=3):
> c:=plot(cos, 0..4*Pi, filled=true, color=red):
> f:=x->if cos(x)>0 and sin(x)>0 then min(cos(x),-sin(x))
  elif cos(x)<0 and sin(x)<0 then max(cos(x),-sin(x))
  else 0 fi:
> b:=plot(f, 0..4*Pi, filled=true, color=white):
> plots[display]([sine, cosine, b, s, c], scaling=constrained);
```

Plot the table of graphs.

```
> with(plots): display(array([sine, cosine])); # row of plots
> display(array(1..2, 1..1, [[sine], [cosine]])); # lines of plots
```

Using the symbol @ for the *composition* of functions $f @ g$, plot iterations of the given function, Fig. 1.9.

```
> n:=8: q:=i->plot((sin@@i)(x), x=0..2*Pi, color=COLOR(
  RGB, 1/i,0,1-1/i)): plots[display]([seq(q(i), i=1..n)]);
```

Taylor decomposition is very useful in studying functions. Compare *Taylor polynomials* of a function with the given function, Fig. 1.10.

```
> x1:=3*Pi: p[0]:=plot(sin(x), x=0..x1, thickness=2):
> for i from 1 to 9 do pp:=convert(taylor(sin(x), x=0,
   2*i-1), polynom): p[i]:=plot(pp, x=0..x1, y=-3..3) od:
> plots[display]([seq(p[i], i=0..9)], scaling=constrained);
```

The domain of points satisfying a set of equations and inequalities is plotted using the command inequal, Fig. 1.11.

```
> plots[inequal]({a+b>3, 2*b-a<6, 3*a+2*b>5, -b+a<=8},
a=-10..30, b=-10..15, optionsfeasible=(color=red),
optionsopen=(color=blue,thickness=2), optionsclosed=
(color=green,thickness=3),optionsexcluded=(color=yellow));
```

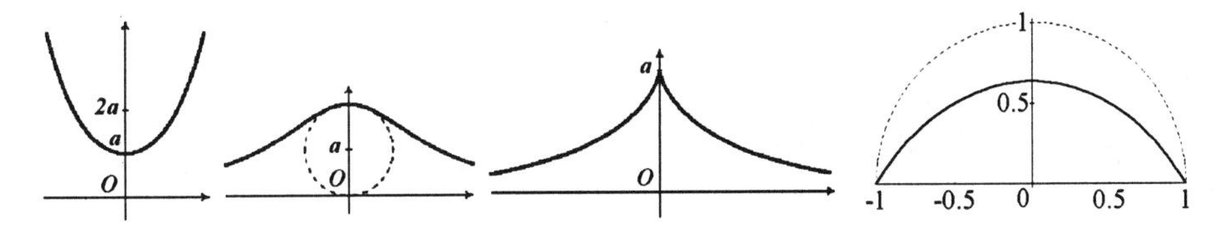

Figs. 1.12–1.15. Catenary, witch of Agnesi, tractrix, and quadratrix

1.2 Graphs of Functions Obtained from Elementary Functions

First plot the graphs of the following functions (set $a = 1$ for functions 1–5).

1. *Catenary* $y = a\cosh(\frac{x}{a})$, Fig. 1.12, is "similar" to a parabola. A catenary is obtained by hanging a cord by its ends between two poles.

2. *Witch of Agnesi (versiera)* $y = \frac{a^3}{a^2+x^2}$, Fig. 1.13.

3. *Cycloid* $x = a\arccos\frac{a-y}{a} - \sqrt{2ay - y^2}$, Fig. 5.2.

4. *Tractrix* $x = a\ln\frac{a-\sqrt{a^2-y^2}}{y} + \sqrt{a^2 - y^2}$, Fig. 1.14.

5. *Dinostratus' quadratrix* $y = x\cot(\frac{\pi x}{2a})$, Fig. 1.15.

6. *Beats.* A complicated situation arises in summing oscillations with different frequencies:

$$y = y_1 + y_2 = A(\cos(\omega_1) + \cos(\omega_2)) = 2A\cos(\tfrac{\omega_1-\omega_2}{2})\cos(\tfrac{\omega_1+\omega_2}{2})$$

If there is a small difference between ω_1 and ω_2, then the multiplier $\cos(\frac{\omega_1-\omega_2}{2})$ changes slowly, and the multiplier $\cos(\frac{\omega_1+\omega_2}{2})$ has almost the same frequency as each of the given oscillations. Beats are sometimes presented (approximately but visually) as harmonic oscillations with slowly changing amplitude.

```
> p:=plot(6*sin(x/4)*sin(4*x), x=-Pi..9*Pi):
> q:=i->plot(6*i*sin(x/4), x=-Pi..9*Pi, linestyle=2):
> plots[display]([p, q(-1), q(1)]); # Fig. 1.16
```

7. *Damped oscillations curve* $y = A\exp(-kx)\sin(\omega x + a)$. For $k = 0$, the oscillations are periodic (without resistance from the medium). Also plot this graph with the option `x=0..infinity`.

```
> p:=plot(exp(1)^(-x/2)*sin(4*x), x=0..3*Pi):
> q:=i->plot(i*exp(1)^(-x/2), x=0..3*Pi, linestyle=2):
> plots[display]([p, q(-1), q(1)]);  # Fig. 1.17
```

8. A *remarkable limit* $y = \frac{\sin(x)}{x}$, Fig. 1.18, (or $y = \frac{\tan(x)}{x}$).

```
> limit(sin(x)/x, x=0); plot(sin(x)/x, x=-10..10);
```

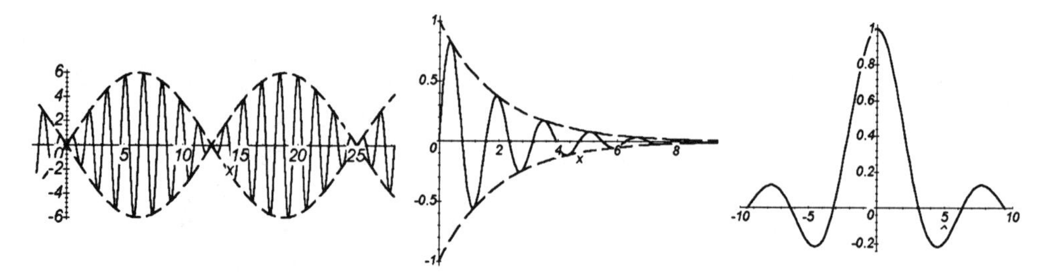

Figs. 1.16–1.18. Beats, dumped oscillations, remarkable limit

1

A sequence f(n) is a function on $\mathbb{N}$. We calculate the sequence $\frac{\sin(x)}{x}$ for n=12..16 and compare with previous results.

```
> f:=n->n*sin(1/n): L:=evalf(seq([n,f(n)],n=12..16));
```

L := [12, 0.998], [13, 0.998], [14, 0.998], [15, 1.00], [16, 1.00]

```
> plot([L], style=point, symbol=circle);
> Digits:=14: evalf(f(1000000));
```

.99999999999983

```
> n:='n': limit(f(n), n=infinity);
```

1

Let us plot a diagram showing the sequence of prime numbers.

```
> PN:=[seq([i, ithprime(i)], i=1..9)]:
> plot(PN, style=point, symbol=box);
```

9. *Growth curves* are curves that describe the laws of development of events over time. The process of analytically smoothing a dynamical line using some functions, i.e., adjusting them to these functions, is in most cases a useful approach to representing empirical data. Plot the following growth curves (1) – (3) for some values of parameters.

(1). *Logarithmic parabola* $y = b^x c^{x^2}$ $(b,\ c > 0)$, Figs. 1.19–1.20.

```
> f:=x->b^x*c^(x^2): plot(subs(b=2, c=2, f(x)), x=-2..1);
```

Write the equation in the form $y = e^{x\ln b + x^2 \ln c}$. For $c > 1$ the branches of the parabola $x\ln b + x^2 \ln c$ are directed upward, and for $0 < c < 1$ they are directed downward. For $c > 1$ and $b > 1$ the curve is displaced to the left, and for $c < b < 1$ it is displaced to the right.

(2). *Logarithmic curve* $y = \frac{k}{1 + b\,e^{-ax}}$ $(k, a, b > 0)$, Fig. 1.21.

```
> f:=x->k/(1+b*exp(-a*x)): plot(subs(k=2, b=0.2, a=1,
```

```
[f(x), k]), x=-7..3, linestyle=[1, 2]);
```

The line $y = k$ is the horizontal asymptote; $x = \ln(b)/a$ is the point of inflection.

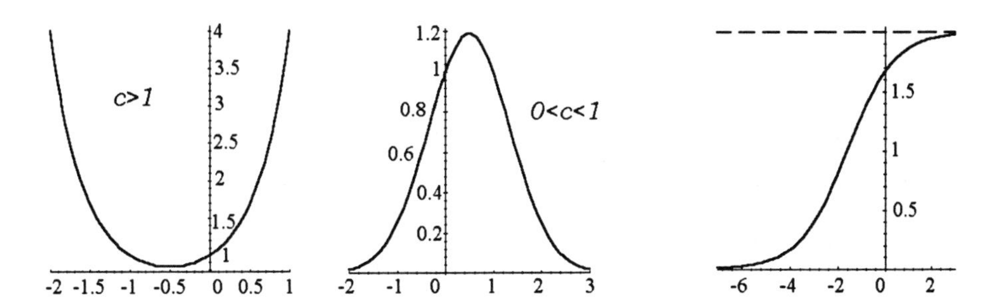

Figs. 1.19–1.21. Logarithmic growth curves

(3). *Gompertz growth curves* $y = a^{b^x}$ $(a, b > 0)$; Fig. 1.22.

```
> f:=x->a^(b^x): plot({subs(b=0.5, a=1.2, f(x)), 1},
  x=-4..1, y=0..18, linestyle=[1, 2]);
```

We write the equation in the form $y = e^{b^x \ln a}$ and consider four cases, Fig. 1.22. One can check that $y = 1$ is the horizontal asymptote.

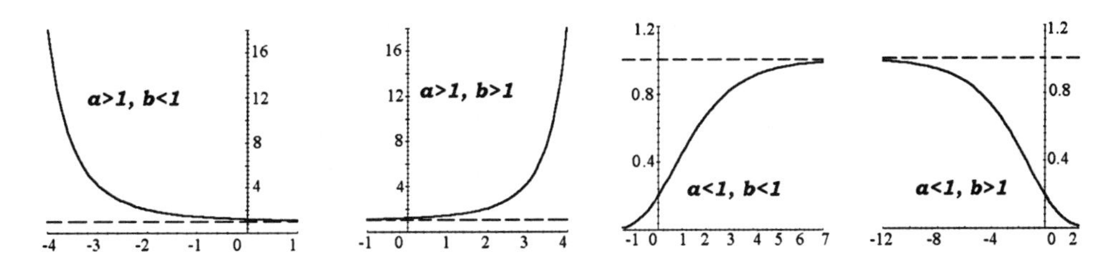

Figs. 1.22 a–d. Gompertz growth curves

1.3 Graphs of Special Functions

The name *special* is usually used for functions that generally cannot be represented in terms of elementary functions. Most of these arise as the solutions of ODEs of special forms and can be expressed using integrals. Let us plot graphs of the following six special functions.

1. The *incomplete Gamma function* $y = \Gamma(x, z) = \int_z^\infty e^{-t} t^{x-1}\, dt$ is one of the most popular special functions.

The *Gamma function* $y = \Gamma(x) = \int_0^\infty e^{-t} t^{x-1}\, dt$ has vertical asymptotes at the points of discontinuity $x = -n$, $(n = 0, 1, 2, \dots)$

```
> plot(GAMMA(x), x=-4..5, y=-10..10);   # Fig. 1.23
```

2. *Integral of probabilities* $y = \mathrm{erf}(x) = \frac{2}{\sqrt{\pi}}\int_0^x e^{-t^2}\,dt$ (*error function*), Fig. 1.24, is derived using Taylor series.

```
> convert(taylor(erf(x), x=0, 31), polynom);
> plot([erf(x), 1, -1], x=-2..2, linestyle=[1, 2, 2]);
```

3. *Exponential integral function* $y = \mathrm{Ei}(n, x) = \int_1^\infty \frac{e^{-xt}}{t^n}\,dt$ $(n \in \mathbb{N},\ x > 0)$. The function is related to the Gamma function by $\mathrm{Ei}(n, x) = x^{n-1}\Gamma(1-n, x)$.

```
> plot([seq(Ei(n,x),n=0..5)],x=-.2..1.6,y=0..1); # Fig. 1.25
```

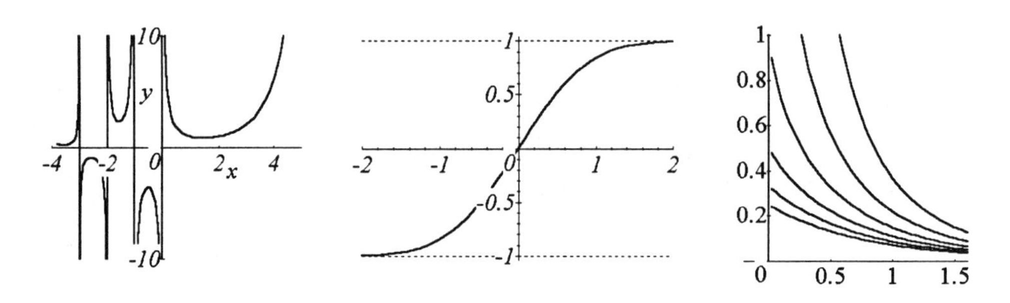

Figs. 1.23–1.25. Graphs of special functions $\Gamma(x)$, $\mathrm{erf}(x)$, $\mathrm{Ei}(n, x)$

4. *Sine integral function* $y = \mathrm{Si}(x) = \int_0^x \frac{\sin(t)}{t}\,dt$. This function is derived using Taylor series; its horizontal asymptote is $y = \frac{\pi}{2}$, Fig. 1.26.

```
> plot([Si(x),Pi/2],x=-1..20,y=0..2,linestyle=[1,2]);
```

Cosine integral function $y = \mathrm{Ci}(x) = \gamma + \ln(x) + \int_0^x \frac{\cos(t)-1}{t}\,dt$. Here $\gamma \approx 0.57721$ is the *Euler constant*; the horizontal asymptote is $y = 0$.

```
> plot(Ci(x), x=-3..21, y=-0.5..0.5);   # Fig. 1.27
```

5. *Elliptic integrals of the first* and *second kind* are given by the formulas

`EllipticF(z,x)=` $\int_{t=0}^{z} 1/(\sqrt{(1-t^2)}\,\sqrt{(1-x^2t^2)})\,dt$, Fig. 1.28

`EllipticE(z,x)=` $\int_{t=0}^{z} \sqrt{(1-x^2t^2)}/\sqrt{(1-t^2)}\,dt$, Fig. 1.29

`EllipticK(x)=EllipticF(1,x), EllipticE(x)=EllipticE(1,x).`

```
> plot([[1,t,t=0..4],EllipticK(x)],x=0..1.2,y=0..4);
> plot([[1,t,t=0..4],EllipticE(x)],x=0..1.2,y=0..4);
```

There are many commands in MAPLE for special functions.

6. For example, let us plot the *spherical Bessel function of the first kind*.

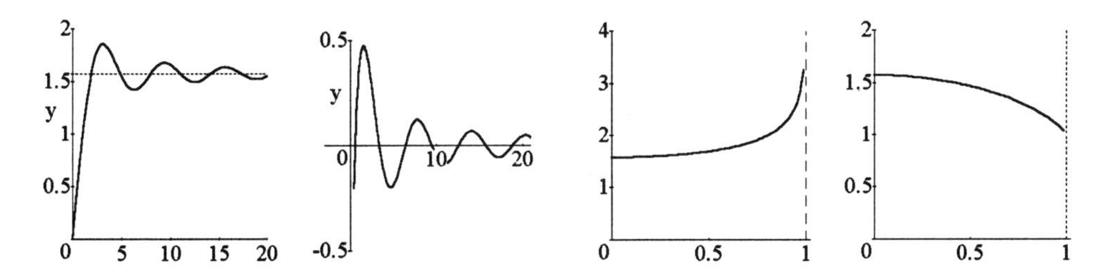

Figs. 1.26–1.29. Graphs of Si(x), Ci(x), and elliptic integrals

```
> graph:=plot(x->J(0,x), 0..20, -0.5..1,
  xtickmarks=8, ytickmarks=4, title=
  `Spherical Bessel function J(0,x) of the first kind`,
  titlefont=[TIMES,ROMAN,16], axesfont=[HELVETICA,BOLD,14]):
> text:=plots[textplot]({[2,0.75,`J(0,x)`],[14,-0.1,`x`]},
  align={ABOVE, RIGHT}, font=[HELVETICA, BOLD, 14]):
> plots[display]({graph, text});
```

1.4 Transformations of Graphs

The animatecurve function (from the library plots) provides support for visualizing the drawing of real functions in one variable.

```
> plots[animatecurve]({x-x^3,sin(x)}, x=0..Pi/2);
```

We invite the reader to play with simple transformations 1–6 of the graph of a function $y = f(x)$ using the command animate.

```
> with(plots): f:=x->x^2;
```

1. *Parallel translation in the direction of an arbitrary vector.*

```
> X:=2: Y:=6: p1:=plottools[arrow]([0,0],[X,Y], .1, .2, .1):
> p2:=animate([x+t, f(x)+3*t, x=-2.5..2.5], t=0..2):
> display([p1, p2]);
```

2. *Contraction along the axis OX.*

```
> p1:=plot([cos(u), sin(u), u=0..2*Pi]):
> p2:=animate([t*cos(u), sin(u), u=0..2*Pi], t=1..2):
> p3:=animate([t*x, f(x)+2, x=-2..2], t=1..2):
> display([p1, p2, p3], scaling=constrained);
```

3. *Contraction along the axis OY.*

```
> p4:=plot([cos(u), sin(u), u=0..Pi]):
```

```
> p5:=animate([cos(u), t*sin(u), u=0..Pi], t=1..2):
> p6:=animate(t*f(x), x=-2..2, t=1..2):
> display([p4, p5, p6], scaling=constrained);
```

4. *Symmetry with respect to the axis OX.*

```
> s1:=animate([t*x, f(x), x=0..2], t=-1..1, color=green):
> s2:=animate([t*x, f(x), x=-2..0], t=-1..1, color=blue):
> s3:=animate([t*x, f(1), x=0..1], t=-1..1, color=green):
> s4:=animate([t*x, f(1), x=-1..0], t=-1..1, color=blue):
> display([s1, s2, s3, s4]);
```

5. *Symmetry with respect to the axis OY.*

```
> s5:=animate([2, t*x*f(2), x=0..1], t=-1..1, color=blue):
> s6:=animate([x, t*f(x), x=-2..2], t=-1..1, color=green):
> display([s5, s6]);
```

6. *Symmetry with respect to the origin of coordinates.*

```
> s7:=animate([t*x,-t*f(x),x=0..2], t=-1..1, color=green):
> s8:=animate([t*x, -t*f(x), x=-2..0], t=-1..1, color=red):
> s9:=animate([2*t*x,t*x*f(2),x=-1..0],t=-1..1,color=red):
> display([s7, s8, s9]);
```

A graph of the *inverse function* f^{-1} can be obtained from the graph Γ_f using symmetry with respect to the bisector $y = x$ (in other words, we change the places of the coordinate axes OX and OY).

Example 1.4.1

1. Plot graphs of the power function $y = x^3$ and its inverse $y = x^{1/3}$ together with the bisector $y = x$; Fig. 1.30.

```
> p1:=plot([surd(x, 3), x^3], x=-2..8, y=-2..8):
> p2:=plots[textplot]({[3, 8, `y=x^3`], [6, 8, `y=x`],
  [7, 3, `y=x^(1/3)`]}):
> p3:=plot(x, x=-2..8, y=-2..8, style=POINT):
> plots[display]([p1, p2, p3]);
```

The command surd makes the choice of the main root:

```
if x>=0 then surd(x,n)=x^(1/n) else surd(x,n)=-(-x)^(1/n) fi
```

2. Find the inverse function for $f(x) = \sin(x)$, $\frac{\pi}{2} \leq x \leq \frac{3}{2}\pi$, and plot both graphs; Fig. 1.31.

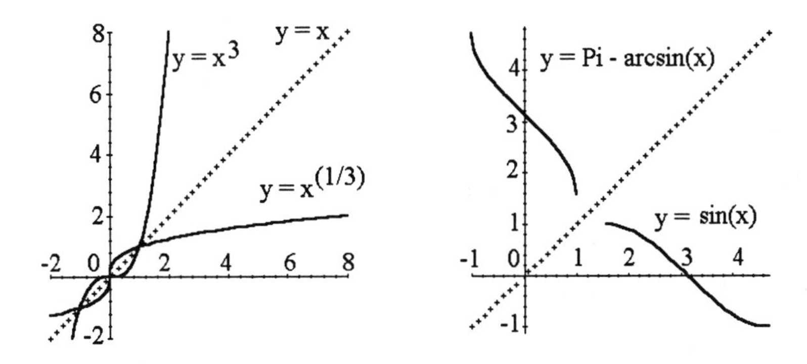

Figs. 1.30–1.31. Graph of the inverse function

Hints. It is clear that `arcsin` requires some modification (it is the inverse function for the condition $-\frac{\pi}{2} \le x \le \frac{\pi}{2}$), because $\arcsin(f(\pi)) = \arcsin(0) = 0$ holds, but 0 does not belong to the domain of definition of f. Here the inverse function is $f^{-1}(x) = \pi - \arcsin(x), \ -1 \le x \le 1$.

```
> p1:=plot(Pi-arcsin(x), x=-1..1):
> p2:=plot(sin(x), x=Pi/2..3*Pi/2):
> p3:=plot(x, x=-1..3*Pi/2, style=point):
> p4:=plots[textplot]({[3, 0.5, `y=sin(x)`],
  [.5, 3.5, `y = Pi-arcsin(x)`]}, align=RIGHT):
> plots[display]([p1, p2, p3, p4]);
```

1.5 Investigation of Functions Using Derivatives

Study the following functions and plot their graphs:

1. $y = \log(x) - \arctan(x)$ 2. $y = (x-3)/\sqrt{(4+x^2)}$ 3. $y = x^3 \log(x)$

4. $y = |1+x|^{3/2}/\sqrt{x}$ 5. $y = x^2 e^{1/x}$ 6. $y = e^x/(4-x^2)$ 7. $y = \sqrt{1-e^{-x^2}}$

8. $y = (1+x)^{1/x}$ 9. $y = e^{1/(1-x^2)}/(1+x^2)$ 10. $y = 2^{\sqrt{x^2+1}-\sqrt{x^2-1}}$

11. $y = \log_2(1-x^2)$ 12. $y = x + \log(x^2-1)$ 13. $y = \tan(x) + \sin(x)$

14. $y = \arcsin((1-x^2)/(1+x^2))$ 15. $y = \arctan((x-3)/(x^2+4))$.

One can plot the graph of $y = f(x)$ and the graphs of the derivatives $f'(x)$ and $f''(x)$ in the same grid. One can then study the functions using the following plan (the graphs that MAPLE plots help us to realize this plan), which is sufficient for most functions in practice.

1. *Find the domain of definition and points of discontinuity; derive the function*

(or its limits) at boundary points of the domain of definition.

2. *Check whether the function is odd or even, periodic or not.*

3. *Find zeros of the function and the intervals on which its sign is constant.*

4. *Find asymptotes of the graph (see Chapter 7).*

5. *Calculate extrema of the function and intervals of monotonicity.*

6. *Find the points of inflection and intervals of convexity and concavity.*

While studying *zeros, asymptotes, boundary* and *discontinuity points*, we calculate asymptotic properties of the function in the neighborhoods of these points.

2

Graphs of Composed Functions

In Chapter 2 we study functions of one variable, defined by different formulas for different values (intervals) of that variable.

Starting in Section 2.1 with the familiar piecewise-continuous functions *Heaviside, signum, floor, ceil,* and *Dirac*, we continue with well-known discrete statistical functions writing our own recursive procedures or using the basic procedure `piecewise`.

In Section 2.2 we create piecewise-differentiable functions using operations *max, min, abs*, then plot graphs of several statistical distributions and manipulate them with recursively defined functions. Finally we play with graphs of limit functions using the command `animate`.

In Chapter 2 the reader will become acquainted with the commands

```
Heaviside, signum, floor, ceil, Dirac, binomial, sqrt, abs
arctan, int, alias, type, discont, rand, piecewise;
readlib(<name of library>), print, interface.
```

2.1 Graphs of Piecewise-Continuous Functions

Definition 2.1.1 A function $f(x)$ given on the interval $I = (a, b)$ is called *piecewise-continuous* (*staircase*) if one can break this interval into a finite number of *intervals of continuity* (resp., *intervals of constancy*) on each of which the function $f(x)$ is continuous (resp., constant).

Values of the function at the points of discontinuity are not always well-defined.

2.1.1 Popular Piecewise-Constant Functions

The *Heaviside (unit step) function* $H(x) = \{$`if x<0 then 0 else 1`$\}$ is most convenient for expressing other piecewise (composed) functions. Let us plot its graph; Fig. 2.7.

```
> plot(Heaviside(x), x=-2..2, ytickmarks=2, discont=true);
```

For an abbreviation of the Heaviside function, use the command

```
> alias(H = Heaviside);
```

The derivative of the Heaviside function is the *Dirac Delta function*, which is equal to 0 everywhere except the singular point $x = 0$. We calculate this function and plot its graph.

```
> diff(H(x), x); plot(%, x=-2..2);
```

$\mathrm{Dirac}(x)$

We plot the graph of another standard function *signum* (Latin for *sign*)

$$y = \mathrm{signum}(x) = \{-1, \text{ if } x < 0, \;\; 1 \text{ if } x > 0, \;\; 0 \text{ if } x = 0\}.$$

```
> plot(signum(x), x=-2..2, discont=true, ytickmarks=2);
```

The variable $\_Envsignum0$ in MAPLE defines the value $signum(0)$ and can be redefined. If we set it equal to 1, we obtain $signum(x) = 2H(x) - 1$. We have the identity $|x| = x(H(x) - H(-x))$ using the standard function `abs(x)`. Using `Heaviside` (or `signum`) we can define the characteristic functions of the segment $[a, b]$, the interval (a, b) and the half-interval, and then plot their graphs.

$\chi(x, [a, b]) = H(x-a) - H(x-b) = (signum(x-a) + signum(x-b))/2,$
$\chi(x, (a, b)) = \chi(x, [a, b]) - \chi(x, [a, a]) - \chi(x, [b, b]),$
$\chi(x, [a, b)) = \chi(x, [a, b]) - \chi(x, [b, b]),$
$\chi(x, (a, b]) = \chi(x, [a, b]) - \chi(x, [a, a]).$

Let us plot the graph of the *integer part of* x: $y = [x]$ using the command `floor(x)` (see also the command `ceil(x)=floor(x)+1`).

```
> plot(floor(x), x=-3..3, discont = true);
```

If you would like to see how some MAPLE procedures and functions are programmed, for example, `floor`, `signum`, `Heaviside`, use the following commands:

```
> interface(verboseproc=2): print(floor);
```

We can return to the usual display mode by using the command

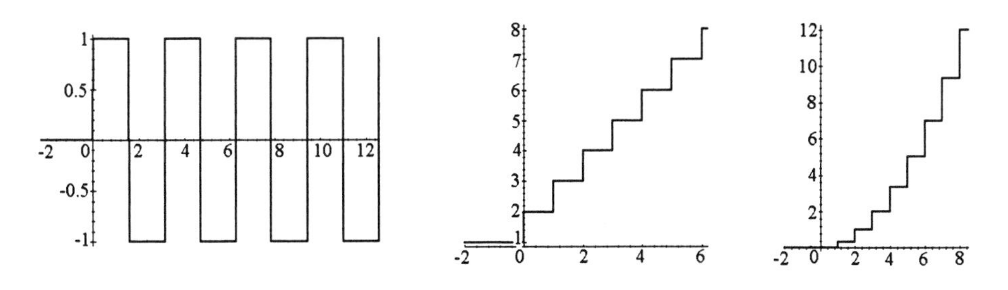

Figs. 2.1–2.3. Impulse, infinite stairs, steep stairs $\{\frac{n}{3}\}$

```
> interface(verboseproc=1);
```

A *periodic impulse* (or triangular wave) $y = H(x) - 2H(x-T) + 2H(x - 2T) - \dots$ can be defined by various methods. We plot its graph for $T = 1$ using the command

```
> plot(H(x)*signum(sin(Pi*x)), x=-1..4);  # Fig. 2.1.
```

Let us plot the *infinite stairs* $y = \sum_{i=0}^{\infty} H(x-i)$ using the command

```
> plot(H(x)*floor(x), x=-2..15); # Fig. 2.2
```

Here is a second method using the *recursive* procedure.

```
> P2:=proc(x) option remember; if type(x, numeric) then
  if x<1 then 1 else P2(x-1)+1 fi: fi end:
> plot(P2, -2..6.2, scaling=constrained);
```

Check that the graph $y = floor(x)$ (without the option `discont=true`) has a similar view. Then plot *steep stairs* with the height of the steps defined by the items of the following sum $S_n = \sum_{n=0}^{m} a_n$.

```
> an:= n -> n/3:  # insert your formula
> P_an:=proc(x) option remember;  if type(x, numeric) then
  if x<0 then 0 else P_an(x-1)+an(floor(x)) fi: fi end:
> P_an(8); plot(P_an, -2..8.2, scaling=constrained);  # Fig. 2.3
```

2.1.2 Discrete Statistical Functions

The main *statistical functions* include the density of a probability function $F(x) = P\{\xi < x\}$ for various distributions. Using the above examples, we plot the graphs $F(x)$ of some *discrete statistical functions*.

1. We start with the *Bernoulli distribution* with the parameter $p \in (0, 1)$.

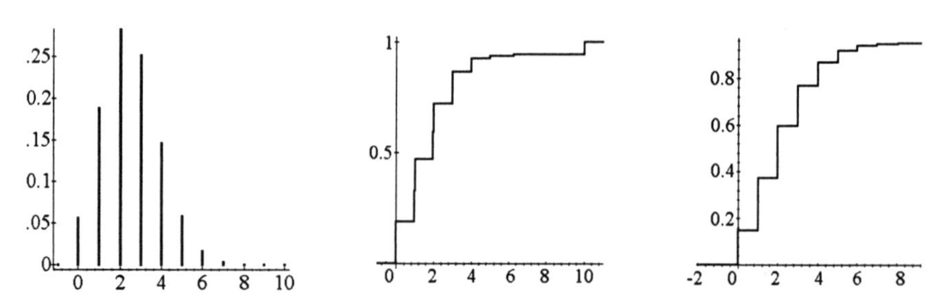

Figs. 2.4–2.6. Graphs of binomial and Poisson distributions

```
> F1:=x -> if x<=0 then 0 elif x<=1 then 1-p else 1 fi;
> plot('subs(p=0.3, F1(x))', x=-1..2);
```

2. *Binomial distribution* $P\{\xi = k\} = C_n^k p^k (1-p)^{n-k}$ $(0 \le k \le n)$ with parameters n, p $(0 < p < 1,\ n \ge 1)$. We define by recursion the distribution function

$$F(x) = \begin{cases} \sum\limits_{k=1}^{m} C_n^k p^k (1-p)^{n-k}, & m < x \le m+1, \\ 1, & x > n, \\ 0, & x \le 0. \end{cases}$$

```
> F2:=proc(x) option remember;
  if type(x, numeric) then if x<=0 then 0 elif x<=n then
  F2(x-1)+binomial(n,ceil(x))*p^ceil(x)*(1-p)^(n-ceil(x))
  else 1 fi: fi end:
> plot(subs(p=0.25, n:=10, F2), -1..n+1); # Fig. 2.5
```

Compare with the following method (plot both graphs):

```
> n:=10: p:=0.25: f2:=x ->
  sum(Dirac(x-k)*binomial(n,k)*p^k*(1-p)^(n-k), k=0..n);
> F2:=int(f2(x), x): plot([F2, 1], x=-1..n, discont=true);
> q:=i -> plot([[i, 0], [i, subs(Dirac(0)=1, f2(i))]]):
  plots[display]([seq(q(i),i=-1..n)],axes=framed);#Fig. 2.4
```

3. *Poisson distribution* $P\{\xi = k\} = \frac{\lambda^k e^{-\lambda}}{k!}$ $(k = 0, 1, 2, \dots)$ with parameter λ $(\lambda > 0)$. We plot, also by recursion, the distribution function

$$F(x) = \begin{cases} \sum\limits_{k=0}^{m} \frac{\lambda^k e^{-\lambda}}{k!}, & m < x \le m+1, \\ 0, & x \le 0. \end{cases}$$

```
> F3:=proc(x) option remember;
  if type(x, numeric) then if x<=0 then 0 else
```

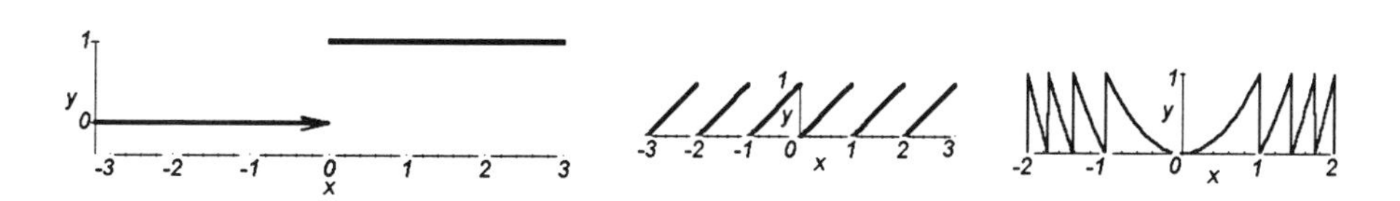

Figs. 2.7–2.9. Graphs of the functions $Heaviside(x)$, $\{x\}$, $\{x^2\}$

```
  F3(x-1)+lambda^ceil(x)*exp(-lambda)/ceil(x)! fi: fi end:
> plot('subs(lambda=0.6, F3(x))', x=-2..6.2); # Fig. 2.6
```

2.1.3 Examples of Piecewise-Continuous Functions

We can also plot the graphs of piecewise-defined functions by the command

$$\texttt{piecewise}(c_1, f_1, \ldots, c_n, f_n, f) = \begin{cases} f_1, & c_1 \text{ true}, \\ \ldots & \\ f_n, & c_n \text{ true}, \\ f, & \text{otherwise}. \end{cases}$$

Load it from the library using the command `readlib(piecewise)` before using it.

Now plot the graph of the *Bernoulli distribution* by the second method.

```
> F1:=x -> piecewise(x<=0, 0, x<=1, 1-p, 1);
> plot(subs(p=0.3, F1(x)), x=-1..2);
```

A function f of the type `piecewise` can be converted into a sum of Heaviside functions using the command `convert(f, Heaviside)`. For example,

```
> convert F1(x), Heaviside);
```

$1 - p\text{Heaviside}(x) + p\text{Heaviside}(x - 1)$

Note that the *characteristic function* of the segment $[a, b]$ (and also of any interval and other sets on the real line) can be written in the short form

```
> assume(a<b): f:=piecewise(x<a, 0, x<=b, 1, 0);
```

One can check the identity `signum(x)=piecewise(x<0, -1, 1)`.

The points of discontinuity of a function $f(x)$ of `piecewise` type are returned by the command

```
> readlib(discont): discont(f, x);
```

The option `discont=true` in the command `plot` for the `piecewise` type function suppresses vertical segments between neighboring parts of the graph.

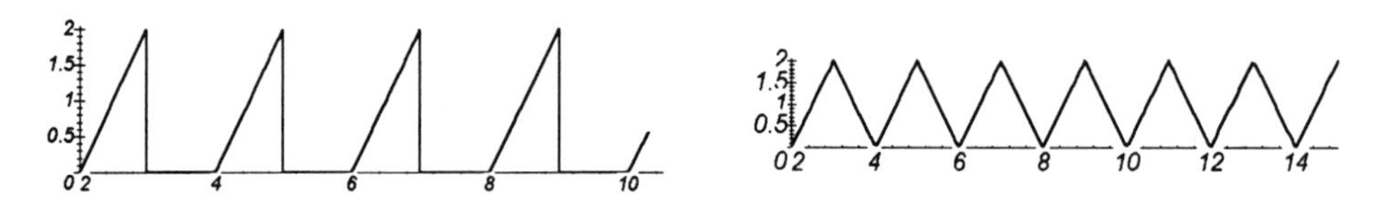

Figs. 2.10–2.11. Saw and triangular waves

Note that the graph $y = \{x\} =$ `x-floor(x)` (the *fractional part*), Fig. 2.8, is similar in form to the graphs $y = \operatorname{arccot}(\cot(x))$ and $y = \arctan(\tan(x))$.

Exercise 2.1.1 Plot the graphs of functions containing operations { }, [ ]:
(1) $y = f([x])$, (2) $y = [f(x)]$, (3) $y = \{f(x)\}$, (4) $y = f(\{x\})$,
where $f = x^2,\ \sqrt{x},\ a^x,\ \cos x$ (see Fig. 2.9).

Let us define the *saw-shaped function* using recursion.

```
> P:=proc(x) option remember; if type(x, numeric) then
  if x<0 then P(x+2) elif x<1 then 2*x elif x<2 then 0
  else P(x-2) fi: fi end:
> plot(P, -2..15, scaling=constrained);   # Fig. 2.10
```

We obtain *triangular waves* (Fig. 2.11) from the previous function by reforming its fragment: "`x<2 then 4-2*x`" or using one of the following formulas: $y = \arccos(\cos(x))$, $y = \arcsin(\sin(x))$.

"Cut out" the function $f(x)$ ($x \in \mathbb{R}$) on the given segment $[a, b]$ and plot its graph using any of the following methods:

```
> f:=x->x^2: # enter your function
> f_ab:=x-> piecewise(x<a, 0, x<=b, f(x), 0);      # first method
> f_ab:=x->f(x)*(H(x-a)-H(x-b));                  # second method
> plot(subs({a=1, b=3}, f_ab(x)), x=0..5);          # Fig. 2.14
```

2.2 Graphs of Piecewise-Differentiable Functions

Definition 2.2.1 A continuous function $f(x)$ given on the interval $I = (a, b)$ is called *piecewise-differentiable* if the interval can be broken into a finite number of segments on each of which $f(x)$ belongs to the class C^1.

In other words, $f'(x)$ is piecewise-continuous.

2.2.1 *The Functions* `max` *and* `min`

The functions *maximum* and *minimum* of some differentiable functions are piecewise-differentiable and can be converted into the type `piecewise`. Consider the following example (Fig. 2.12).

```
> g:=min(x^2,x+3): convert(g,piecewise); plot(g,x=-2..5);
```

$$\begin{cases} x+3, & x \leq (1-\sqrt{13})/2, \\ x^2, & x < (1+\sqrt{13})/2, \\ x+3, & (1+\sqrt{13})/2 \leq x. \end{cases}$$

The graph of max (or min) of several linear functions is a convex (or concave) polygon.

```
> n:=9: k:=seq(rand(-5..5)(),i=1..n);
  b:=seq(rand(1..30)(),i=1..n);
```

$k := -1, -3, 5, -3, 1, 3, 5, -4, 4 \quad b := 16, 14, 10, 28, 8, 30, 25, 19, 16$

```
> fmax:=max(seq(k[i]*x+b[i], i=1..n)):
  fmin:=min(seq(k[i]*x+b[i], i=1..n)):
> Pmax:=plot(fmax(x), x=-10..10, thickness=2):
  Pmin:=plot(fmin(x), x=-10..10, thickness=2):
  PP:=plot([seq(k[i]*x+b[i],i=1..n)], x=-10..10):
> plots[display]([Pmax, Pmin, PP], axes=framed); # Fig. 2.15
```

2.2.2 *Functions Containing the Operation* abs

1. We recommend that the reader consider some examples and formulate a plan for plotting the graphs of functions whose analytical expression contains the operation abs: $y = f(|x|), \quad y = |f(x)|, \quad y = |f(|x|)|$.

2. For a given increasing sequence of points $\{a_i\}$ and weights $\{b_i\}$, where $1 \leq i \leq n$, find the minimum of the function $f_n(x) = \sum_{i=1}^{n} b_n\,|x - a_i|$.

```
> a:=i -> 2+sqrt(i): b:=i -> 1: n:=6: # enter your data
> f:=x -> sum(b(i)*abs(x-a(i)),i=1..n); f(x);
> plot(f,a(1)-1..a(n)+1);
```

An example is the *problem of machine tools* with weight coefficients: Place the machine tool at the point x, such that the sum of the distances from x to the machines tools at the given points -3, -1, 2, and 6 with given weights 2, 1, 3, 1 is minimal.

```
> f:=x -> 2*abs(x-(-3))+abs(x-(-1))+3*abs(x-2)+abs(x-6);
> f1:=convert(f(x), piecewise); plot(f1, x=-5..6); # Fig. 2.13
```

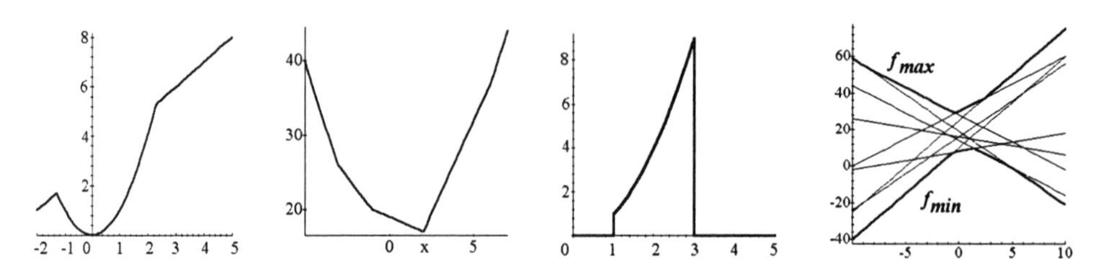

Figs. 2.12–2.15. Graphs of piecewise-differentiable functions

For unit weights $b_i = 1$, explain the following rule for calculating the *optimum* point x (where the function $f_n(x)$ takes its minimum).
Hints. For odd n write down the numbers $a(1), \ldots, a(n)$ in increasing order, then take the optimum point x equal to the middle number of this sequence. What does the graph of $f_n(x)$ look like, and what is the optimum point x for even n?

3. The primitive of a function of the class C^k, where $k \geq 0$, belongs to the class C^{k+1}. Direct symbolic integration of piecewise-differentiable functions in MAPLE V, Release V does not always give the best result. We recommend converting such functions to the `piecewise` type before integrating them.

For example, if we integrate the function $|2 - |x||$ by the standard method, we obtain a function that is discontinuous at the points -2 and 2 (check this). Converting this function first to the `piecewise` type allows us to find a continuous primitive function.

```
> int(abs(2-abs(x)),x); plot(int(abs(2-abs(x)),x),x=-5..5);
```

$$\frac{1}{2}\,\frac{|2-|x||\cdot|x(-4+|x|)}{-2+|x|}$$

```
> f:=convert(abs(2-abs(x)), piecewise); g:=int(f, x);
  plot(g,x=-5..5);
```

$$f := \begin{cases} -2-x, & x \leq -2, \\ 2+x, & x \leq 0, \\ 2-x, & x \leq 2, \\ x-2, & 2 < x, \end{cases} \qquad g := \begin{cases} -2x - \frac{1}{2}x^2, & x \leq -2, \\ 2x + 4 + \frac{1}{2}x^2, & x \leq 0, \\ 2x + 4 - \frac{1}{2}x^2, & x \leq 2, \\ -2x + 8 + \frac{1}{2}x^2, & 2 < x. \end{cases}$$

2.2.3 Piecewise-Differentiable Statistical Functions

Using the above examples, plot graphs of piecewise-differentiable statistical functions with continuous density.

The *uniform distribution* $f(x) = \begin{cases} \frac{1}{b-a}, & x \in [a,b], \\ 0, & x \notin [a,b] \end{cases}$ on the segment $[a, b]$ has the continuous distribution $F(x)$. Let us plot its graphs.

```
> f:=x -> piecewise(x<=a, 0, x<=b, 1/(b-a), 0);
> plot(subs({a=1, b=3}, f(x)), x=-1..4);                # Fig. 2.16
> F:=int(subs({a=1,b=3}, f(x)), x); plot(F, x=-1..4); #Fig. 2.17
```

The *Simpson distribution* (triangle) on the segment $[a, b]$.

$$f(x) = \begin{cases} \frac{2}{b-a} - \frac{2}{(b-a)^2}|a+b-2x|, & x \in [a,b], \\ 0, & x \notin [a,b]. \end{cases}$$

```
> f:= x -> piecewise(
  x<a, 0, x<=b, 2/(b-a)-2/(b-a)^2*abs(a+b-2*x), x>b, 0):
> plot(subs({a=1, b=3}, f(x)), x=-2..4);                # Fig. 2.18
> F:=int(subs({a=1,b=3}, f(x)), x); plot(F, x=-1..4);# Fig. 2.19
```

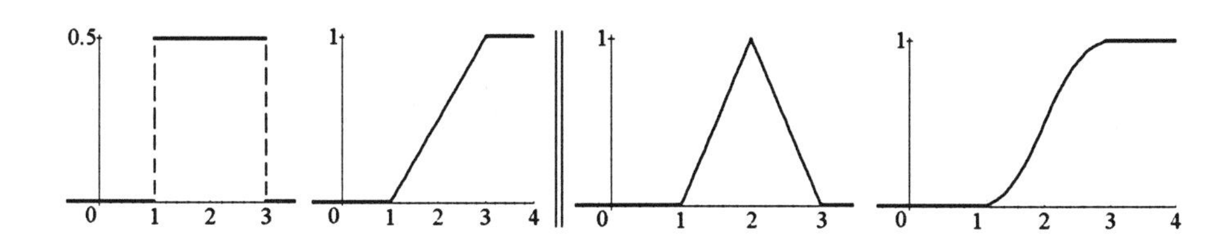

Figs. 2.16–2.19. Uniform and Simpson distributions

The χ^2 *distribution* $f(x) = \begin{cases} \frac{1}{2^{a/2}\Gamma(\frac{a}{2})} x^{\frac{a}{2}-1} \exp(-\frac{x}{2}), & x > 0, \\ 0, & x \le 0 \end{cases}$ with a degrees of freedom; Fig. 2.20.

```
> f:=x->piecewise(
  x<=0, 0, 1/(2^(a/2)*GAMMA(a/2))*x^(a/2-1)*exp(-x/2));
> plot([seq(subs(a=i, f(x)), i=2..5)], x=-1..10, y=0..0.5);
```

The *exponential distribution* $f(x) = \begin{cases} \lambda \cdot \exp(-\lambda x), & x \ge 0, \\ 0, & x < 0 \end{cases}$ with parameter $\lambda > 0$; Fig. 2.21.

```
> f(x):=x->piecewise(x<0, 0, lambda*exp(-lambda*x));
> plot([seq(subs(lambda=i/4,f(x)),i=1..4)],-1..4, y=0..1);
```

The *F (Fisher) distribution* with m_1, m_2 degrees of freedom

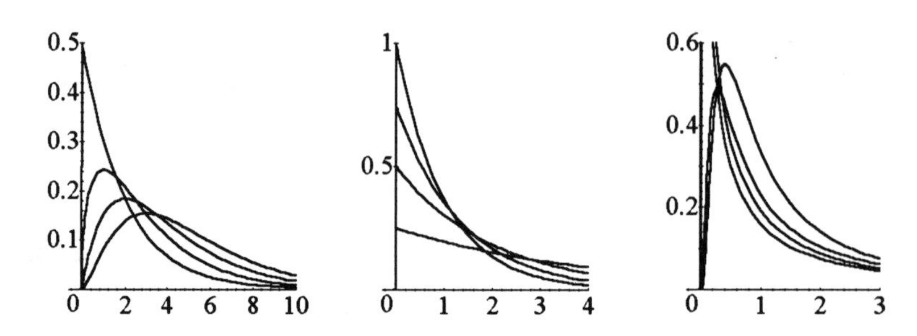

Figs. 2.20–2.22. Graphs of χ^2, exponential, and Fisher distributions

$$f(x) = \begin{cases} \dfrac{\Gamma(\frac{m_1+m_2}{2}) m_1^{\frac{m_1}{2}} m_2^{\frac{m_2}{2}} x^{\frac{m_1}{2}-1}}{\Gamma(\frac{m_1}{2})\Gamma(\frac{m_2}{2})} (m_2 + m_1 x)^{-\frac{m_1+m_2}{2}}, & x > 0, \\ 0, & x \leq 0. \end{cases}$$

```
> f(x) := x -> piecewise(x<=0, 0,
  (GAMMA((m1+m2)/2)*m1^(m1/2)*m2^(m2/2)*x^(m1/2-1))/
  (GAMMA(m1/2)*GAMMA(m2/2))*(m2+m1*x)^(-(m1+m2)/2));
> plot([seq(seq(subs(m1=10^i,m2=j,f(x)),i=0..1),j=1..2)],
  -1..3, y=0..0.6);  # Fig. 2.22
```

2.2.4 *Recursively Defined Functions*

Plot the graphs of *recursively defined functions*, using recursive procedures.

1. $f(x+1) = 2f(x)$ and $f(x) = x(1-x) \quad (0 \leq x \leq 1)$.

```
> f1:=proc(x) option remember; if type(x, numeric) then
  if x<0 then f1(x+1)/2 elif x<1 then x*(x-1)
  else 2*f1(x-1) fi: fi end:
> plot(f1, -2..3, scaling=constrained);  # Fig. 2.23
```

2. $f(x+\pi) = f(x) + \sin(x)$ and $f(x) = 0 \quad (0 \leq x \leq \pi)$.

```
> f2:=proc(x) option remember; if type(x, numeric) then
  if x<0 then f2(x+Pi)+sin(x+Pi) elif x<Pi then 0
  else f2(x-Pi)+sin(x-Pi) fi: fi end:
> plot(f2, -2*Pi..4*Pi, scaling=constrained);  # Fig. 2.24
```

2.2.5 *Functions That Are Defined Using* `limit`

We can approach plotting graphs of functions defined in terms of limits using the command `animate`.

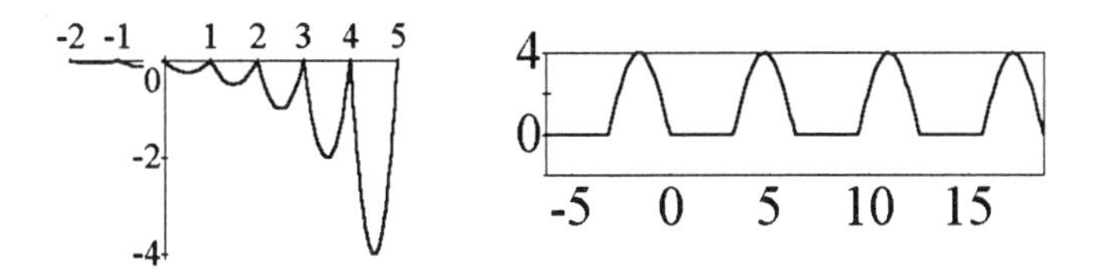

Figs. 2.23–2.24. Graphs of recursively defined functions

1. $y = \lim_{n\to\infty} \frac{x^n}{1+x^n} \quad (x \geq 0).$

```
> animate(x^n/(1+x^n), x=0..3, n=1..50, thickness=2);
```

Then we plot graphs of y_n together with graphs of their limits.

2. $y = \lim_{n\to\infty} [(x-1)\arctan x^n] \quad (x > -1).$

```
> f:=(x-1)*arctan(x^n):
> p:=plot(limit(f, n=infinity), x=0..3, thickness=1):
> q:=animate(f, x=0..3, n=1..50, thickness=2):
> display([p, q], axes=framed);  # Fig. 2.25
```

3. $y = \lim_{n\to\infty} \frac{|x|^n-1}{|x|^n+1} \quad (x \neq 0),$

```
> f:=(abs(x)^n-1)/(abs(x)^n+1):
> p:=plot(limit(f, n=infinity), x=-3..3, thickness=1):
> q:=animate(f, x=-3..3, n=1..50, thickness=2):
> display([p, q], axes=framed);  # Fig. 2.26
```

4. $y = \lim_{n\to\infty} [x \arctan(n \cot(x))],$

```
> f:=x*arctan(n*cot(x)):
> p:=plot(limit(f, n=infinity), x=-3*Pi/2..3*Pi/2):
> q:=animate(f, x=-3*Pi/2..3*Pi/2, n=1..50, thickness=2):
> display([p, q], axes=framed);  # Fig. 2.27
```

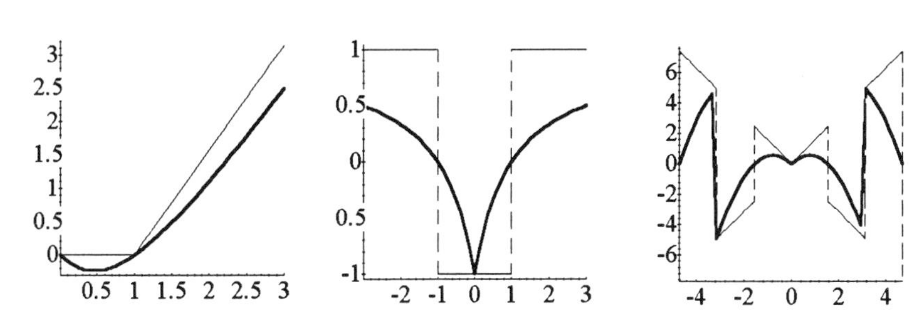

Figs. 2.25–2.27. Animation (1 still) of limits of functions 2–4 above

Exercise 2.2.1 Plot graphs of some other functions:

5. $y = \lim_{n\to\infty} (1-x)^{2n} \quad (|x| \le 1)$

6. $y = \lim_{n\to\infty} \sqrt[n]{1+e^{n(x+1)}}$

7. $y = \lim_{n\to\infty} \frac{\log(2^n+x^n)}{n} \quad (x \ge 0)$

8. $y = \lim_{n\to\infty} \frac{x+x^2 e^{nx}}{1+e^{nx}}$

9. $y = \lim_{n\to\infty} \frac{x+e^{nx}}{1+e^{nx}}$

10. $y = \lim_{n\to\infty} \frac{x^{2n}\sin(\pi x/2)+x^2}{1+x^{2n}}$.

3

Interpolation of Functions

In mathematics and its applications one often needs to find a continuous function at a finite (or countable) set of points taken from its domain of definition.

The problem of *interpolation* is to find a continuous function $y = f(x)$ ($x \in I$), such that $f(x_i) = y_i$ for all $i = 0, 1, \ldots, n$. Below, all real $x_0, x_1, \ldots, x_n$ (*nodes of a net*) are assumed different and presented in increasing order. The choice of function depends on additional boundary conditions. For example, after connecting the successive points (x_i, y_i), one obtains a piecewise-linear function.

In Sections 3.1 and 3.2 we study some MAPLE commands that carry out *polynomial* and *spline interpolation*. In Section 3.3 a concrete application of spline functions for generating plane and space curves (in particular knots) is given. This theme is developed in Chapter 14.

In Chapter 3 the reader will become acquainted with the commands

`delta, interp, spline, tubeplot.`

3.1 Polynomial Interpolation of Functions

Given vectors xx=$[x_1, x_2, \ldots, x_{n+1}]$ and yy=$[y_1, y_2, \ldots, y_{n+1}]$, the *interpolation polynomial of Lagrange* $L_n(x)$ in the variable x is given by the formula

$$L_n(x) = \sum_{i=1}^{n+1} y_i \frac{\varphi_n(x)}{(x-x_i)\,\varphi_n'(x_i)},$$

where $\varphi_n(x) = \prod_{i=1}^{n+1}(x - x_i)$. This is calculated using the command `interp( xx, yy, x)`. The *Lagrange base* (coefficients of $L_n(x)$ in front of y_i) is convenient when one needs to calculate a large number of interpolation poly-

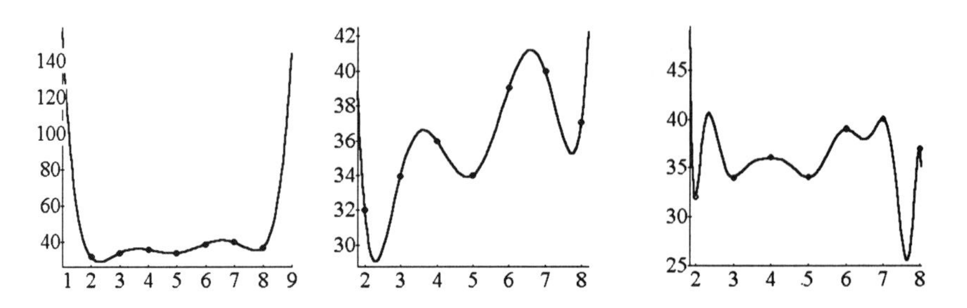

Figs. 3.1–3.3. Lagrange and Hermite interpolation polynomials

nomials for different initially given $\{y_i\}$ at the same nodes $\{x_i\}$ of the net. In this case one needs to calculate the base only once.

Example 3.1.1 First we plot $L_6(x)$ over the range `x=1..9` and look at its strange behavior outside the segment containing the given nodes, Fig. 3.1. We can conclude from this that interpolation polynomials cannot be used for the *extrapolation* of functions.

```
> xx:=[2,3,4,5,6,7,8]: yy:=[32,34,36,34,39,40,37]:
> f:=interp(xx, yy, x); INTERPOL:=plot(f, x=1..9):
```

$$f := \frac{7}{72}x^6 - \frac{44}{15}x^5 + \frac{1277}{36}x^4 - \frac{2629}{12}x^3 + \frac{52231}{72}x^2 - \frac{72839}{60}x + 831$$

```
> points:=[seq([xx[j], yy[j]], j=1..nops(xx))]:
> Points:=plot(points, style=point, symbol=circle):
> plots[display]([Points, INTERPOL]);     # Fig. 3.1
```

Then we plot the graph of $L_6(x)$ on the segment `x=1.8..8.2` (Fig. 3.2) and look at its oscillations. This tells us something about weak *approximate properties* of interpolation polynomials.

Despite the ease of deriving the Lagrange interpolation polynomial, its tendency for unlimited growth of the net requires careful attention. It is known (K. Weierstrass, 1885) that every continuous function defined on a segment can be approximated as closely as desired on this segment by a polynomial. In particular, *for given n and $f(x)$, let* $B(f, n) = \sum_{i=0}^{n} C_n^i x^i (1-x)^{n-i} f(\frac{i}{n})$. *Then the sequence of polynomials* $\{B(f, n)\}$, $n \in N$ *converges uniformly to* $f(x)$ *on* $[0, 1]$.

In approximating a continuous function by $L_n(x)$, the interpolation polynomial of Lagrange, its graph is not necessarily close to the graph of the function $f(x)$ at each point of the segment $[a, b]$; it can vary from this function by an arbitrarily large amount.

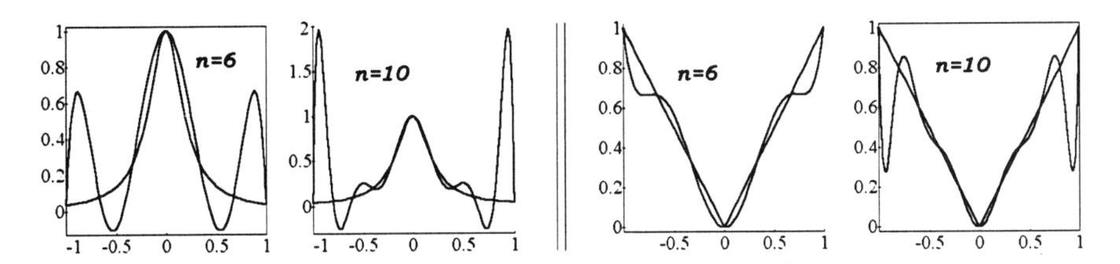

Figs. 3.4–3.7. Examples by Runge ($n = 6, 10$) and Bernstein ($n = 6, 10$)

Example 3.1.2 [K. Runge, 1901] Check experimentally that for an unbounded increase in the number of nodes on the segment $[a, b]$ for the function $f(x) = \frac{1}{1+25x^2}$, we have $\lim\limits_{n\to\infty} \max\limits_{-1\le x\le 1} |f(x) - L_n(x)| = \infty$.

```
> f:=x->1/(1+25*x^2): x:=i-> -1+2*i/n: y:=i->f(x(i)):
> n:=6: g:=interp([seq(x(i),i=0..n)], [seq(y(i),i=0..n)],x);
```

$$g := -\frac{1265625}{96356}x^6 + \frac{2019375}{96356}x^4 - \frac{211600}{24089}x^2 + 1$$

```
> plot([f(x), g], x=-1..1);        # Figs. 3.4–3.5
```

Note that for an analytic function $f(x)$ on the segment $[a, b]$, the equality $\lim\limits_{n\to\infty} \max\limits_{a\le x\le b} |f(x) - L_n(x)| = 0$ holds.

Example 3.1.3 [S. Bernstein, 1912] Check using calculations that the sequence of Lagrange interpolation polynomials $L_n(x)$ on uniform nets for the continuous function $f(x) = |x|$ on the segment $[-1, 1]$, with an increasing number of nodes n, does not converge to the function $f(x)$. For plotting using the previous program (Figs. 3.6–3.7), we are replacing `f:=x->abs(x):`

The generalization of Lagrange interpolation polynomials are *Hermite interpolation polynomials.* These arise when the derivatives at the points in addition to the values of the function are known. Hence, the number of conditions is doubled, and the interpolation polynomial has degree $2n + 1$. Given vectors `xx=`$[x_1, x_2, \ldots, x_{n+1}]$, `yy=`$[y_1, y_2, \ldots, y_{n+1}]$ and derivatives $dy = [y'_1, y'_2, \ldots, y'_{n+1}]$, the *Hermite interpolation polynomial* $\mathrm{Herm}_n(x)$ in the variable x is defined by the formula

$$\mathrm{Herm}_n(x) = \sum_{i=1}^{n+1} y_i[1 - 2(x - x_i)\tfrac{d}{dx}\bar{L}_i(x_i)]\bar{L}_i(x)^2 + \textstyle\sum_{i=1}^{n+1} y'_i(x - x_i)\bar{L}_i(x)^2,$$

where $\bar{L}_i(x)$ is the Lagrange polynomial with conditions $\bar{L}_i(x_j) = \delta_{i,j}$.

To find the Hermite polynomial use the command `interp(xx, yy, x)`. In the program the derivatives at nodes of the net are for simplicity assumed to be zero, i.e., tangent lines to the graph at the nodes are horizontal (Fig. 3.3).

```
> xx:=[2,3,4,5,6,7,8]: n:=nops(xx)-1:
> yy:=[32,34,36,34,39,40,37]: zz:=[0,0,0,0,0,0,0]
  POINTS:=plot([seq([xx[j], yy[j]], j=1..nops(xx))],style=
  point,symbol=circle): delta:=(i,j)->if i=j then 1 else 0 fi:
> for k from 1 to n+1 do
  L[k]:=unapply(interp(xx,[seq(delta(k,j),j=1..n+1)],x),x);
  Lx[k]:=unapply(diff(L[k](x),x),x) od:
> HERM:=sum(yy[i]*(1-2*(x-xx[i])*Lx[i](xx[i]))*L[i](x)^2,
  i=1..n+1)+ sum(zz[i]*(x-xx[i])*L[i](x)^2, i=1..n+1):
> HERMITE:=plot(HERM, x=1.9..8.1):
> plots[display]([Points, HERMITE]);   # Fig. 3.3
```

Conclusions: Polynomial interpolation works succesfully only for a small number of points (not more than 15) because if the number of points is increased, then the degree of the polynomial increases too, and large oscillations take place on intervals between the given points.

3.2 Spline Interpolation of Functions

If the requirement for the smoothness of an interpolation function is weakened, then without greatly complicating the calculations, one can obtain nice approximate properties of the spline. The idea is to derive polynomials independent of one another (or other standard functions) on each segment $[x_i, x_{i+1}]$ and then glue them at their endpoints.

Before the first work on *splines* (Shenberg, 1946) there was the earlier work of Leibniz and Euler using the calculus of variations (for example, the Euler *polygon* for the numerical solution of ODEs).

The notion of a *spline* came from practice. The word means a "flexible ruling" used in drawing smooth curves. Modern computers allow automatic processing in designing the shape of manufactured articles. Today, splines are used in various graphical packages and systems for design in industry, for example in `AutoCad`.

A polynomial *spline function* in the variable `var`, defined by the vectors `xx=`$[x_1, \ldots, x_{n+1}]$ and `yy=`$[y_1, \ldots, y_{n+1}]$, is derived using the command `spline(xx, yy, var, d)`. Here the parameter d defines the order of the spline function, which can be `linear` (in which case we obtain a polygon through the given points), `quadratic`, `cubic` (by assumption), and `quartic`. Before using the command we load it from the library by using the command `readlib(spline)`. The result is the function $f(x) \in C^2$ of the type `piecewise` composed of n polynomials P_i of degree d:

$$S(x) = \begin{cases} P_1(x), & \text{if} \quad x \le x_2, \\ P_i(x), & \text{if} \quad x_i \le x \le x_{i+1} \text{ and } i = 2, 3, \ldots, n-1, \\ P_n(x), & \text{if} \quad x_n \le x. \end{cases}$$

For example:

```
> readlib(spline): cubic1:=spline([0,1,2,3], [0,1,4,3],x);
```

$$\text{cubic}_1 := \begin{cases} \frac{1}{5}x + \frac{4}{5}x^3 & \text{if} \quad x < 1, \\ \frac{14}{5} - \frac{41}{5}x + \frac{42}{5}x^2 - 2x^3 & \text{if} \quad x < 2, \\ -\frac{114}{5} + \frac{151}{5}x - \frac{54}{5}x^2 + \frac{6}{5}x^3 & \text{otherwise.} \end{cases}$$

In case of the cubic spline ($d = 3$) we derive $4n$ real coefficients of n polynomials $P_i(x) = a_{i,0} + a_{i,1}x + a_{2,0}x^2 + a_{i,3}x^3$. The equalities $S(x_i) = y_i$ for $1 \le i \le n+1$ give us $n+1$ equations. The condition $S(x) \in C^2$ means that the function $S(x)$ and its derivatives $S'(x)$ and $S''(x)$ at all $n-1$ inner nodes of the net are continuous. This gives additionally $3(n-1)$ equations for finding all the coefficients. Finally, we obtain $3(n-1) + (n+1) = 4n-2$ equations for the coefficients $a_{i,j}$. Two missing conditions come from the restrictions on values of the spline and/or its derivatives at the endpoints of the segment $[a, b]$. For example, $S'(a) = S'(b) = 0$ (*first kind*) or $S''(a) = S''(b) = 0$ (*second kind*) or $S'(a) = S'(b)$, $S''(a) = S''(b)$ (*periodic*).

In contrast to Lagrange interpolation polynomials, the sequence of interpolating cubic splines on a uniform net always converges to a continuous interpolation function. Moreover, the rate of convergence is increased when the differential properties of the function are reinforced.

Example 3.2.1

1. Plot the "triangular waves" using a linear spline and then smooth them by a cubic spline.

```
> v:=spline([seq(i,i=0..9)],[seq((-1)^i,i=0..9)],x,linear):
> c:=spline([seq(i, i=0..9)], [seq((-1)^i, i=0..9)], x):
> plot([v, c], x=0..9);  # Fig. 3.10
```

2. Check that for the function $f(x) = \frac{1}{1+25x^2}$, the cubic spline with $n = 6$ nodes contains an error of approximation of the same order as that of the interpolation polynomial $L_5(x)$, and on the net with $n = 21$ nodes, this error just cannot be seen on the scale of usual figures in a book (in this case $L_{20}(x)$ gives an error of about 10000 %).

3. Compare the behavior of Lagrange interpolation polynomials and polynomial splines derived by the measuring air temperature over 10 days given in Section 1.1.

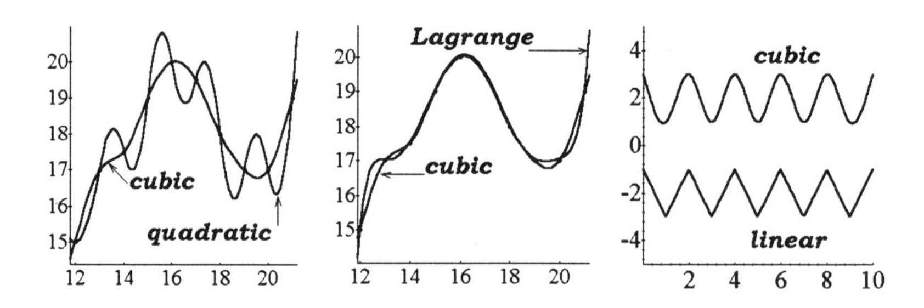

Figs. 3.8–3.10. Interpolating data t^0 *over* 10 *days*

```
> Days:=[12,13,14,15,16,17,18,19,20,21]:
> TT:=[15,17,17.5,19,20,19.5,18,17,17,19]:
> f2:=spline(Days,TT,x,quadratic): f3:=spline(Days,TT,x):
> h:=unapply(interp(Days, TT, x), x):
> a:=plot([f2, f3], x=11.8..21.2): b:=plot(h, 11.8..21.2):
> c:=plot(seq[Days[i], TT[i]],i=1..10) style=point, symbol=diamond:
> plots[display]([a, b, c]); # Figs. 3.8–3.9
```

Remark 3.2.1 Cubic spline interpolation is the unique solution to the following extremal problem. *Derive the function* $f(x)$ $(x \in [a, b])$ *minimizing the functional* $J(f) = \int_a^b (f''(x))^2\,dx$ *among all functions of the class* C^2 *whose graphs contain the given points* $\{(x_i, y_i)\}$ $(i = 1, 2, \ldots, n+1)$. Moreover, the following additional conditions are satisfied: $f''(a) = f''(b) = 0$. The choice of the functional $J(f)$ is explained by the fact that, according to the Bernoulli–Euler physical law, the linearized equation of bended ruling has the form Ei $f''(x) = -M(x)$, where $f(x)$ is a bend function, $M(x)$ varies linearly from bearing to bearing bending moment, and Ei is the rigidity of ruling.

3.3 Constructing Curves Using Spline Functions

We considered above arrays of points whose abscissas form an increasing sequence of real numbers. However, the above method allows us to obtain interpolation or smooth curves for arbitrary given points in the plane or in space. The class of modeled curves is essentially enlarged: now they are defined by parametric equations $x = x(t)$, $y = y(t)$, $z = z(t)$. The plan for plotting the spline curve is the following.

1. *For an arbitrary given segment* $[a, b]$ *of parameter* t*, one defines an auxiliary net* $a = t_1 < \cdots < t_{n+1} = b$ *whose number of nodes coincides with the number of given control points.*

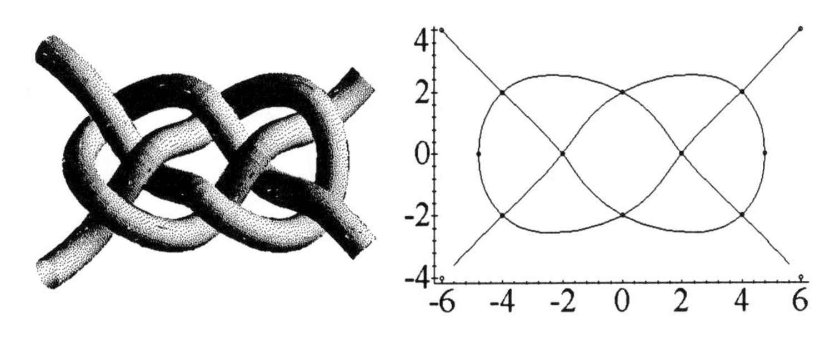

Figs. 3.11–3.12. Well-known knot to sailors

2. *Given an array* $\mathbf{P} = \{P_i(x_i, y_i, z_i),\ i = 1, \ldots, n+1\}$ *of control points, one defines three auxiliary arrays* $\mathbf{X} = \{(t_i, x_i),\ i = 1, \ldots, n+1\}$, $\mathbf{Y} = \{(t_i, y_i),\ i = 1, \ldots, n+1\}$, $\mathbf{Z} = \{(t_i, z_i),\ i = 1, \ldots, n+1\}$.

3. *For each array* **X**, **Y**, *and* **Z** *one derives the corresponding spline function* $x(t)$, $y(t)$, *and* $z(t)$, *where* $t \in [a, b]$.

Remark 3.3.1 Such a parametrized curve may not be C^1-regular, because one cannot exclude the possibility of simultaneous vanishing (for some value t_0) of the derivatives: $x'(t_0) = y'(t_0) = z'(t_0) = 0$. In Chapter 14 we study a more powerful alternative for plotting spline curves.

Example 3.3.1 As a concrete application of spline functions, try to tie the following *knot* (Fig. 3.11), well known to sailors. (For producing "thick" curves we use the command `tubeplot`, studied in Section 8.1.1).

```
> tt:=[0,1/3,2/3,1,4/3,3/2,5/3,2,7/3,8/3,3.1]:
  xx:=[-6,-4,-2,0,4,4.8,4,0,-2,-4,-6]:
  yy:=[4,2,0,-2,-2,0,2,2,0,-2,-4]:
  zz:=[1,1,-1,1,-1,0,1,-1,1,-1,-1]:
> readlib(spline): X:=spline(tt, xx, t):
  Y:=spline(tt, yy, t): Z:=spline(tt, zz, t):
> plots[tubeplot]({[X(t),Y(t),Z(t)], [-X(t),Y(t),-Z(t)]},
  t=0..3, radius=.6, tubepoints=30, numpoints=190,
  orientation=[90, 26], color=[.8, .6, .4],
  projection=.5, style=patchnogrid, scaling=constrained,
  ambientlight=[.6, .6, .5], light=[75, 50, 1, .9, .7]);
```

To understand the key idea, we plot the figure in the plane.

```
> XY:=plot([[X(t),Y(t),t=0..3], [-X(t),Y(t),t=0..3]],
  color=[red, blue]):
> Points:=plot([seq([op(i,xx),op(i,yy)],i=1..nops(xx)),
```

```
  seq([-op(i,xx), op(i,yy)], i=1..nops(xx))], style=point):
> plots[display](XY, Points, scaling=constrained); # Fig. 3.12
```

4
Approximation of Functions

Assuming that the values $\{y_i = f(x_i),\ i = 1, \ldots, n+1\}$ are given up to a certain error, it is better to use *smoothing polynomials*, or *spline functions*. In this way the graph of the derived function $y = f(x)$ lies near the given points (x_i, y_i).

In Section 4.1 we briefly consider the classical method of least squares. Then in Sections 4.2 and 4.3, we construct Bezier curves (graphs). One application of rational Bezier curves is in plotting all conic sections without using trigonometric functions.

In Chapter 4 the reader will become acquainted with the commands

`evalm, expand, leastsquare.`

4.1 Method of Least Squares

The *method of least squares* is often used for smoothing purposes. In MAPLE it is called by the command `leastsquare` from the subpackage `stats[fit, ... ]`. Solving smoothness problems by the least squares method consists in finding the polynomial $P(x)$ of degree not more than n that minimizes the value $\sum_{i=1}^{n+1}(P(x_i) - y_i)^2$.

Example 4.1.1 Consider the data from the Example 3.1.1.

```
> xx:=[2,3,4,5,6,7,8]: yy:=[32,34,36,34,39,40,37]:
> Points:=plot([seq([xx[j],yy[j]],j=1..nops(xx))],
  style=point, symbol=circle):
> with(stats): P1:=op(2,fit[leastsquare[[x,y]]]([xx,yy]));
```

$P_1 := \frac{429}{14} + \frac{15}{14}x$

```
> P2:=op(2,fit[leastsquare[[x,y],y=a2*x^>2+a1*x+a0]]([xx,yy]));
```

$P_2 := -\frac{4}{21}x^2 + \frac{125}{42}x + \frac{373}{14}$

```
> P3:=op(2,fit[leastsquare[[x,y],y=a3*x^3+a2*x^2+a1*x+a0]]
([xx, yy]));
```

$P_3 := -\frac{1}{9}x^3 + \frac{31}{21}x^2 - \frac{577}{126}x + \frac{513}{14}$

```
> PL:=plot([P1,P2,P3], x=0..10, color=[blue,green,red]):
> plots[display]([Points, PL]); # Fig. 4.1
```

4.2 Bezier Curves

Another alternative for smoothing is provided by Bezier curves.

Definition 4.2.1 The *Bezier curve*, defined by control points $P_1, \ldots, P_{n+1}$, is given by the formula $B(x) = \sum_{i=0}^{n} C_i^n x^i (1-x)^{n-i} P_{i+1}$. The multipliers $C_i^n x^i (1-x)^{n-i}$ at the points in the above equation are called *Bernstein functions*.

A cubic Bezier curve is derived by four control points and the base functions $(1-x)^3$, $3x(1-x)^2$, $3x^2(1-x)$, x^3. Gluing many such segments (see Section 14.2) leads to long Bezier curves. These cubic segments play a role in designing computer fonts.

```
> xx:=[2,3,4,5,6,7,8]: yy:=[32, 34, 36, 34, 39, 40, 37]:
> n:=nops(xx)-1: Points:=plot([seq([xx[j],yy[j]],j=1..n+1)],
  style=point, symbol=circle):
> bez:=x->evalm(sum(binomial(n,i)*x^>i*(1-x)^(n-i)*[xx[i+1],
  yy[i+1]],i=0..n)):PBEZ:=plot([bez(x)[1],bez(x)[2],x=0..1]):
> plots[display]([Points, PBEZ], axes=framed);
```

Note that the function `bez(x)[1]` is linear when the step of its net is constant.

```
> expand(bez(t)[1]);
```

$2 + 6t$

Fig. 4.2 illustrates some standard properties of Bezier curves. For instance, these curves always lie in the convex hulls of their control points since base Bernstein functions are nonnegative with sum equal to 1.

```
> simplify(sum(binomial(n,i)*x^i*(1-x)^(n-i), i=0..n));
```

1

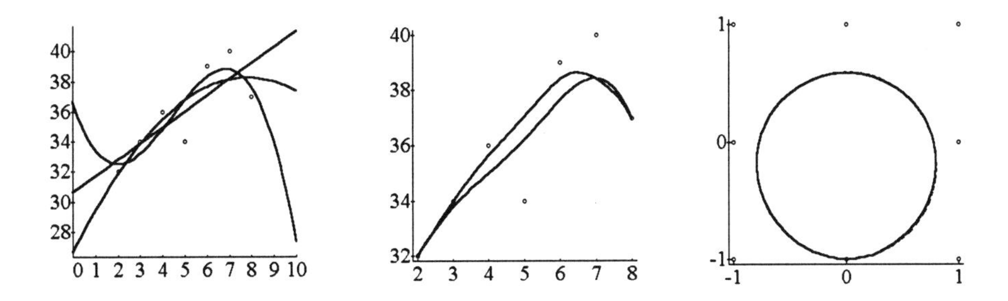

Figs. 4.1–4.3. Least squares method. Bezier curves

A Bezier curve goes through initial and end control points. Its nice approximate properties are explained by the Weierstrass theorem. Note that the addition of a point increases the degree of a Bezier curve. Another defect is that the whole Bezier curve changes when one of the control points moves.

4.3 Rational Bezier Curves

We can extend this approach to use quotients of polynomials.

Definition 4.3.1 The *rational Bezier curve* associated with control points $P_1, \dots, P_{n+1}$ and corresponding *weights* $w_1, w_2, \dots, w_{n+1}$ is given by the formula

$$B(x) = \frac{\sum_{i=0}^{n} C_i^n x^i (1-x)^{n-i} w_i \, P_{i+1}}{\sum_{i=0}^{n} C_i^n x^i (1-x)^{n-i} w_{i+1}}.$$

To extend the earlier program for a Bezier curve to the case of a rational curve, change the last four lines as shown here. (In Fig. 4.2, $w_4 = w_6 = 5$, and other weights are equal to 1.)

```
> ...
> w:=[1,1,5,1,5,1,1]: rbez:=x->evalm(sum(binomial(n,i)*
  x^i*(1-x)^(n-i)*w[i+1]*[xx[i+1], yy[i+1]], i=0..n)/
  sum(binomial(n,i)*x^i*(1-x)^(n-i)*w[i+1], i=0..n));
> PRBEZ:=plot([rbez(x)[1], rbez(x)[2], x=0..1]):
> plots[display]([Points, PRBEZ], axes=framed);
```

If all w_i are the same, then we obtain a usual Bezier curve.

One application of rational Bezier curves is in plotting all conic sections (circle, ellipse, hyperbola, parabola) without using trigonometric and hyperbolic functions.

Example 4.3.1 Compare the curve PRBEZ given by 8 control points on the perimeter of the square with center at the origin, and the circle CIRC of radius

$R = 0.8$ with center at $C(0, -0.2)$; Fig. 4.3.

```
> xx:=[0,1,1,1,0,-1,-1,-1,0]: yy:=[-1,-1,0,1,1,1,0,-1,-1]:
> w:=[1,1,1.6,1,1.2,1,1.6,1,1]:
> CIRC:=plot([0.8*cos(t), 0.8*sin(t)-0.2, t=0..2*Pi]):
```

Solution

```
> n:=nops(xx)-1: PNTS:=plot([seq([xx[j],yy[j]],j=1..n+1)],
  style=point, symbol=circle):

> rbez:=x-> evalm(sum(binomial(n,i)*x ^i*(1-x) ^(n-i)
  *w[i+1]*[xx[i+1], yy[i+1], i=0..n)/
  sum(binomial(n,i)*x^i*(1-x)^(n-i)*w[i+1], i=0..n)):

> PRBEZ: = plot([rbez(x)[1], rbez(x)[2], x=0..1]):
> plots[display]([PNTS, PRBEZ, CIRC], axes=framed,
  scaling=constrained);
```

Remark 4.3.1 In generalizing Section 3.2 on cubic splines, one comes to the notion of the smoothing polynomial spline. A *smoothing cubic spline* is a function $f(x)$, $(x \in [a, b])$ of the class C^2, which

- on each segment $[x_i, x_{i+1}]$ is a polynomial of degree 3,
- minimizes the functional $J(f) = \int_a^b (f''(x))^2\,dx + \sum_{i=1}^{n+1} \rho_i (f(x_i) - y_i)^2$, where y_i and $\rho_i > 0$ are given numbers.

Boundary conditions, for example, have the form $f''(a) = f''(b) = 0$. The choice of weight coefficients ρ_i contained in the functional allows one in some sense to control the properties of smoothing splines. If all $\rho_i \to \infty$, then the spline is actually an interpolation spline, and for small values of ρ_i, the spline is close to the straight line obtained by the least squares method.

Some generalizations of these smoothing splines are known for arbitrary degree $d > 3$.

Part II

Curves with MAPLE

5

Plane Curves in Rectangular Coordinates

There are many ways to classify curves. One of them is to think of curves as either *algebraic* or *transcendental*. An algebraic (plane) curve is given by a polynomial equation $P(x, y) = 0$. Its degree $n = \deg P$ is called the *order of the curve*. *Curves of order* $n = 2$ are studied in analytic geometry. The first classification of *curves of order* $n = 3$ was obtained by Newton. The case $n > 3$ is more difficult. But among easily obtained curves, there are many that are nonalgebraic, for example, the cycloid and spiral of Archimedes; we study them using parametrized or implicit equations (Chapter 5) or polar coordinates (see Chapter 6).

The names of curves, like the names of geographical objects, contain much interesting information. Let us group some curves by the meaning of their names:

By the name of a scientist: Dinostratus' quadratrix, conchoid of Nicomedes, Pascal's limaçon, lemniscate of Bernoulli, cissoid of Diocles.

By the method of construction: caustics, equidistant (parallel), pedal, evolutes, evolvents.

By an important property: trisectrices, quadratrices, tractrices.

By an essential property of their shape: astroid (star-shaped), deltoid (Greek letter Δ), cardioid (heart-shaped), nephroid (kidney-shaped).

By historical factors: ellipse, parabola, hyperbola.

By the structure of the formula: (semi)cubical parabola, logarithmic spiral, exponential and logarithmic curves, integral sine (cosine).

In Section 5.1 we consider the basic notion of a regular curve (for analogous definitions for surfaces, see Sections 19.1 and 19.2).

In Sections 5.2–5.4 we investigate cycloidal curves and other remarkable parametrized curves, and we provide tables with interesting figures.

In Section 5.5 we study implicitly given curves (appearing as level curves of functions in two variables); another family of curves is given by trajectories of vector fields (continued in Section 8.4 for trajectories in space).

In Section 5.6 level sets are used in solving problems with conditional extrema. Three approaches are compared: experimental, geometrical and analytical (using Lagrange multipliers).

In Chapter 5 the reader will become acquainted with the commands

```
solve, dsolve, numer, Diff, grad, allvalues;
implicitplot, contourplot, densityplot, fieldplot,
gradplot, odeplot, DEplot,  disk, ellipse, arc, line.
```

5.1 What Is a Curve?

Intuitively, from Euclid, a *curve* is the trajectory of a moving point, a boundary of a surface, a one-dimensional figure. The mathematically correct definition of a curve is based on notions from *topology*, but it starts with the key notion of an *elementary curve*, which can be imagined as an *interval* $I = (a, b)$ (or a *segment* $\bar{I} = [a, b]$) of the straight line after continuous deformation.

Definition 5.1.1 A set γ in $\mathbb{R}^2$ is called a *curve* if it can be covered by a finite or countable number of elementary curves.

One can distinguish *self-intersecting, simple* (i.e., without self-intersections; for example, graphs), *closed, connected* curves.

If we fix rectangular coordinates (with orthonormal base $\{\mathbf{i}, \mathbf{j}\}$) in the plane $\mathbb{R}^2$ with origin O, then the coordinates $x,\ y$ of a point on the elementary curve γ given by the map $\vec{\mathbf{r}}(t) = [x(t),\ y(t)]$ (a *parametrization*) are functions of $t \in I$:

$$x = x(t), \quad y = y(t). \tag{5.1}$$

In other words, the position vector $\vec{\mathbf{r}} = \vec{OP}$ of the point $P(x, y)$ on the curve γ is the following vector-valued function:

$$\vec{\mathbf{r}} = \vec{\mathbf{r}}(t) = x(t)\,\mathbf{i} + y(t)\,\mathbf{j}, \qquad t \in I. \tag{5.1a}$$

We define functions $x(t),\ y(t)$ and then use the following command:

```
> r:=[x(t), y(t)];
```

For example, a straight line in a plane is given by a linear vector-valued function in the variable t: $\vec{\mathbf{r}}(t) = [a_1 t + b_1,\ a_2 t + b_2]$.

Remark 5.1.1 We can derive in MAPLE some calculus type of operations on vector-valued functions $\vec{\mathbf{r}}(t) = [x(t),\ y(t)]$ of the class C^k, $(k \geq 0)$.

The *limit* $\vec{\mathbf{r}}(t_0) = \lim\limits_{t\to t_0} \vec{\mathbf{r}}(t)$,

```
> r0:=[subs({t=t0}, x(t)), subs({t=t0}, y(t))];
```

The *first derivative* $\vec{\mathbf{r}}\,'(t) = \lim\limits_{\Delta t\to 0} \frac{\vec{\mathbf{r}}(t+\Delta t)-\vec{\mathbf{r}}(t)}{\Delta t}$.

```
> rt:=[Diff(x(t),t), Diff(y(t),t)];
```

Analogously, we calculate higher order derivatives $\vec{\mathbf{r}}\,''$, $\vec{\mathbf{r}}\,^{(3)}$, etc. and *Taylor series* (see Section 1.1).

Definition 5.1.2 A point P on a simple curve $\gamma \in \mathbb{R}^2$ is called *regular of the class* C^k (C^∞ or C^ω) if some neighborhood of this point admits a parametrization $\vec{\mathbf{r}}(t) = [x(t),\ y(t)]$, $(t \in I)$, where the functions $x(t)$ and $y(t)$ belong to the class C^k (C^∞ or C^ω) and the vector $\vec{\mathbf{r}}\,'(t)$ is nonzero at P. For the opposite case, a point P on a curve γ is called a *singular point* (see Fig. 1.14 with a tractrix). A curve consisting of regular points is called a *regular curve of the class* C^k (C^∞ or C^ω) (*smooth* when $k = 1$).

Definitions 5.1.1 and 5.1.2 can be generalized for the space curves studied in Chapter 8.

5.2 Plotting Cycloidal Curves

The trajectory of a point on the circle of radius R traveling (without sliding) along another circle of radius R' (or along a straight line) is called a *cycloidal curve* (Greek, $\varkappa\upsilon\varkappa\lambda o\ \varepsilon\iota\delta\eta\varsigma$, circle-shaped). If the circle drives along and inside of motionless circle, then such a curve is called a *hypocycloid*; if outside, then the curve is called an *epicycloid*. If we follow a point not on the border of the wheel but on its spoke or on the spoke's continuation, then we obtain a *curtate* or *prolate trochoid* (Greek, $\tau\rho o\chi o\ \varepsilon\iota\delta\eta\varsigma$, wheel-shaped).

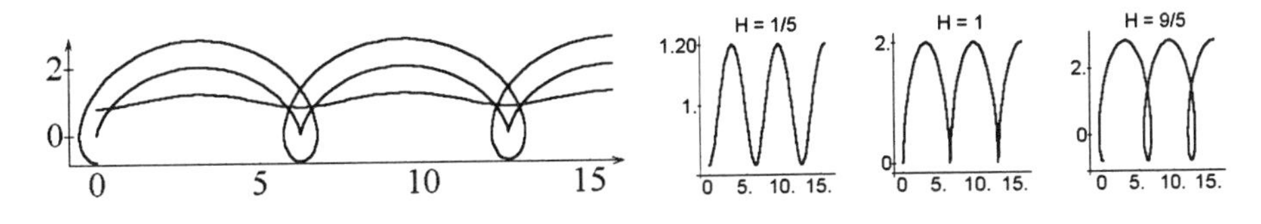

Figs. 5.1–5.2. Common cycloid, curtate and prolate cycloids

Example 5.2.1

1. Using the equations $x = Rt - hR\sin(t)$, $y = R - hR\cos(t)$, let us plot in the interval $t \in [-3\pi, 3\pi]$ the *cycloid* for $h = 1$ and its *trochoids*: *prolate* for $h > 1$ and *curtate* for $0 < h < 1$, Figs. 5.1–5.2.

```
> p:=array(-1..1): for i from -1 to 1 do  h:= 1+i*8/10:
  p[i]:=plot([t-h*sin(t), 1-h*cos(t), t=0..5*Pi],
  title=convert(H=h, string)) od:
> plots[display](p,tickmarks=[3,2],thickness=2);   # Fig. 5.1
> plots[display]([seq(p[i],i=-1..1)],scaling=constrained);
```

2. Parametric equations of *trochoids* (*prolate* for $h > 1$ and *curtate* for $0 < h < 1$), where $m = \frac{R}{R'}$ is the *modulus*, are the following:

$$\begin{cases} x = (R + mR)\cos(mt) - hmR\cos(t + mt), \\ y = (R + mR)\sin(mt) - hmR\sin(t + mt). \end{cases}$$

In the case $h = 1$ we have cycloidal curves: *epicycloids* for $m > 0$ and *hypocycloids* for $m < 0$; see Figs. 5.3–5.4 below.

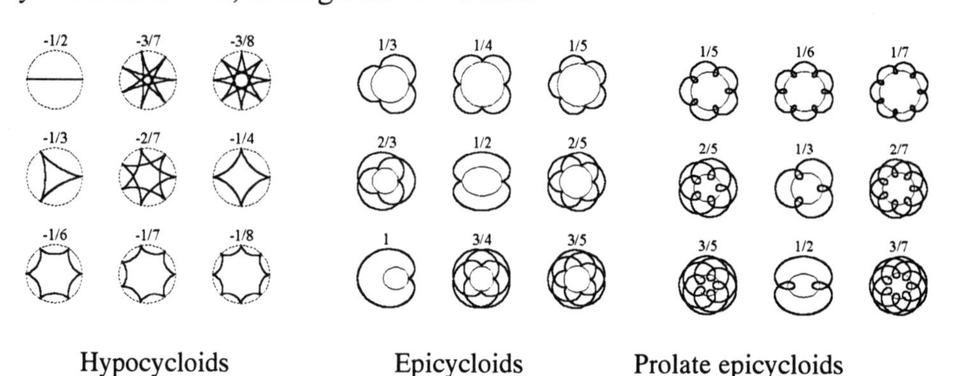

Hypocycloids Epicycloids Prolate epicycloids

Figs. 5.3–5.5. Cycloidal curves

If m is represented in the form of an irreducible fraction $\frac{a}{b}$, then the cycloidal curve is closed with b branches; its period with respect to the parameter t is $2\pi\, b$. If the modulus m is irrational, then the curve consists of an infinite number of branches.

We plot the tables of cycloidal curves (Figures 5.3–5.8) with changing parameters and modulus $m = a/b$.

For an *epicycloid* the program is the following:

```
> p:=array(1..3, 3..5): h:=1:
> for a from 1 to 3 do for b from 3 to 5 do
  p[a,b]:=plot([[(1+a/b)*cos(a/b*t)-h*a/b*cos(t+a/b*t),
```

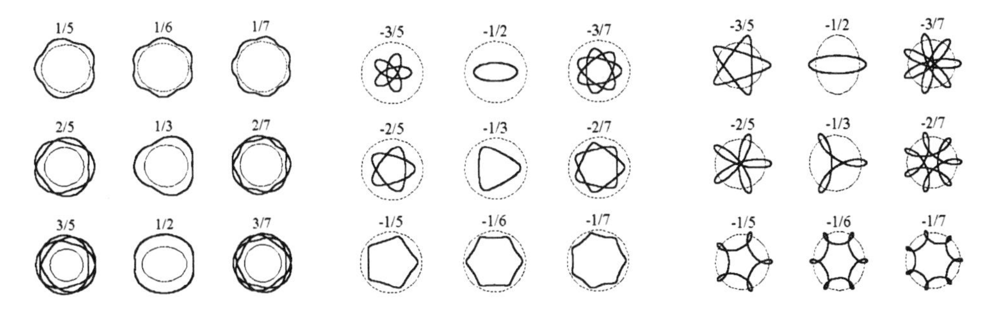

Curtate epicycloids Curtate hypocycloids Prolate hypocycloids

Figs. 5.6-5.8. Cycloidal curves

```
  (1+a/b)*sin(a/b*t)-h*a/b*sin(t+a/b*t), t=0..2*b*Pi],
  [cos(t), sin(t), t=0..2*Pi]], thickness=[2, 1],
  linestyle=[1, 2], title=convert(a/b, string)) od od:
> plots[display](p, axes=none);
```

For modeling *hypocycloids* we put `p:=array(-3..-1, 3..5)` in the above program and use the cycle `for a from -3 to -1 do for b from 6 to 8 do`. For modeling *prolate* and *curtate trochoids* we assume $h := 1.5$ and then $h := 0.5$.

3. Let us study separately the following interesting cycloidal curves.

(a) *Cardioid* (Greek, $\varkappa\alpha\rho\delta\iota\alpha$, heart), the epicycloid with modulua $m = 1$, Figs. 5.4, 9.19, the curve of 4th degree $(x^2+y^2+2Rx)^2 = 4R^2(x^2+y^2)$. Let us substitute $x^2+y^2 = \rho^2$, $x = \rho\cos\varphi$ in the polynomial equation and reduce the fraction by ρ^2 to obtain the equation in polar coordinates $\rho = 2R\,(1 - \cos\varphi)$. Slightly changing the equation of the cardioid, we obtain *Pascal's limaçon*, $\rho = l - 2R\cos\varphi$ with the loop and self-intersection for $l < 2R$, without a loop but with the singular point O for $l > 2R$, Figs. 6.10, 9.19.

(b) The epicycloid with modulus $m = \frac{1}{2}$ (in Fig. 5.4 of epicycloids above: $a = 3$, $b = 6$) is called a *nephroid* (*nephros* means kidney-shaped; Proctor, 1878); this curve has two cuspidal points.

(c) *Astroid* (Greek, $\alpha\sigma\tau\rho o\,\nu$, star; $\varepsilon\iota\delta o\,\varsigma$, view), the hypocycloid with modulus $m = -\frac{1}{4}$. It is the sixth degree curve $(x^2 + y^2 - R^2)^3 + 27R^2x^2y^2 = 0$. Its implicit (second method of plotting) and parametrized equations are

$$x^{\frac{2}{3}} + y^{\frac{2}{3}} = R^{\frac{2}{3}} \iff \vec{\mathbf{r}}(t) = [R\cos^3(\tfrac{t}{4}),\ R\sin^3(\tfrac{t}{4})]. \quad \text{(see Fig. 5.3)}$$

(d) The *deltoid* (*Steiner curve*) is the hypocycloid with $m = -\frac{1}{3}$. It is the fourth degree curve $(x^2 + y^2)^2 + 8Rx(3y^2 - x^2) + 18R^2(x^2 + y^2) - 27R^4 = 0$.

For plotting *cycloidal curves* separately, we fix $h := 1$ and, for example, $R := 1$. We plot (by dotted line) the stationary circle. Then, changing values of the modulus $m := 1, \frac{1}{2}, \frac{1}{3}$, we obtain the *epicycloids*: *cardioid, nephroid*, and so on. For plotting *hypocycloids* we plot the negative values of the variable $m := -\frac{1}{3}, -\frac{1}{4}$, etc., and plot the *deltoid, astroid*, etc. In both cases for $m := \frac{1}{b}$ do not forget to extend the domain of the parameter t to the segment $[-b\pi, b\pi]$.

5.3 Experiment with Polar Coordinates

We will produce a number of interesting figures in the formulas relating *polar* and *rectangular* coordinates $[\rho\cos(t), \rho\sin(t)]$ (see Section 6.1) by inserting two additional parameters A, B.

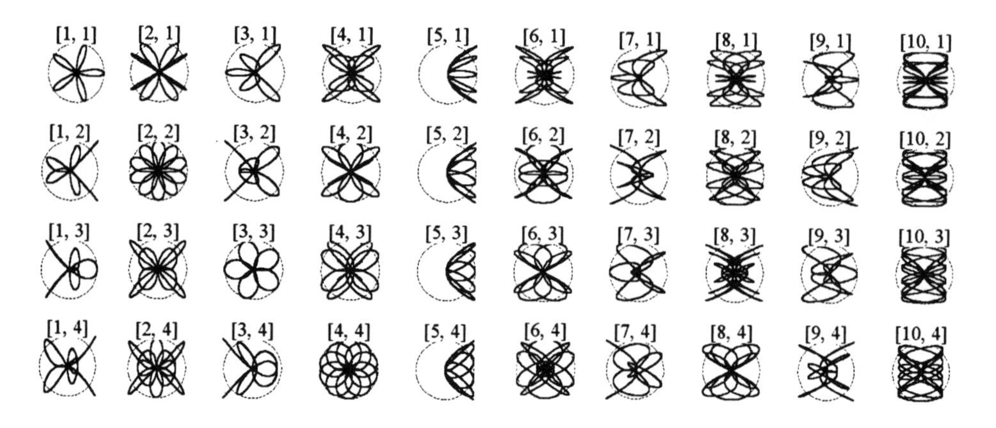

Fig. 5.9. Experiment with polar coordinates: $k = 5$

Now substituting in the equations $[\rho\cos(At), \rho\sin(Bt)]$ the relationship $\rho = \sin(kt)$ (for *roses*, see Section 6.3) for $k = 5$, and then changing A and B, we obtain a splendid view; Fig. 5.9.

Let us continue this experiment *up and down, around and near* [Kos] for some other values $k = 6, 7, \ldots$ or another relationship $\rho = f(t)$.

```
> p:=array(1..4, 1..10):
> for a from 1 to 4 do for b from 1 to 10 do
> p[a,b]:=plot([[sin(5*t)*cos(a*t), sin(5*t)*sin(b*t),
  t=0..2*Pi], [cos(t),sin(t),t=0..2*Pi]],thickness=[2,1],
  linestyle=[1, 2], title=convert([a,b], string)) od od:
> plots[display](p, scaling=constrained, axes=none);
```

5.4 Some Other Remarkable Curves

1. The *Lissajous curve*, Fig. 21.9, is defined by $\vec{\mathbf{r}}(t) = [a\sin(nt+\varphi),\ b\sin(t)]$.

```
> n:=2: phi:=0: a:=1: b:=1:        # enter your data (Fig. 21.4)
> plot([a*sin(n*t+phi), b*sin(t), t=0..4*Pi]);
```

The Lissajous curve helps us to study the mechanical problem–*small oscillations with two degrees of freedom*, i.e., the system of differential equations $x''(t) = -y(t),\quad y''(t) = -n^2 x(t)$, where $n > 0$. We meet this curve again in Sections 21.2 and 21.3.

Let us plot the graph $y = \sin(nt+\varphi)$ to obtain this curve in a geometrical way, and then wind the strip $\{|y| \le 1,\ x \in \mathbb{R}\}$ with the graph onto the circular cylinder with the axis OY and radius 1. The orthogonal projection of the cylinder with the sine curve onto the plane XY is the Lissajous curve. For $n = 1$ this curve is an ellipse (line segment when $\varphi = 1$).

Exercise 5.4.1 Plot tables of the Lissajous curves for several values of n as φ changes. Deduce that the curve is closed and algebraic when n is a rational number. What figure is filled by the Lissajous curve with n irrational? For $n = 2$, find values of φ for which the Lissajous curve is a parabola and a figure eight; Fig. 21.4.

2. The *clothoid (Cornu spiral)* is defined as the curve whose *curvature* (see Chapter 12) is proportional to its arc length: $k(s) = \frac{s}{a}$, and hence the angle between the tangent line and the axis OX is given by $\alpha(s) = \frac{s^2}{2a} + C$. Let us fix rectangular coordinates in such a way that the curve is tangent to the axis OX at the origin O, and start from this point to read the arc length (the constant C vanishes). We deduce from the above formulas that $x'(s) = \cos(\frac{s^2}{2a}),\ y'(s) = \sin(\frac{s^2}{2a})$. Integration leads to the equations of the clothoid: $x(s) = \int_0^s \cos(\frac{s^2}{2a})\,ds,\ y(s) = \int_0^s \sin(\frac{s^2}{2a})\,ds$. The clothoid makes an infinite number of revolutions around the points $M(\pm x_0, \pm y_0)$, where $x_0 = \int_0^\infty \cos(\frac{s^2}{2a})ds = \frac{\sqrt{\pi a}}{2},\ y_0 = \int_0^\infty \sin(\frac{s^2}{2a})ds = \frac{\sqrt{\pi a}}{2}$.

```
> p:=[int(cos(s*s/2), s=0..t), int(sin(s*s/2), s=0..t)];
```

$$[\text{FresnelC}(\tfrac{t}{\sqrt{\pi}})\sqrt{\pi},\ \text{FresnelS}(\tfrac{t}{\sqrt{\pi}})\sqrt{\pi}]$$

```
> plot([p[1], p[2],t=-7..7],scaling=constrained); # Fig. 5.1
```

The first clothoid was investigated by the physicist Cornu in 1874 with respect to the diffraction of light. The clothoid is also used in railway design, since it is the ideal transitional curve.

3. The natural equations of both the logarithmic spiral and the clothoid are particular cases of the equation $k(s) = \frac{s^m}{a}$, which defines the family of curves called *pseudospirals* (aside from the clothoid $m = 1$ and logarithmic spiral $m = -1$, they include the evolvent of the circle $m = -\frac{1}{2}$; see Section 9.4). Evolutes of pseudospirals again belong to this class of curves, for example, the evolute of the clothoid for $m = 3$.

Generalizing this, one can study the relationship $k(s) = f(s)$. The curves having natural equations of the form $k(s) = \frac{1}{a}\cos(s)$ are called *patterned curves* (in mechanics these curves are called *elastics*; Euler was the first to derive their equation). We plot them when $a = \frac{\pi}{2}, \pi, 2\pi$; Figs. 5.10–5.12.

```
> a:=2*Pi:
  p:=[int(cos(a*sin(s)),s=0..t),int(sin(a*sin(s)),s=0..t)]:
> plot([p[1], p[2], t=-2*Pi..2*Pi], scaling=constrained);
```

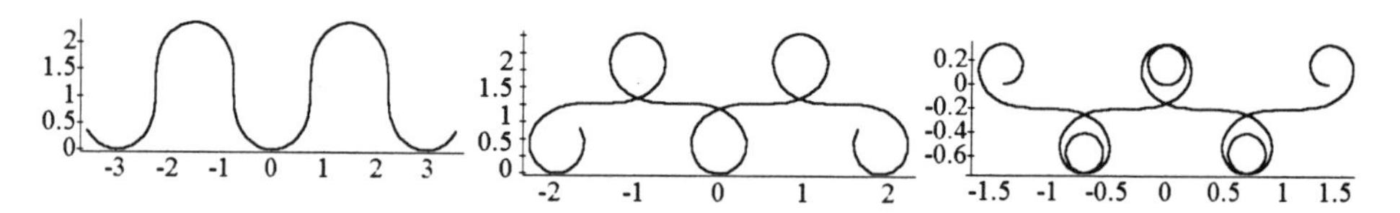

Figs. 5.10–5.12. Patterned curves (pseudospirals): $a = \frac{\pi}{2}, \pi, 2\pi$

It is not difficult to plot the *SICI spiral*, the plane curve whose equation in rectangular coordinates takes the form $\vec{\mathbf{r}}(t) = [\mathrm{Ci}(t), \mathrm{Si}(t)]$, where Ci is the integral cosine and Si is the integral sine. One can check that the arc length of the curve from the point $t = 0$ to the point t is equal to $\ln(t)$, and the curvature is $k(t) = t$ (see Section 12.1.2).

```
> plot([Ci(t),Si(t),t=1.5..30],axes=framed);     # Fig. 5.14
```

5.5 Level Curves, Vector Fields, and Trajectories

An implicitly given curve $f(x, y) = 0$ can be plotted using the command `implicitplot` from the library `plots`. One increases the number of points in the option `grid` (by assumption, it is `[25, 25]`) to improve the quality of figures.

```
> plots[implicitplot](x^3+y^3-3*x*y, x=-2..2, y=-2..2,
  grid=[40,40],tickmarks=[3,3]); # folium of Descartes,; Fig. 5.15
```

Using the function `x^3+y^3-3*x*y+0.01` one can plot the disconnected curve (perturbed folium of Descartes).

The algebraic curves package algcurves for MAPLE Release 5 helps us to study these curves. (An analysis of a collection of polynomials is possible using routines from the Groebner) package. For implicit plotting of given *algebraic curves*, the following method is sometimes used.

One can solve equations (of degree 2–4), then enter the command plot to obtain the graph of each branch of the function, and finally enter the command display to collect the pieces into the whole curve. The expressions for roots are often complicated, but their graphs are plotted exactly, and the obtained curve is glued together from different colored branches. We recommend the reader plot some curves of third and fourth degrees by both methods: *folium of Descartes, cissoid, strophoid, trisectrix of Maclaurin, cardioid, Nicomedian conchoid, lemniscate of Bernoulli, kappa*. Some of these curves also are successfully plotted in polar coordinates; see Chapter 6, where they are graphed.

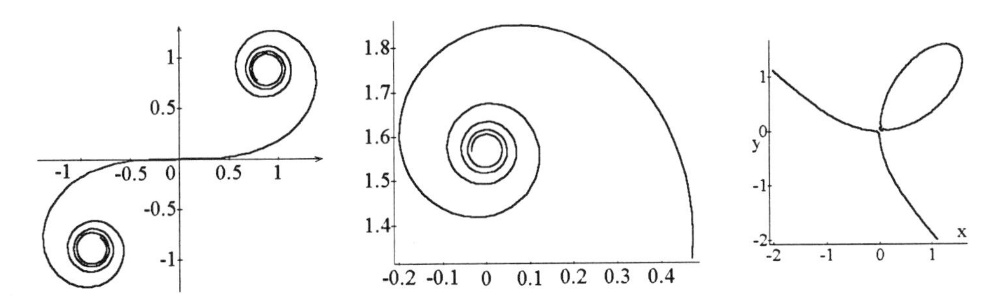

Figs. 5.13–5.15. Clothoid. *SICI* spiral. Folium of Descartes

The command contourplot allows one to obtain implicitly defined curves $f(x, y) = c$ for the values $c = c_1, \ldots, c_n$. For example,

```
> plots[contourplot](sin(x*y), x=-Pi..Pi,y=-Pi..Pi,
  grid=[50,50], contours=[-0.8,-1/2,0,1/2,0.8]);   # Fig.5.16
```

Using the command densityplot, the graph of the density of level curves of the function (darker areas correspond to greater density) is obtained in Fig. 5.17.

```
> plots[densityplot](sin(x*y), x=-Pi..Pi, y=-Pi..Pi,
  grid=[30,30], axes=boxed);
```

The graph of the gradient vector field for the above function (in two variables) is obtained in Fig. 5.18 using the command gradplot

```
> plots[gradplot](sin(x*y),x=-Pi..Pi,y=-Pi..Pi,arrows=SLIM);
```

Several procedures in MAPLE help us to study vector fields. The graph of an arbitrary two-dimensional vector field can be plotted using the command `fieldplot`; see Fig. 5.18 again.

```
> plots[fieldplot]([y*cos(x),x*cos(y)],x=-Pi..Pi,y=-Pi..Pi,
  arrows=SLIM);
```

For obtaining the integral curves for the given vector field, we solve an ODE using the command `dsolve` and then plot the graphs of the solutions. The option `type` has values `exact`, `series`, or `numeric`; the option `method` has values `rkf45`,`dverk78`, or `classical`, and it allows us to choose the method of solution.

The command `DEplot(deqns,vars,trange,inits,xrange,yrange, eqns)` from the library `DEtools` plots the solution (curve) of ODEs.

```
> DEtools[DEplot](diff(y(x),x,x,x)+x*sqrt(abs(diff(y(x),
  x)))+x^2*y(x),{y(x)},x=-4..5,[[y(0)=0,D(y)(0)=1,
  (D @@ 2)(y)(0)=1]],stepsize=.1);   # Fig. 5.20
```

The command `odeplot` from the library `plots` plots the solution (curve) of ODEs. It is analogous to the command `DEtools[DEplot]`. For example,

```
> f1:=diff(y(x),x,x,x)+x*sqrt(abs(diff(y(x),x)))+x^2*y(x);
  F1:=dsolve({f1, y(0)=0, D(y)(0)=1, D(D(y))(0)=1}, y(x),
  type=numeric); plots[odeplot](F1,[x,y(x)],-4..5); # Fig. 5.20
```

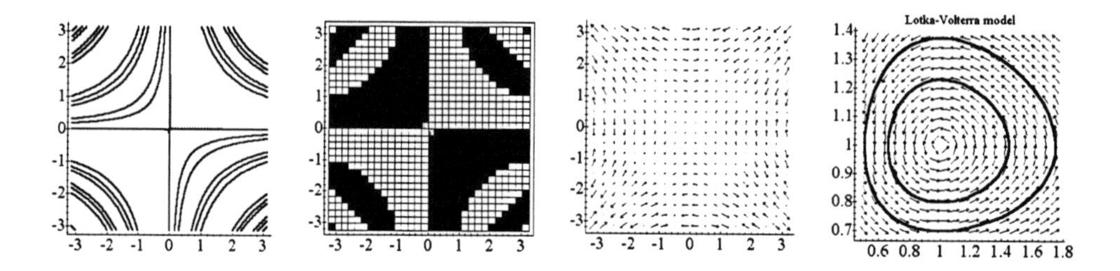

Figs. 5.16–5.19. Level curves: their density and vector fields

Continuing the examples, we plot two integral curves and also the direction field for the following system of ODEs; Fig. 5.19.

```
> DEtools[DEplot]({diff(x(t),t)=x(t)*(1-y(t)),
  diff(y(t),t)=.3*y(t)*(x(t)-1)}, [x(t),y(t)], t=-7..7,
  [[x(0)=1.2,y(0)=1.2], [x(0)=1,y(0)=.7]], stepsize=.2,
  title=`Lotka-Volterra model`,arrows=MEDIUM,method=rkf45);
```

Let us plot the trajectory of the system of three ODEs in the plane of the variables z and x.

```
> DEtools[DEplot]({D(x)(t)=y(t)-z(t), D(y)(t)=z(t)-x(t),
```

```
D(z)(t)=x(t)-2*y(t)}, {x(t), y(t), z(t)}, t=-2..2,
[[x(0)=1,y(0)=0,z(0)=2]], stepsize=.05, scene=[z(t),x(t)],
arrows=MEDIUM, method=classical[foreuler]);  # Fig. 5.21
```

The package `DEtools` also includes the commands `dfieldplot` and `phaserportrait`, but the command `DEplot` covers their functionality.

Example 5.5.1 Implicitly defined curves can be plotted in MAPLE by a number of commands.

1. The following example shows how to use the library `plottools`.

```
> with(plottools): disk([-0.4,0.4],0.1, color=blue):
  head:=ellipse([0,0.5], 0.7, 0.9, color=black):
  mouth:=arc([0,0.1], 0.35, 5/4*Pi..7/4*Pi, color=red,
  thickness=7): eye1:=disk([0.4,0.4],0.1, color=blue):
  eye2:=disk([-0.4, 0.4], 0.1, color=blue):
  nose:=line([0,0.35], [0,-0.1], color=black, thickness=5):
> plots[display]({head, eye1, eye2, nose, mouth},
  scaling=constrained, axes=none); # Fig. 5.22, happy face
```

2. One can write an analogous program using the library `geometry`.

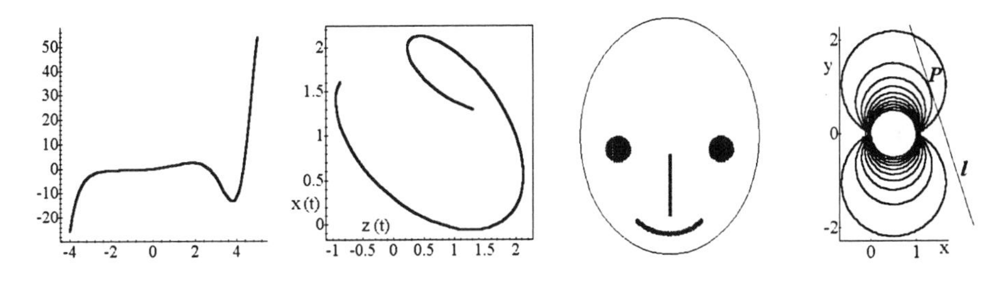

Figs. 5.20–5.23. Trajectories of ODEs in the plane.
Happy face. Level curves and extremum

5.6 Level Curves of Functions and Extremal Problems

Level sets of functions are useful in solving problems with extrema. The problem of *relative extrema* is formulated as follows:

What is the maximal (minimal) value of the function f defined in the plane along the given curve γ ?

Theorem 5.6.1 *Let the differentiable function F be given in the domain $G \subset \mathbb{R}^2$) and let the gradient ∇F be nonzero along γ_c : $F(x, y) = c$ (i.e., the curve γ_c of level c for F is a smooth curve). Assume that $g(x, y)$ is a smooth function* *on G and that $P \in \gamma_c$ is an extremal point of g on γ_c. Then the tangent line (see Chapter 9) to γ_c at the point P is orthogonal to the gradient $\nabla g(P)$, i.e., there exists a real number λ such that* $\nabla g(P) = \lambda \, \nabla F(P)$ (λ is called a *Lagrange multiplier*).

Remark 5.6.1 The equation in Theorem 5.6.1 is equivalent to the system

$$\begin{cases} F(x, y) & = c, \\ F'_x(x, y) & = \lambda \, g'_x(x, y), \\ F'_y(x, y) & = \lambda \, g'_y(x, y). \end{cases}$$

Example 5.6.1 An excursion bus travels along a highway (the straight line l). There is a castle (the line segment $P_1 P_2$ of length d) at some distance and elevated at some angle from the highway. What is the optimal stopping point of the bus on the highway (the point $P \in l$) from which the front of the castle can be viewed from P under the maximal angle?

Let us use this example to study and compare different ways of using *computer algebra* in MAPLE to solve problems of conditional extremum:

1. *Experimental*: We plot some level curves of the *objective function* $G = \cos(\angle P_1 P P_2)$ and choose visually the best from among them.

```
> with(plots): f:=sqrt(x^2+y^2): g:=sqrt((x-d)^2+y^2):
> G:=simplify((f^2+g^2-d^2)/(2*f*g)); P:=i -> implicitplot
  (subs(d=1,G)=i/10,x=-3..3,y=-3..3,grid=[70,70]):
> display([seq(P(i),i=0..10)],scaling=constrained); # Fig. 5.23
```

2. *Geometrical*: We analyze the configuration of level curves together with the given curve and use elementary geometry to find the solution.

3. *Analytical*:

(a) We reduce the problem to the case of an unconditional extremum.

(b) We apply the method of Lagrange multipliers.

Hints. **Method 2**. Let the function $f(P) = \angle P_1 P P_2$ reach its maximum at the point $P \in l$. The value of the angle of view φ of the segment $P_1 P_2$ is constant on arcs of circles through the end points of the segment, i.e., level curves of the function $f(P)$. A local maximum is reached at a point on the straight line l that is *tangent* to one of arcs of the family of level curves.

The problem has two solutions. Let us fix orthogonal coordinates in the plane such that $P_1 = (-\frac{d}{2}, 0)$ and $P_2 = (\frac{d}{2}, 0)$; then write down the equation of the straight line $l: \ ax + by = 1$. Find the center $O'(0, y)$ of the arc tangent to l from the equation $|O'P_1| = d(O', l)$: $\sqrt{y^2 + \frac{1}{4}d^2} = \frac{|b\,y-1|}{\sqrt{a^2+b^2}}$ (the second root corresponds to the best view of the back side of the castle). We find the optimal angle φ from the equation $\tan(\varphi) = \frac{d}{2|y|}$.

```
> f1:=sqrt(y^2+d^2/4): f2:=(b*y-1)/sqrt(a^2+b^2):
> solve(f1^2=f2^2, y);
```

$$\frac{-4\,b+2\,\sqrt{4\,b^2-a^4d^2-a^2d^2b^2+4\,a^2}}{4\,a^2},\quad \frac{-4\,b-2\,\sqrt{4\,b^2-a^4d^2-a^2d^2b^2+4\,a^2}}{4\,a^2}$$

Method 3 a. Let us fix orthogonal coordinates such that $P_1 = (0,0)$, $P_2 = (d,0)$, and write down the equation of the highway $l : ax + by = 1$. Represent the point on the straight line l in the form $P = (x = \frac{1-bt}{a}, y = t)$. Let us find $F := \cos(\angle P_1 P P_2)$ by the cosine theorem and substitute instead of variables f, g, h, the sides of $\Delta(P_1 P P_2)$ expressed in terms of coordinates. We differentiate the function $F(t)$ and find three roots of the numerator.

```
> f:=sqrt(t^2+((1-b*t)/a)^2): g:=sqrt(t^2+(d-(1-b*t)/a)^2):
  h:=d: F:=simplify((f^2+g^2-h^2)/(2*f*g));
```

$$F := \frac{t^2a^2+b^2t^2-2\,bt+1+dabt-da}{a^2\sqrt{\frac{t^2a^2+b^2t^2-2\,bt+1}{a^2}}\sqrt{\frac{t^2a^2+d^2a^2+2\,dabt-2\,da+b^2t^2-2\,bt+1}{a^2}}}$$

```
> dF:=simplify(diff(F, t));
```

$$dF := \frac{d^2a^2t\,(b^2t^2+t^2a^2+da-1)}{(t^2a^2+b^2t^2-2\,bt+1)\sqrt{\frac{t^2a^2+b^2t^2-2\,bt+1}{a^2}}}\cdot$$

$$\cdot\frac{1}{(t^2a^2+d^2a^2+2\,dabt-2\,da+b^2t^2-2\,bt+1)\sqrt{\frac{t^2a^2+d^2a^2+2\,dabt-2\,da+b^2t^2-2\,bt+1}{a^2}}}$$

```
> T:=[solve(numer(dF)=0,t)];P:=simplify([(b*T[2]-1)/a,T[2]]);
```

$$T := [0, \frac{\sqrt{(b^2+a^2)(1-da)}}{b^2+a^2}, -\frac{\sqrt{(b^2+a^2)(1-da)}}{b^2+a^2}]$$

$$P := [-\frac{-b\sqrt{(a^2+b^2)(1-da)}+a^2+b^2}{(a^2+b^2)a}, \frac{\sqrt{(a^2+b^2)(1-da)}}{a^2+b^2}]$$

Method 3 b. For deriving the maximum of the function

$$G := \cos(\angle P_1\, P\, P_2) = \frac{f^2+g^2-d^2}{2\,f\,g},$$

where $f = \sqrt{x^2+y^2}$, $g = \sqrt{(x-d)^2+y^2}$ with the restriction $F := ax + by - 1 = 0$, it is sufficient to solve the system grad $(G) = \lambda$ grad (F) with respect to $x,\ y$, and λ under the given restriction. Since the line segment $[P_1,\ P_2]$

does not intersect l, the function F has the same sign at the points P_1 and P_2, that is, $ad < 1$.

```
> f:=sqrt(x^2+y^2): g:=sqrt((x-d)^2+y^2): F:=a*x+b*y-1:
> G:=simplify((f^2+g^2-d^2)/(2*f*g));
```

$$\frac{x^2+y^2-xd}{\sqrt{x^2+y^2}\sqrt{x^2-2xd+d^2+y^2}}$$

```
> with(linalg): GradG:=grad(G, [x,y]): GradF:=grad(F, [x,y]):
```

```
> s:=[solve({F, seq(GradG[i]=lambda*GradF[i], i=1..2)},
  {x, y, lambda})];
```

$$[\{y = 0, \lambda = 0, x = a^{-1}\}, \{x = -\frac{b\mathit{RootOf}((b^2+a^2)\_Z^2-1+da)-1}{a},$$
$$y = \mathit{RootOf}((b^2 + a^2)\_Z^2 - 1 + da),\ \lambda = \ldots$$

```
> x0:=allvalues(-(b* RootOf((b^2+a^2)* _Z^2-1+da)-1)/a)[1];
  y0:=allvalues( RootOf((b^2+a^2)* _Z^2-1+da))[1];
```

$$x_0 := \left(\frac{b\sqrt{-(b^2+a^2)(-1+da)}}{b^2+a^2} - 1\right)a^{-1} \quad y_0 := \frac{\sqrt{-(b^2+a^2)(-1+da)}}{b^2+a^2}$$

```
> GO:=subs({x=x0, y=y0}, G);
```

Conclusion: The point of the best view with $y > 0$ has coordinates

$$x_0 = (\frac{b\sqrt{(b^2+a^2)(1-da)}}{b^2+a^2} - 1)a^{-1},\ y_0 = \frac{\sqrt{(b^2+a^2)(1-da)}}{b^2+a^2}.$$

Exercise 5.6.1 Given a circle with center O and a point A inside it, find the point P on the circle such that the angle $\angle APO$ is maximal.

6
Curves in Polar Coordinates

Section 6.1 is a short introduction to using polar coordinates with MAPLE. In Sections 6.2–6.5 we plot the polar graphs of some remarkable curves (in particular, spirals, roses and crosses) and use inversion transformation.

In Chapter 6 the reader will become acquainted with the commands

`polarplot, pieslice, conformal, coordplot, cot, inversion.`

6.1 Basic Plots in Polar Coordinates

The location of a point M in the plane with origin O is uniquely defined by the *distance* $|OM| = \rho$ and the *angle* $\varphi \in [-\pi, \pi]$ between the segment OM and the *polar axis*. The real pair (ρ, φ) is called the *polar coordinates* of the point M.

The relation between polar coordinates (ρ, φ) and rectangular coordinates (x, y) when the axis OX plays the role of the polar axis is the following:

$$x = \rho \cos \varphi, \quad y = \rho \sin \varphi$$

These formulas were given by Isaac Newton in 1670. If $\rho = \rho(\varphi)$ is the polar equation of a curve, then its equations in rectangular coordinates are

$$x = \rho(t) \cos(t), \quad y = \rho(t) \sin(t).$$

The conversion of a complex number $z = a + ib$ into polar form is possible using the command `polar(`*number*`)`. The graph of $\rho = f(t)$ in polar coordinates can be plotted using the command `polarplot` from the library `plots`. By assumption in MAPLE the parameter is given by `t=-Pi..Pi`.

```
> plots[polarplot](1, t=0..Pi);  # upper half-circle
> plots[polarplot](t, t=0..4*Pi); # two coils of Archimedes' spiral
```

The *spiral of Archimedes* $\rho = a\varphi$ (plotted above) was studied by Archimedes in the third century B.C. in relation to the problem of the trisection of an angle.

Also, plotting in polar coordinates is possible using the basic command `plot` with the additional option `coords=polar`. Hence the polygons and graphs of functions (see Figs. 6.1–6.2) in polar coordinates can be plotted analogously to the case of rectangular coordinates.

Example 6.1.1

1. We plot the polygon through some points of Archimedes' spiral.

```
> t:=i -> i*Pi/6: p1:=plot([seq([t(i), t(i)], i=0..40)],
  coords=polar, style=point, symbol=circle):
> p2:=plot([seq([t(i), t(i)], i=0..40)], coords=polar):
> plots[display]([p1, p2], scaling=constrained);  # Fig. 6.1
```

2. We plot a regular *star* (m, n)*-gon* (convex for $m = 1$) with relatively prime m and n (see also Example 17.3.1).

```
> n:=5: m:=2:
  plot([seq([1,m*i*2*Pi/n], i=0..n)],coords=polar); # Fig. 6.2
```

Using an easy generalization of this program, we plot the *disconnected star* (m, n)*-gon.*

```
> n:=8: m:=2: t:=i -> i*2*Pi/n:
  plot([seq([[1,t(i)], [1,t(i+m)]], i=1..n)], coords=polar);
```

3. We plot the *circular diagram* of Fig. 6.3.

```
> with(plots): A:=[0, 10,30,40,20]: # enter A[2], A[3]... in %
   B:=i-> sum((A[j]/100)*2*Pi, j=2..i):
  P:=polarplot([1, seq([[0,0], [1,B(i)]], i=1..nops(A)-1)]):
  T:=textplot({seq([.5,B(i),convert('A'[i],string)],i=2..nops(A))},
  coords=polar): display([P, T]);
```

Another method is based on the command `pieslice` from the library `plottools`.

```
> with(plots): P:=seq(display(plottools[pieslice]([0,0], 5,
  Pi*i/10..Pi*(i+1)/10, color=COLOR(HUE,evalf(i/20))),
  scaling=constrained), i=0..20): display({P}, axes=none);
```

4. Let us plot the stopwatch with a *moving* arrow.

```
> with(plots): n:=60:     # 60 seconds in a minute
```

```
> q:=k -> polarplot([[0,0], [0.9,-k*2*Pi/n]], thickness=3):
> p:=polarplot([1,.1], color=blue):
> text:=textplot([seq([sin(Pi*i/6), cos(Pi*i/6), `i`],
  i=1..12), [-.1,-.3, `Cosmos`]], font=[TIMES,BOLD,18]):
> display([seq(display([p, text, q(k)]), k=1..n)],
  insequence=true, axes=none);  # Fig. 6.4
```

Then we plot a more natural arrow for the stopwatch (replacing the line in the program).

```
> q:=k -> plottools[arrow]([0, 0], [0.8*sin(k*2*Pi/n),
  0.8*cos(k*2*Pi/n)], .05, .15, .2, color=green):
```

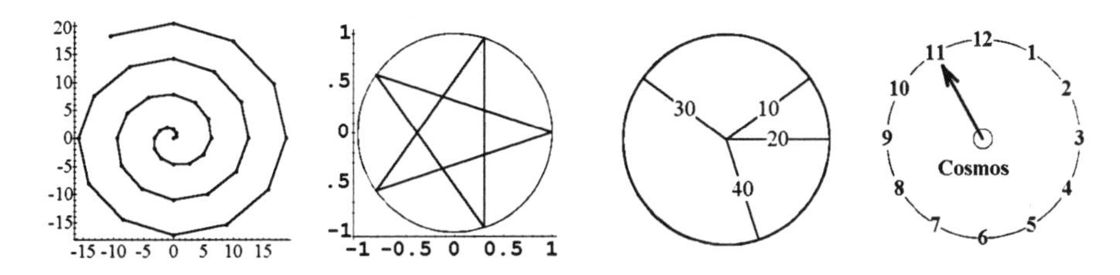

Figs. 6.1–6.4. Polar coordinates: polygon, star n-gon, circular diagram, a working stopwatch

5. Let us plot the *butterfly* and the cochleoid.

```
> f:=exp(1)^cos(t)-2*cos(4*t)+sin(t/12)^5; # butterfly, Fig. 6.5
> plot(f, t=-12*Pi..12*Pi, numpoints=999, coords=polar);
> plot(sin(t)/t, t=-6*Pi..6*Pi, coords=polar); # cochleoid
```

The *cochleoid* $\rho = \frac{\sin(\varphi)}{\varphi}$ (Greek, $\varkappa o\,\chi\lambda o\,\varsigma$, snail) of Fig. 6.6, with $|\varphi| < 720°$ belongs to a class of quadratrics, well-known since the 17th century.

6. The equations $\rho = 2a\cos\varphi$, and $\rho = \frac{a}{\cos(\varphi-\varphi_0)}$ define in polar coordinates the circle and the straight line, respectively.

What is the geometrical meaning of the parameters a and φ_0 ?

Both equations are special cases of the equation $\rho^m = a^m \cos(m\varphi)$, which defines in polar coordinates the family of curves called *sinusoidal spirals*. One can check that, except for the straight line and the circle, this family includes the rectangular hyperbola ($m = -2$), the parabola ($m = -\frac{1}{2}$), the cardioid ($m = \frac{1}{2}$), and the lemniscate of Bernoulli ($m = 2$).

Write a program for plotting the sinusoidal spiral with $m = \pm 4$.

7. If, for an arbitrary *plane conic section*, we fix the point O not at a focus and if the axis of the section through the foci plays the role of the polar axis,

then the equation of the conic section in polar coordinates will take the form $\rho = \frac{p}{1-e\cos\varphi}$, where $e < 1$ for an *ellipse*, $e = 1$ for a *parabola*, and for a *hyperbola* $e > 1$. To obtain the second branch of the curve, we replace 1 by -1 in the denominator.

We plot conic sections of different types using the following program, where (p1 are ellipses, p2 are parabolas, p3 are hyperbolas); Fig. 6.7.

```
> p:=1: f:=e->p/(1-e*cos(t)):
> p1:=plot([seq(f(i/15),i=9..14)],t=-Pi..Pi,coords=polar):
> v:=1.2: p2:=plot(f(1),t=Pi/2-v..3*Pi/2+v,coords=polar):
> v:=0.5: p3:=plot([seq(f(1+i/5), i=1..3)],
  t=Pi/2-v..3*Pi/2+v, coords=polar):
> plots[display]([p1, p2, p3]); # all types of conic sections
```

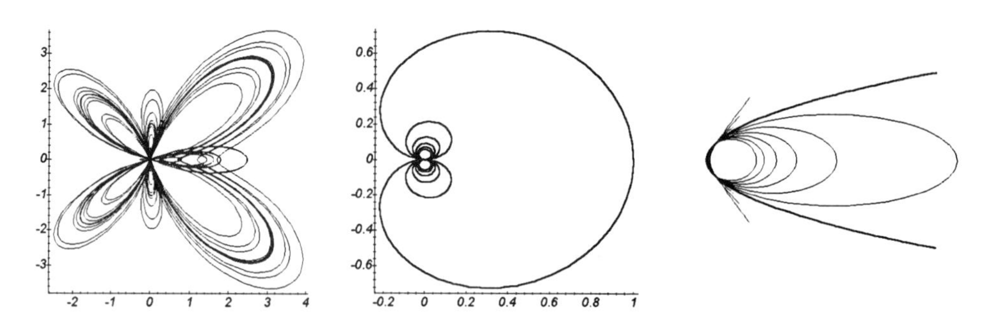

Figs. 6.5–6.7. Butterfly and cochleoid. Conic sections

8. Let us plot the *sunflower* $\rho = 3 + |\cos(n\,\varphi)|$, Fig. 6.8.

```
> n:=7: f:=3+abs(cos(n*t)):
> plot([f,3],t=0..2*Pi,coords=polar,color=[yellow,black]);
```

9. Plot the *loop coupling* $\rho = 2\cos(2\varphi) + 1$

```
> plot(2*cos(2*t)+1, t=0..2*Pi, coords=polar);  # Fig. 6.9
```

10. Let us plot the *Pascal limaçon* $\rho = 2a\cos(\varphi) - b$ $(a, b > 0)$ (with a loop for $b < 2a$, without a loop for $b > 2a$, and the *cardioid* for $b = 2a$).

```
> a:=1: f:=b->2*a*cos(t)+b: plot([f(3*a),f(2*a),f(a)],
  t=0..2*Pi, coords=polar, color=[red,blue,black]); # Fig. 6.10
```

One can give a generalization $\rho = 2a\cos(n\varphi) - b$ $(a, b > 0)$ of the cardioid and limaçon and plot them, for example, for $n = 3$, $a = 0.5$ and $b = 2$.

Remark 6.1.1 For plotting the parametrized curves $\rho = f(t)$, $\varphi = g(t)$ $(a \le t \le b)$ in polar coordinates, we use the command polarplot (or plot with the additional option `coords=polar`), analogous to the case of rectangular coordinates.

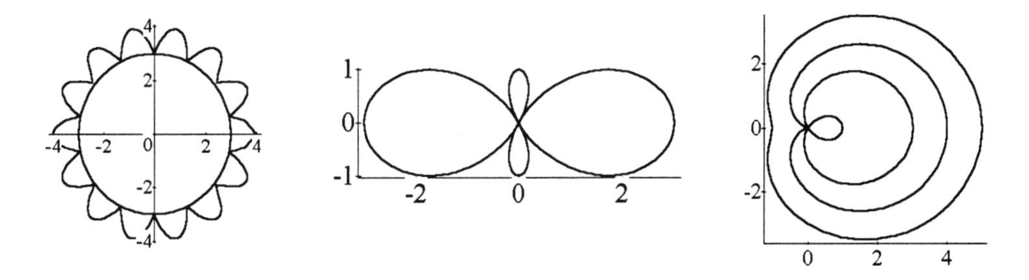

Figs. 6.8–6.10. Sunflower, loop coupling, Pascal's limaçon and cardioid

11. Let us plot the *leaf of a Japanese maple* (*Acer palmatum*) by the equation $\rho = (1 + \sin(t))\,(1 + 0.3\cos(8\,t))\,(1 + 0.1\cos(24\,t))$.

```
> R:=(1+sin(t))*(1+.3*cos(8*t))*(1+.1*cos(24*t)):
> plot([R,t,t=0..2*Pi], coords=polar);   # Fig. 6.11
```

Figs. 6.11–6.13. A leaf of a Japanese maple and another leaf. Coordinate nets under conformal transformation of the plane

One can plot another *leaf* by the following equation in polar coordinates:

```
> g:=100/(100+(t-Pi/2)^8): # for scaling
> R:=g*(2-sin(7*t)-cos(30*t)/2): # Fig. 6.12
> plot([R,t,t=-Pi/2..3/2*Pi], coords=polar, numpoints=999);
```

Remark 6.1.2 a) The option coords=*name* is used for plotting a number of curvilinear coordinates in MAPLE. For example, we plot domains in the plane with nets of curvilinear coordinates such as *polar, parabolic, elliptic, bipolar, hyperbolic* (Fig. 6.14)

```
> coordplot(polar, title=` Polar , scaling=constrained` );
```

by replacing polar with the name of the corresponding coordinates parabolic, elliptic, bipolar, hyperbolic.

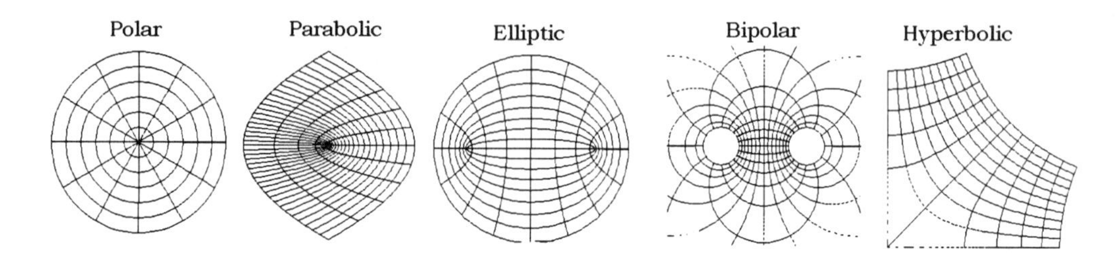

Fig. 6.14. Nets of curvilinear coordinates in the plane

b) Another way of obtaining figures with orthogonal families of curves is to use conformal transformations (defined by complex functions) of the plane; Fig. 6.13.

```
> conformal(1/z, z=-1-I..1+I, grid=[25,25], numxy=[100,100],
  axes=framed, scaling=constrained, view=[-6..6, -6..6]);
```

6.2 Remarkable Curves in Polar Coordinates

Implicit equations of (algebraic) curves are often simplified when we use polar coordinates instead of rectangular ones. Consider several curves given by their geometrical or kinematic properties. Determining their equations is more convenient using polar coordinates.

1. Let us fix the circle with diameter $|OA| = 2a$ and the tangent line l at the point A. On the ray emanating from O and rotating around the point O, one places the line segment OM of length equal to the segment of the ray between the circle and the line l. The set of all points M_1, M_2 (two directions on l for placing M) is called the *cissoid* (Greek, $\varkappa\iota\sigma\sigma o\,\varepsilon\iota\delta\eta\varsigma$, ivy-shaped, studied by Diocles in the third century B.C. in relation to the *problem of doubling a cube*). Its equation is $\rho = 2a\frac{\sin^2\varphi}{\cos\varphi}$, Fig. 6.15 a.

```
> f:=2*sin(t)^2/cos(t): h:=2/cos(t): plot([f, h],
  t=-Pi/2+.2..Pi/2-.2, coords=polar, linestyle=[1, 2]);
```

Remark 6.2.1 The geometrical construction of the cissoid can be applied to two arbitrary curves, given in polar coordinates by equations $\rho = \rho_1(\varphi)$ and $\rho = \rho_2(\varphi)$, with the aim of plotting the curve $\rho = \rho_1(\varphi) - \rho_2(\varphi)$, called the *cissoidal transformation*. The analogous transformation in rectangular coordinates is known as the *difference of graphs* $f(x) = f_1(x) - f_2(x)$ of two functions.

2. Let us fix the point A and the straight line l at distance a from this point; AC is the perpendicular from A onto the line l. On a ray rotating around the point A, place the segments BM_1 and BM_2 from the point B of its intersection with the line l. Moreover, $BM_1 = BM_2 = CB$ is assumed. The set of all points M_1, M_2 is called the *strophoid* (Greek, $\sigma\tau\rho o\,\varphi o\,\varsigma$, a twisted strip). It was studied by Torricelli in 1645, and its equation is $\rho = a\frac{1\pm\sin\varphi}{\cos\varphi}$, Fig. 6.15 b.

```
> f:=(1+sin(t))/cos(t): h:=2/cos(t):
> T:=[-Pi/2..Pi/2-.5, Pi/2+.5..3*Pi/2]:
> p:=i->plot(f, t=T[i], coords=polar, thickness=2):
> q:=plot(h,t=-Pi/2+.5..Pi/2-.5, coords=polar,linestyle=2):
> plots[display]([p(1), p(2), q], tickmarks=[2,3]);
```

3. Let us fix the straight line g parallel to the polar axis at distance a from it. On the ray emanating from O and rotating around the point O, place the segments BM_1 and BM_2 of given length l on both sides of the point B of intersection of the ray with g. The set of all points M_1, M_2 is called the *conchoid* (Greek, $\varkappa o\nu\chi o\,\varepsilon\iota\delta\eta\varsigma$, shell-shaped). It was studied by Nicomedes in the third century B.C. in relation to the *problem of the trisection of an angle*). Its equation is $\rho = \frac{a}{\sin(\varphi)} \pm l$. For $l > a$ the conchoid has a loop; for $0 < l < a$ it has a cuspidal point of the first kind. In Fig. 6.15c, the curve is rotated by a 90° angle.

```
> a:=5: l:=4*a: f1:=a/sin(t)+l: f2:=a/sin(t)-l: h:=a/sin(t):
> plot([f1, f2, h], t=.15..Pi-.15, coords=polar,
  thickness=[2,2,1], linestyle=[1,1,2]);
```

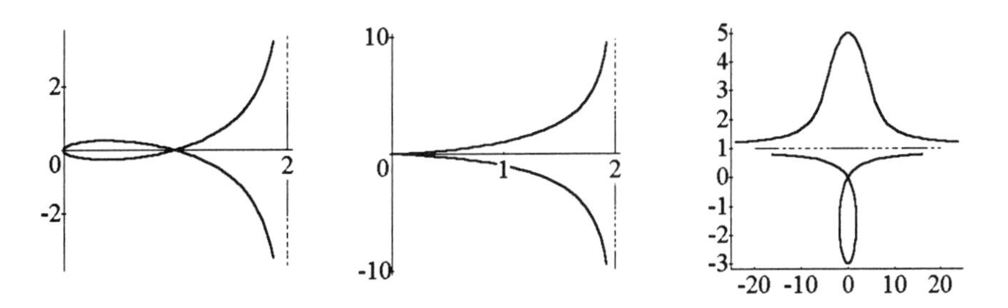

Figs. 6.15 a–c. Cissoid, strophoid, and conchoid of a line

Remark 6.2.2 An obvious generalization of this construction is the *conchoid of a plane curve,* which can be obtained by increasing or decreasing the position vector at each point on the given curve on the constant segment l.

If $\rho = f(\varphi)$ is the equation of the given curve in polar coordinates, then the equation of its conchoid is $\rho = f(\varphi) \pm l$. In other words, Nicomedes studied the conchoid of a line. One can prove that the conchoid of a circle is Pascal's limaçon; Fig. 6.10.

4. *Kappa* is the curve consisting of all points on the tangent lines from the origin to a fixed circle of radius a with moving center along the axis OX. The shape of *kappa* is similar to the Greek letter $\varkappa$ of that name. Its equation is $\rho = a \cot \varphi$.

```
> a:=1: e:=.4: f1:=a*cot(t): f3:=a/sin(t):
> plot([f1, -f1, f3, -f3], t=e..Pi-e,
  coords=polar, linestyle=[1,1,2,2]);  # Fig. 6.16
```

The straight lines $y = \pm a$ are horizontal asymptotes, the point O is the node.

Kappa is a member of the family of curves $\rho = a\cot(k\varphi)$, called *nodal curves*. This family includes the strophoid ($k = \frac{1}{2}$) and the *windmill*, ($k = 2$).

```
> a:=1: e:=0.2:      # windmill, Fig. 6.17
> f1:=a*cot(2*t): f3:=(a/2)/sin(t): f4:=(a/2)/cos(t):
> p1:=plot([f1, -f1], t=e..Pi/2-e, coords=polar):
  p2:=plot([f1, -f1], t=-Pi/2+e..-e, coords=polar):
  p3:=plot([f3,-f3], t=e..Pi-e, coords=polar, linestyle=2):
  p4:=plot([f4, -f4], t=-Pi/2+e..Pi/2-e, coords=polar,
  linestyle=2):  plots[display]([p1, p2, p3, p4]);
```

5. *Cassini ovals* are defined to be the sets of points in the plane for which the product of the distances to two fixed points are constant. It is not difficult to deduce the following polynomial equation of Cassini ovals:
$(x^2 + y^2)^2 - 2c^2(x^2 - y^2) = a^4 - c^4$ or in polar coordinates,

$$\rho^2 = c^2\{\cos(2\varphi) \pm \sqrt{\cos(2\varphi)^2 + (\tfrac{a^4}{c^4} - 1)}\}.$$

For $0 < a < c$, the curve consists of two simple closed components (for $a = 0$ it degenerates into two points); for $c < a < c\sqrt{2}$, the curve is closed with waist that disappears for $a \geq c\sqrt{2}$. All points on the ovals with the tangent line parallel to the axis OX lie on the circle $x^2 + y^2 = c^2$. The problem of plotting these curves on the computer display is fascinating.

```
> f1:=a-> 3*sqrt(abs(cos(2*t)+sqrt(cos(2*t)^2-cos(2*a)^2))):
  f2:=a-> 3*sqrt(abs(cos(2*t)-sqrt(cos(2*t)^2-cos(2*a)^2))):
  f3:=b-> 3*sqrt(cos(2*t)+sqrt((cos(2*t)^2+b))):
> p[0]:=plot(3, coords=polar, linestyle=2):
> n:=24: for i from 1 to 6 do ti:=i*Pi/n: p[i]:=plot(
```

```
  [f1(ti),f2(ti),-f1(ti),-f2(ti)],t=-ti..ti,coords=polar):
  p[i+6]:=plot(f3(i), t=-Pi..Pi, coords=polar)  od:
> plots[display]([seq(p[i], i=0..12)]);  # Fig. 6.18
```

For $a = c$ (p[6] in the above program) we obtain the *lemniscate of Bernoulli* $\rho = c\sqrt{2\cos(2\varphi)}$ (in Latin *lemniscate* means "adorned with ribbons"), which has a point of self-intersection.

```
> c:=1: f1:=c*sqrt(2*abs(cos(2*t))):
> plot([f1,-f1], t=-Pi/4..Pi/4, coords=polar);
```

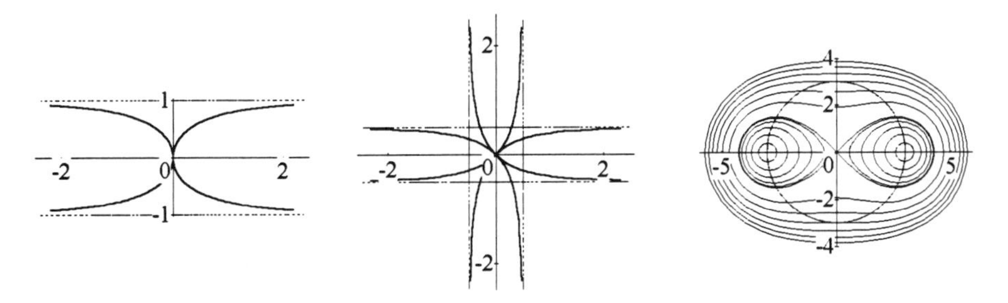

Figs. 6.16–6.18. Kappa, windmill, Cassini ovals

Exercise 6.2.1 Consider some other examples of curves whose polar equations contain trigonometric functions.

6.3 Inversion of Curves

Definition 6.3.1 *Inversion* (symmetry with respect to the circle of radius R with center O) is the transformation of the plane that maps any point M onto the point M' on the same ray OM, and the product of distances OM and OM' is constant and equal to R^2, the *degree of inversion.*

For example, an inversion maps a straight line and a circle again onto a straight line and a circle. (The command `inversion(Q, P, C)` from the library `geometry` plots the inversion (the object `Q`) of a point, line, or circle `P` with respect to the circle `C`). If the curve is given by the equation $\rho = f(\varphi)$, then the inverse curve with respect to the circle of radius R with the center at the pole has the equation $\rho = \frac{R^2}{f(\varphi)}$ (see examples below). There are different mechanisms (*inversors*) that help us avoid long calculations, and there are standard instruments to plot the curve, which is the image under inversion of a given plane curve.

Example 6.3.1 1. The reflections of roses $\rho = R\cos(k\varphi)$ under *inversion* with respect to the base circle of radius R are the curves *ears* $\rho = \frac{R}{\cos(k\varphi)}$ outside this circle. For example, the inversion of a *three-leafed rose* $\rho = R\cos(3\varphi)$ (see Section 6.5.1) is the *trisectrix of Longchamps* $\rho = \frac{R}{\cos(3\varphi)}$; Fig. 6.20.

```
> a:=1: e:=.1: f:=a/cos(3*t): p:=i ->
  plot(f, t=(2*i-1)*Pi/6+e..(2*i+1)*Pi/6-e, coords=polar):
> q:=i -> plot([t, i*Pi/3-Pi/6, t=-3..3], coords=polar,
  linestyle=2): h:=plot(1, coords=polar, linestyle=2):
> plots[display]([seq(p(i),i=0..2), seq(q(i),i=0..2), h]);
```

2. A similar equation $\rho = \frac{R}{\cos(\varphi/3)}$ defines the *trisectrix of Maclaurin*; Fig. 6.19. It was studied by C. Maclaurin in 1742 in relation to the *problem of the trisection of an angle.*

```
> p1:=plot(1/cos(t/3), t=-3*Pi/2+e..3*Pi/2-e, coords=polar);
> p2:=plot([cos(t/3),1], t=-3*Pi/2..3*Pi/2, coords=polar,
  linestyle=[1,2]):
> p3:=plot(-3/cos(t), t=-Pi/2+e..Pi/2-e, coords=polar,
  linestyle=2): e:=.7: plots[display]([p1, p2, p3]);
```

3. Inversion of the four-leafed rose $\rho = R\cos(4\varphi)$ leads to the *cross-shaped curve* $\rho = \frac{R}{\cos(4\varphi)}$ (see other *crosses* below in Section 6.5.2).

```
> a:=1: f:=a/sin(2*t):
> p:=i->plot(f, t=(2*i-1)*Pi/4+Pi/4+e..(2*i+1)*Pi/4+Pi/4-e,
  coords=polar): h:=plot(1, coords=polar, linestyle=2):
e:=.15: plots[display]([seq(p(i), i=0..3),h]);  # Fig. 6.21
```

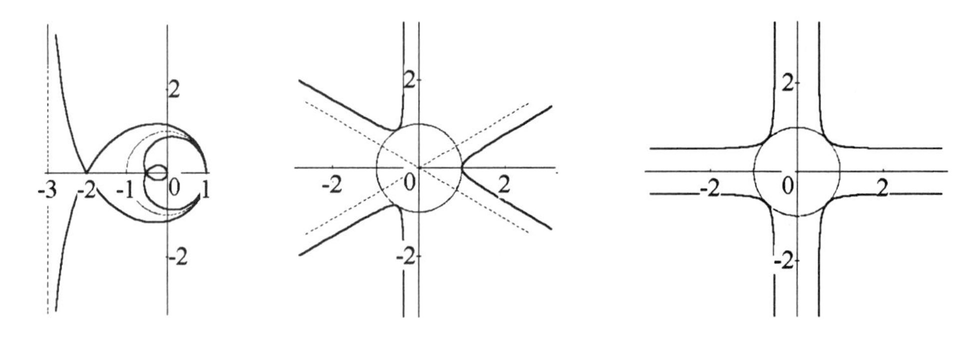

Figs. 6.19–6.21. Trisectrices of Maclaurin and Longchamps, cross-shaped curve

Other mutually inverse curves are Archimedes' spiral and the hyperbolic spiral, the rectangular hyperbola and the lemniscate of Bernoulli, the parabola and the cardioid, the cochleoid and the Dinostratus' quadratrix.

6.4 Spirals

The *spirals* $\rho = f(\varphi)$ (where f is monotone function) have especially simple equations in polar coordinates. By contrast, their equations in rectangular coordinates are complicated. The spiral of Archimedes (considered in Section 6.1) is a member of the family of *algebraic spirals*, which are given by polynomial equations $F(\rho, \varphi) = 0$. We plot the spirals using globally the option `scaling=constrained`.

1. The *neoid* $\rho = a\varphi + l$ (the conchoid of Archimedes' spiral) is used in the construction of a spinning machine.

```
> plot(0.2*t+0.5, t=0..6*Pi, coords=polar); # Fig. 6.22
```

2. *Galileo's spiral* $\rho = a\varphi^2 - l$ has been known since the 17th century from the problem of the trajectory of a point falling near the earth's equator, starting with initial velocity from the usual rotation of the earth. The *inverse Galileo spiral* for $l = 0$ has the equation $\rho = \frac{a}{\varphi^2}$.

```
> plot(0.01*t^2-0.02, t=0..6*Pi, coords=polar); # Fig. 6.23
> plot(100/t^2, t=6..10*Pi, coords=polar); # Fig. 6.24
```

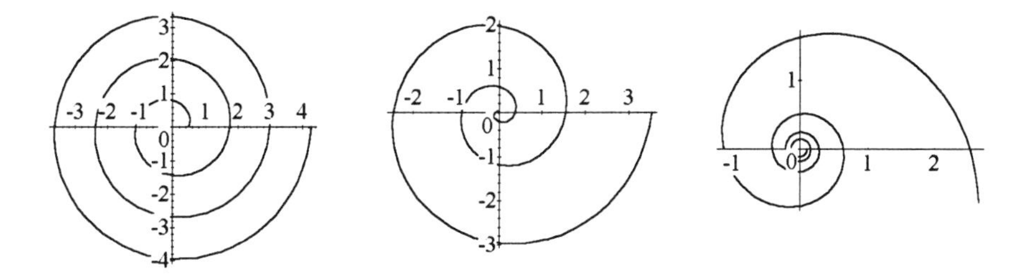

Figs. 6.22–6.24. Spirals: neoid, Galileo's and its inversion

3. *Fermat's spiral* $\rho = a\sqrt{\varphi}$ was discovered in 1636.

```
> plot(sqrt(t), t=0..4*Pi, coords=polar); # Fig. 6.25
```

Its conchoid is the *parabolic spiral* $\rho = a\sqrt{\varphi} + l,\ l > 0$, Fig. 6.26

```
> plot([sqrt(t)+.5,-sqrt(t)+.5], t=0..4*Pi, coords=polar);
```

4. The *logarithmic spiral* $\rho = ae^{k\varphi}$ ($\varphi \in \mathbb{R}$), first studied by Descartes, is the curve in which the angle between the polar radius and a tangent line at its endpoint is constant. In view of this property, the curve is used extensively in applications. In nature, some shells have the shape of a logarithmic spiral.

```
> plot(1.1^t, t=-6*Pi..4*Pi, coords=polar); # Fig. 6.27
```

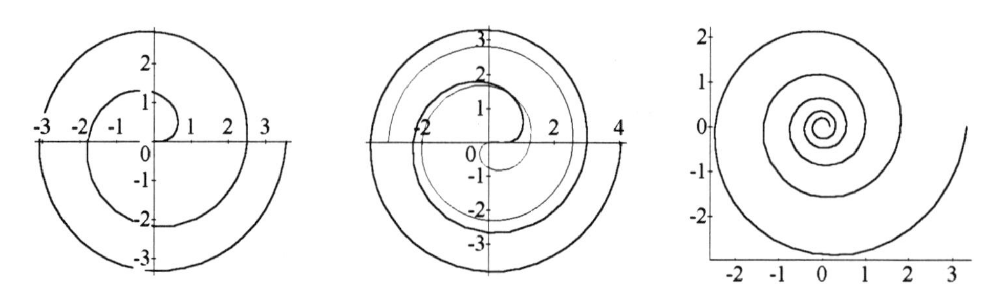

Figs. 6.25–6.27. Spirals: Fermat's, parabolic, logarithmic

5. The *hyperbolic spiral* $\rho = \frac{a}{\varphi}$ was studied by P. Varignon in 1704. The pole plays the role of an asymptotic (limit) point. The asymptote is the straight line parallel to the polar axis at a distance a from it. (Its conchoid is plotted analogously.)

```
> p1:=plot(2/t, t=0..6*Pi, coords=polar):
> p2:=plot(2/sin(t), t=.4..1.5, coords=polar, linestyle=2):
> plots[display]([p1, p2]); # Figs. 6.28–6.29
```

6. The *lituus* (from 1714) is the spiral curve defined by the equation $\rho = \frac{a}{\sqrt{\varphi}}$; the polar axis is its asymptote.

```
> plot(2/sqrt(t), t=0.1..6*Pi, coords=polar); # Fig. 6.30
```

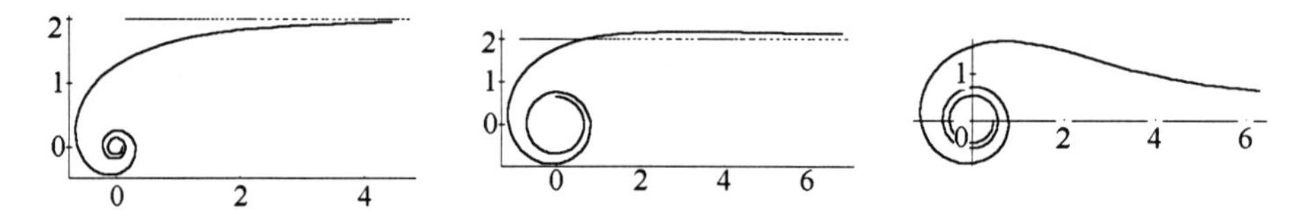

Figs. 6.28–6.30. Spirals: hyperbolic and its conchoid, lituus

6.5 Roses and Crosses

6.5.1 Roses

The curves $\rho = R\sin(k\varphi)$ are called *roses*. Since the function $\sin(k\varphi)$ is periodic, a rose consists of equal *leaves*; Fig. 6.31.

If the modulus k is a rational number, then the rose is a closed algebraic curve of even order. For an integer k, the rose consists of k leaves for odd k and of $2k$ leaves for even k.

If the modulus $k = \frac{p}{q}$ $(q > 0)$ is a rational number and p, q are relatively prime, then the rose consists of p leaves when both p, q are odd and of $2p$

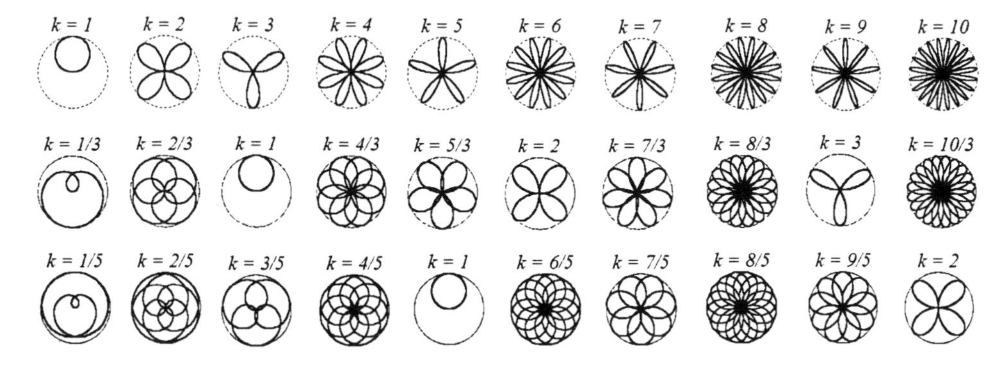

Fig. 6.31. Table with roses for $\rho = \sin(k\varphi)$.

leaves when one of these integers is even; moreover, in contrast to the first case, each subsequent leaf partially covers the previous one.

If the modulus k is irrational, then the rose consists of an infinite number of leaves, partially covering each other.

Plot the roses in Fig. 6.31 using the following program:

```
> p:=array(1..3, 1..10):
> for a from 1 to 10 do for c from 1 to 3 do b:=2*c-1:
  p[c,a]:=polarplot([sin(a/b*t), 1], t=0..2*b*Pi,
  linestyle=[1,2], title=convert(k=a/b, string)) od od:
> plots[display](p, axes=none);  # Fig. 6.31
```

Exercise 6.5.1 Check that roses belong to the family of cycloidal curves: epitrochoids for $k > 1$ and hypotrochoids for $k < 1$.
Hints. Substitute $h = R + mR$ in the parametric equations of trochoids (see Section 5.1) and obtain the *trochoidal roses* $\rho = 2R(1+m)\sin(\frac{\varphi+\pi/2}{2m+1})$.

6.5.2 Crosses

Let us plot (approximately) the following crosses by their polar equations.

1. *Leaf cross.*

```
> f:=max(2*cos(2*t)^2,0.3): plot([f, 0.25, 2.05],
  t=0..2*Pi, coords=polar, thickness=2);
```

2. *St. Andrew's cross.*

```
> f:=min(1/(2*abs(cos(2*t))),2): plot([f,2.05], t=0..2*Pi,
  coords=polar, thickness=[2, 1], linestyle=[1, 2]);
```

3. *Catacomb cross.*

Fig. 6.32. Crosses: 1, 2, 3, 4a, 4b, 5, 6 and the graph of $1/f_1(t)$

```
> f:=min(2/(3*abs(sin(2*t))),2): plot([f,2.07],t=-.1..2*Pi,
  coords=polar, thickness=[2, 1], linestyle=[1, 2]);
```

4. *St. George's cross*, two examples.

```
> f:=min((9/(10*abs(sin(2*t))))^5,2): plot([f,2.05], t=-.1
  ..2*Pi, coords=polar, thickness=[2,1], linestyle=[1,2]);
> g:=min(4/((2*sin(2*t)))^4,2): plot([g,2.05],t=-.1..2*Pi,
  coords=polar, thickness=[2, 1], linestyle=[1, 2]);
```

5. *On-Bread cross.*

```
> f:=min(1/(10*abs(sin(2*t))),1): plot([f, 1.02, 1.2],
  t=-.01..2*Pi, coords=polar, thickness=[2,1,1]);
```

6. *St. George's cross (sharp).* Explain the program below.

```
> f1:=4/(2*sin(2*t))^4:
> sol:=solve(f1=2, t): t0:=sol[1];
> plot([1/f1, 1/2], t=-t0..2*Pi-t0); f2:=t -> a*t^2-a*t0^2+2:
```

$$t_0 := \frac{1}{2}\arcsin\left(\frac{1}{2}\,2^{1/4}\right)$$

```
> a:=2: f:=piecewise(t<t0, f2(t), t<Pi/2-t0, f1,
  t<Pi/2+t0, f2(t-Pi/2),
  t<Pi-t0,f1,t<Pi+t0,f2(t-Pi),t<3*Pi/2-t0,f1,

> plot([f, 2.02], t=-t0..2*Pi-t0, coords=polar,
  thickness=[2, 1], linestyle=[1, 2]);
```

7
Asymptotes of Curves

One can distinguish *vertical, horizontal*, and *oblique* asymptotes of plane curves with respect to rectangular coordinates (see Fig. 7.1 a). An oblique asymptote to a graph can be *left* (if $x \to -\infty$), *right* (if $x \to \infty$), and *two-sided* if the graph approaches the asymptote for both $x \to -\infty$ and $x \to \infty$.

Curves sometimes have *asymptotic points*, for example, the clothoid, lituus, hyperbolic and logarithmic spirals, and the circular tractrix.

In this short chapter four situations for plotting asymptotes are considered: curves are graphs in Cartesian and polar coordinates, parametrized curves in plane and in space.

In Chapter 7 the reader will become acquainted with the commands

`union, spacecurve.`

1. We derive the coefficients in the equation $y = kx + b$ of the asymptote of the graph of $y = f(x)$ for $x \to \pm\infty$ as follows:

$$k_+ = \lim_{t\to\infty} \tfrac{f(x)}{x}, \qquad b_+ = \lim_{t\to\infty} [f(x) - k_+ x].$$

Analogously we derive coefficients k_- and b_- when $x \to -\infty$.

Example 7.0.1 Let us derive and plot asymptotes of the graph

a) $y = \sqrt[3]{x^3 + 2}$, b) $y = \sqrt[3]{6x^2 - x^3}$, c) $y = \frac{x^3}{2(x+1)^2}$.

a) *Answer*: two-sided asymptote $y = x$; Fig. 7.1 b.

```
> f:=surd(x^3+2, 3): x1:=2:
> k:=limit(f/x, x=infinity); b:=limit(f-k*x, x=infinity);
```

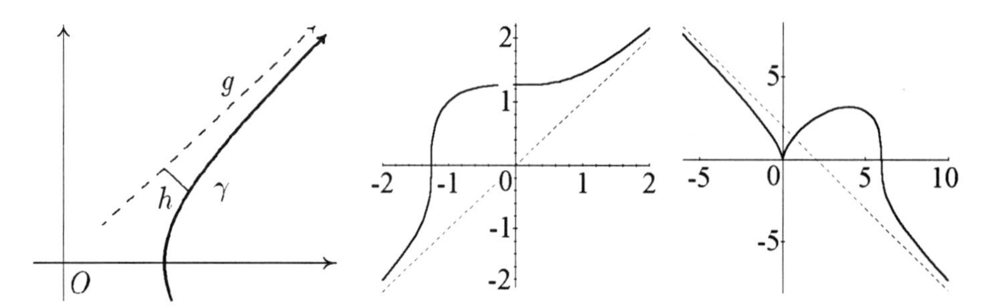

Figs. 7.1 a–c. Asymptotes of the graphs

$k := 1 \qquad b := 0$

```
> plot([f,k*x+b],x=-x1..x1,thickness=[2,1],linestyle=[1,2]);
```

b) *Answer*: the two-sided asymptote $y = -x + 2$, Fig. 7.1 c.

```
> f:=surd(6*x^2-x^3,3):
> k:=limit(f/x, x=-infinity); b:=limit(f-k*x, x=-infinity);
```

$k := -1 \qquad b := 2$

```
> plot([f,k*x+b],x=-6..10,thickness=[2,1],linestyle=[1,2]);
```

c) *Answer*: vertical and oblique asymptotes are the following:

$$x + 1 = 0, \quad x - 2y - 2 = 0.$$

```
> f:=x^3/(2*(x+1)^2):
  k:=limit(f/x, x=infinity); b:=limit(f-k*x, x=infinity);
```

$k := \frac{1}{2} \qquad b := -1$

```
> p.1:=plot(f, x=-0.84..5): p.2:=plot(f, x=-6..-1.3):
  p.3:=plot(k*x+b, x=-6..5, linestyle=2):
  p.4:=plot([-1, x, x=-12..1.8], linestyle=2):
> plots[display]([seq(p.i, i=1..4)]); # Fig. 7.2
```

Remark 7.0.1 We plot the asymptotes of the curve $\rho = \rho(\varphi)$ in *polar coordinates* (see Sections 6.2, 6.4 and 6.5) as follows:

(1) *Derive the values* φ_i, *for which* $\lim_{\varphi \to \varphi_i} |\rho(\varphi)| = \infty$.

(2) *Let* $p_i = \lim_{\varphi \to \varphi_i} \frac{\rho^2(\varphi)}{|\rho'(\varphi)|}$. *If this limit exists, then the infinite branch of the curve has the asymptote* $\rho_i = \frac{p_i}{\cos(\varphi - \varphi_i)}$, *and if the limit does not exist, then the infinite branch has no asymptotes.*

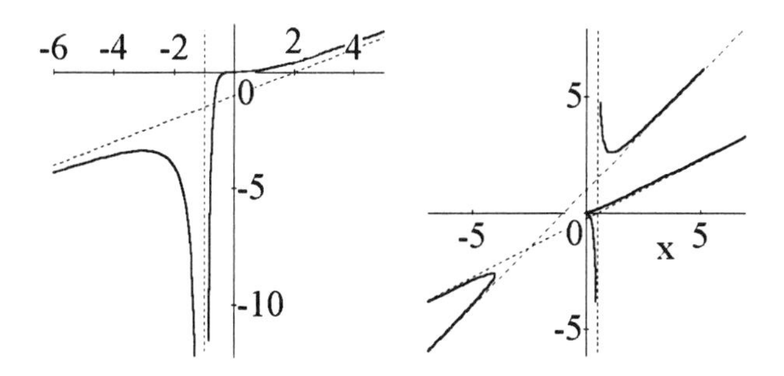

Figs. 7.2–7.3. Curves with asymptotes

If $p_i > 0$, then the asymptote lies on the right of the position vector of the curve, which runs to infinity. If $p_i < 0$, then the asymptote lies on the left side.

2. We calculate and plot the asymptotes of the parametrized plane curve $[x(t), y(t)]$ by the following plan (for convenience we write the results of our calculations in a table).

(1) *Find points* $\{a_i\}$ *where the curve runs to* ∞, *i.e.,* $\lim_{t\to a_i}[x^2(t)+y^2(t)]=\infty$.

(2) *For each* $t = a_i$ *derive the coefficient of inclination* $k_i = \lim_{t\to a_i}\frac{y(t)}{x(t)}$.

(3) *For each finite* k_i *derive* $b_i = \lim_{t\to a_i}[y(t) - k_i x(t)]$ *and plot the oblique asymptote.*

(4) *For each infinite* k_i *derive* $x_i = \lim_{t\to a_i} x(t)$ *and plot the vertical asymptote* $x = x_i$.

Example 7.0.2 Let us derive and plot asymptotes of the parametrized curve $[\frac{t^2}{1-t}, \frac{t^3}{1-t^2}]$.

Answer: The curve has four points running to infinity. Its three asymptotes are $x = \frac{1}{2}$, $y = \frac{1}{2}x - \frac{1}{4}$, $y = x + 1$; Fig. 7.3.

```
> X:=t^2/(1-t): Y:=t^3/(1-t^2);
> N:=`union` (-6..-1.15,-0.9..0.9,1.15..6):
> p1:=plot([seq([X, Y, t=N[i]], i=1..3)], thickness=2):
> A:=[-infinity,-1,1,infinity]: for i from 1 to nops(A) do
  k.i:=limit(Y/X, t=A[i]): b.i:=limit(Y-k.i*X, t=A[i]) od;
```

$k_1 := 1 \quad b_1 := 1 \quad k_2 := \text{undefined} \quad b_2 := \text{undefined}$

$k_3 := \frac{1}{2} \quad b_3 := -\frac{1}{4} \quad k_4 := 1 \quad b_4 := 1$

```
> p2:=plot([seq(k.i*x+b.i, i=[1,3,4]), [subs(t=A[2], X),
```

```
  x, x=-6..8]], x=-7..7, linestyle=2): %;
> plots[display]([p1, p2], scaling=constrained);
```

3. We plot the asymptotes $y = B + Mx$, $z = C + Nx$ of the space curve $\vec{\mathbf{r}}(t) = [x(t),\ y(t),\ z(t)]$ as follows:

(**1**) *Find the points* $\{a_i\}$ *where the curve runs to* ∞*, i.e.,* $\lim_{t\to a_i}[x^2(t) + y^2(t) + z^2(t)] = \infty$.

(**2**) *For each* $t = a_i$ *derive the coefficients of inclination* $M_i = \lim_{t\to a_i} \frac{y(t)}{x(t)}$ *and* $N_i = \lim_{t\to a_i} \frac{z(t)}{x(t)}$.

(**3**) *For each finite* $\{M_i,\ N_i\}$ *derive* $B_i = \lim_{t\to a_i}[y(t) - M_i x(t)]$ *and* $C_i = \lim_{t\to a_i}[z(t) - N_i x(t)]$*; then plot the oblique asymptote.*

(**4**) *For each infinite* M_i *derive* $x_i = \lim_{t\to a_i} x(t)$*; for each infinite* N_i *derive* $x_i = \lim_{t\to a_i} x(t)$*, and then plt the vertical asymptotes.*

Example 7.0.3 Let us plot the following space curve and find its three asymptotes (coordinate axes):

```
> f:=exp(-t^2):
> r:=[f*(t-2)*(t-1)/t, f*(t-2)*t/(t-1), f*t*(t-1)/(t-2)];
> plots[spacecurve](r, t=-1...2, thickness=2);
```

The command `spacecurve` is studied in Section 8.1.1.

8
Space Curves

Section 8.1 is a short introduction to the basic MAPLE support for plotting space curves using various coordinate systems. Section 8.2 is devoted to helix type curves lying on a cylinder, a sphere, a torus and other surfaces of revolution. In Section 8.3 we study remarkable MAPLE strategies for producing splendid views (of three-dimensional objects) with coloring, shadows and moving. We plot curves that are intersections of two surfaces: the Viviani curve, bicylindrical curves, and solutions of Euler's equations from mechanics. In the final Section 8.4, we deal with vector fields and their trajectories in space (see Section 5.4 for two-dimensional case).

In Chapter 8 the reader will become acquainted with the commands

`spacecurve, tubeplot, plot3d, display3d, DEplot3d, textplot3d, fieldplot3d, gradplot3d, polygonplot3d, PLOT3D, ANIMATE;`
`det, cosh, sinh, JacobiCN, JacobiSN, JacobiDN, transform.`

We begin using the options of 3D plotting:

`light, ambientlight, projection, orientation.`

8.1 Introduction

8.1.1 *Commands* `spacecurve` *and* `tubeplot`

A space curve $\vec{\mathbf{r}}(t) = [x(t),\ y(t),\ z(t)]$ is plotted using one of the following commands from the library `plots`:

```
spacecurve(one or several curves, options);
```

```
tubeplot(one or several curves, options);
```

Their syntax is similar; the difference is that `tubeplot` plots a *tubular surface* around the given curve, so the options with lighting are effective.

Example 8.1.1

1. Let us show that the curve $\vec{\mathbf{r}}\,(t) = [1+3t+2t^2,\ 2-2t-4t^2,\ 1-t^2]$ lies on a plane and describe it.

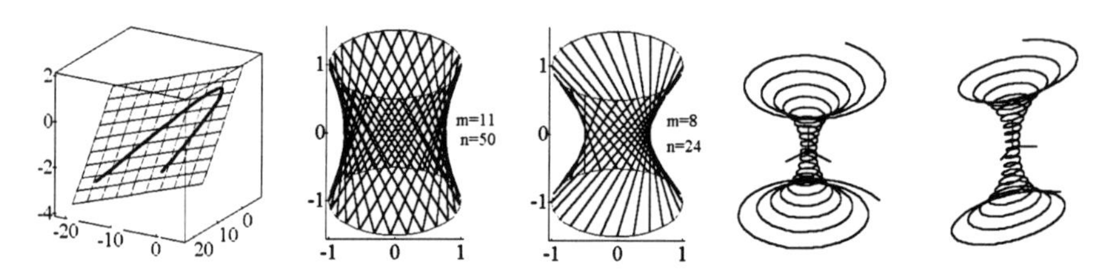

Figs. 8.1–8.5. Plane curve in space. Hyperboloid. Curve on a catenoid

Hints. Note that the curve $\vec{\mathbf{r}}\,(t) = \vec{\mathbf{r}}_1 + t\,\vec{\mathbf{r}}_2 + t^2\,\vec{\mathbf{r}}_3$ lies in a plane if the three vectors $\vec{\mathbf{r}}_1, \vec{\mathbf{r}}_2, \vec{\mathbf{r}}_3$ are coplanar, i.e., the determinant consisting of their coordinates vanishes. Any two (linearly independent) vectors from among them define the plane. In our case

```
> r1:=[1, 2, 1]: r2:=[3, -2, 0]: r3:=[2, -4, -1]:
> linalg[det]([r1, r2, r3]); with(plots):
```

0

```
> A:=spacecurve(evalm(r1+t*r2+t^2*r3),t=-2..2,thickness=3):
> B:=plot3d(evalm(u*r1+v*r2), u=-4..2, v=-1..7):
> display([A, B], orientation=[30,70]);  # Fig. 8.1
```

2. Using cycloidal curves (close plane curves) we obtain three-dimensional curves: astroid ($m = -\frac{1}{4}$), deltoid ($m = -\frac{1}{3}$), nephroid ($m = \frac{1}{2}$), and three-dimensional roses.

```
> m:=-1/4: h:=1:  # 3—dimensional astroid, then enter your data
  > r3:=[(1+m)*cos(m*t)-h*m*cos(t+m*t),
  (1+m)*sin(m*t)-h*m*sin(t+m*t),cos(t)]:
  plots[spacecurve](r3,t=0..2*Pi/abs(m),axes=boxed);
```

3. We plot *two engaged circles* (tori).

```
> plots[tubeplot]({[cos(t),sin(t),0],[0,cos(t)+.8,sin(t)]},
  t=0..2*Pi,radius=.25,tubepoints=40,orientation=[0, 60],
  lightmodel=light2, style=patchnogrid);# Fig. 8.6
```

Then we generalize this for a *chain* containing $2n$ circles; Fig. 8.7.

Figs. 8.6–8.8. Two engaged circles. Short chain. Chain armor

```
> n:=3: P:=i -> plots[tubeplot]({[2*cos(t),2*sin(t)+6*i,0],
  [0,2*cos(t)+3+6*i,2*sin(t)]}, t=0..2*Pi, radius=.4,
  orientation=[0,60],style=patchnogrid, lightmodel=light4):
> plots[display]([seq(P(i),i=0..n-1)],scaling=constrained);
```

Finally we generalize it for the *chain armor* with m lines; Fig. 8.8.

```
> with(plots): n:=4: m:=4: P:=(i,j)-> tubeplot({[2*cos(t)+6*j,
  2*sin(t)+6*i,0],[6*j,2*cos(t)+3+6*i,2*sin(t)], [2*cos(t)-
  3+6*i,0+6*j,2*sin(t)]},t=0..2*Pi,radius=.4, orientation=
  [40,50],style=patchnogrid,lightmodel=light4): display([(
  seq(seq(P(i,j), i=0..n), j=0..m))], scaling=constrained);
```

8.1.2 Intuitive Projection

How is a space curve $\vec{\mathbf{r}}(t) = [x(t), y(t), z(t)]$ (or more generally, any space figure) plotted on a planar computer display? Sometimes one creates the image by intuition, as in the following *method of plotting the hyperboloid of one sheet*, using our knowledge of a *star-shaped polygon*.

The coordinates of the sequence of vertices of the star-shaped (n, m) polygon are calculated by the formulas $[R\cos(2\pi i\frac{m}{n}),\ R\sin(2\pi i\frac{m}{n})]$ $(1 \le i \le n)$; see Section 6.1. We transform the circumscribed circle of this polygon to the ellipse and raise and lower it (in the plane) a distance of ± 1 and obtain two star-shaped polygons. Now we connect the ith vertex of one of them with the $(i+1)$th vertex of another, and obtain a part of the hyperboloid of one sheet. We plot it by two methods.

```
> m:=11: n:=50: T:=i -> i*2*m*Pi/n: # first method; Fig. 8.2
> p1:=plot([seq([cos(T(i)),.5*sin(T(i))+(-1)^i],i=1..m*n)]):
  p2:=plot([[cos(t), 0.5*sin(t)+1, t=0..2*Pi],
  [cos(t), 0.5*sin(t)-1, t=0..2*Pi]]):
  plots[display]([p1, p2], scaling=constrained);
```

```
> m:=8: n:=3*m: T:=i-> i*2*Pi/n: # second method; Fig. 8.3
> p1:=plot([seq([[cos(T(i)), 0.5*sin(T(i))-1],
  [cos(T(i+m)), 0.5*sin(T(i+m))+1]], i=1..n)]):
  p2:=plot([[cos(t),.5*sin(t)+1,t=0..2*Pi],
  [cos(t),.5*sin(t)-1,t=0..2*Pi]]):
  plots[display]([p1, p2], scaling=constrained);
```

8.1.3 Standard Coordinate Systems in Space

Other than `rectangular` coordinates, there are two especially popular methods for fixing the place of the point P in the space.

(a) Consider the *height* z as the projection of the segment OP onto the axis OZ; the *distance* $|OP'| = \rho$, where P' is the projection of the point P onto the plane XY; and the *angle* $\varphi \in [-\pi, \pi]$ between the segment OP' and the axis OX. The triple of numbers $(\rho,\ \varphi,\ z)$ is called the *cylindrical coordinates* of the point P. The relation with the rectangular coordinates (x, y, z) is analogous to the case of the polar coordinates:

$$x = \rho \cos\varphi, \quad y = \rho \sin\varphi, \quad z = z.$$

Coordinate surfaces are cylinders $\rho = const$, half-planes $\varphi =$ const through the axis OZ, and also horizontal planes $z =$ const.

(b) Consider the *distance* $|OP| = \rho$, the (*vertical*) *angle* $\phi \in [-\pi, \pi]$ between the segment OP and axis OZ and the (*horizontal*) *angle* $\theta \in [0, \pi]$ between the segment OP' and the axis OX. The triple of numbers $(\rho,\ \theta,\ \phi)$ is called the *spherical coordinates* of the point P. The relation with rectangular coordinates is the following:

$$x = \rho \cos\theta \sin\phi,\ y = \rho \sin\theta \sin\phi,\ z = \rho \cos\phi.$$

Coordinate surfaces are spheres $\rho =$ const with centers at O, half-planes $\theta =$ const through the axis OZ, and circular cones $\phi =$ const with the common axis OZ.

Plotting curves in cylindrical or spherical coordinates is possible by the standard command `spacecurve` with the additional option `coords= cylindrical` or `coords= spherical`.

In addition to the above three, in `MAPLE` there are about thirty systems of curvilinear coordinates in $\mathbb{R}^3$ (see Section 19.3.2), and we can define a number of new coordinate systems using the command `addcoords`.

8.2 Knitting on Surfaces of Revolution

8.2.1 Simplified Model of Projection

For a clearer understanding of how to plot images using *parallel projection*, consider a simplified model that reduces the problem to the basic 2-dimensional command `plot`. We also project the coordinate system $\{O;\ \vec{\mathbf{i}},\ \vec{\mathbf{j}},\ \vec{\mathbf{k}}\}$ of space for visibility. Let O' be the origin of rectangular coordinates $\{O';\ \vec{\mathbf{e}}_1, \vec{\mathbf{e}}_2\}$ of the 2-dimensional window, and let $O'X'$, $O'Y'$ be the horizontal and vertical axes.

We consider two examples of projections $p: \mathbb{R}^3 \to \mathbb{R}^2$ that are similar to the standard *isometric* and *dimetric orthogonal projections* (*axonometric projections* are studied in greater detail in *descriptive geometry*). By the *Pohlke–Schwarz Theorem*, the system $\{O;\ \vec{\mathbf{i}}, \vec{\mathbf{j}}, \vec{\mathbf{k}}\}$ can be projected (within the transformation of similarity in the plane of the image) onto an arbitrary triple of vectors through the point O'.

Assume that the projections of the axes OX, OY form the angles $\pm(90°+\arctan(\frac{1}{2})) \approx \pm 117°$ with the axis $O'Y'$, and the projection of OZ coincides with the axis $O'X'$: $p(O) = O'$, $p(\vec{\mathbf{i}}) = -\vec{\mathbf{e}}_1 - \frac{1}{2}\vec{\mathbf{e}}_2$, $p(\vec{\mathbf{j}}) = \vec{\mathbf{e}}_1 - \frac{1}{2}\vec{\mathbf{e}}_2$, $p(\vec{\mathbf{k}}) = \vec{\mathbf{e}}_2$. In view of the *linearity* of projections, we obtain

$$x'\vec{\mathbf{e}}_1 + y'\vec{\mathbf{e}}_2 = p(x\vec{\mathbf{i}} + y\vec{\mathbf{j}} + z\vec{\mathbf{k}}) = xp(\vec{\mathbf{i}}) + yp(\vec{\mathbf{j}}) + zp(\vec{\mathbf{k}})$$
$$= x(-\vec{\mathbf{e}}_1 - \tfrac{1}{2}\vec{\mathbf{e}}_2) + y(\vec{\mathbf{e}}_1 - \tfrac{1}{2}\vec{\mathbf{e}}_2) + z\vec{\mathbf{e}}_2 = (y-x)\vec{\mathbf{e}}_1 + (z - \tfrac{x+y}{2})\vec{\mathbf{e}}_2.$$

Hence $\boxed{x' = y - x,\ y' = z - \frac{1}{2}(x+y).}$ The point $P(x, y, z)$ projects onto the point $P'(y - x,\ z - (x + y)/2)$ of the display. Let us call this projection `isometric`.

`Dimetric` projection is defined analogously: the axes OY, OZ are projected onto the axes $O'X'$, $O'Z'$ of the 2-dimensional window, and the projection of the axis OX is parallel to the bisector of the 3-dimensional coordinate angle, namely,

$$\boxed{x' = y - \frac{1}{2}x,\ y' = z - \frac{1}{2}x.}$$

There are well-known surfaces obtained by revolution of a plane curve γ (the *meridian*) around a straight line (the *axis of revolution*) in space.

Trajectories of points of the curve γ (*parallels*) are circles contained in planes orthogonal to the axis of revolution. Let the curve γ lie in the plane XOZ and be given by parametric equations $\gamma: x = f(t),\ y = 0,\ z = g(t)$. Denoting by $u := t$ and $v \in [0, 2\pi]$ the *angle of revolution* of the initial meridian, we obtain the following *equations of the surface of revolution*:

$$\vec{\mathbf{r}(u, v)} = f(u)\cos(v)\vec{\mathbf{i}} + f(u)\sin(v)\vec{\mathbf{j}} + g(u)\vec{\mathbf{k}}$$
$$= [\, f(u)\cos(v),\ f(u)\sin(v),\ g(u)\,], \qquad \text{see Section 20.2.}$$

The idea of *knitting* on the surface of revolution on the display is as follows. If one winds a thread on a transparent surface of revolution, then the shape of the surface would be visible. This effect exists for a dense winding of a surface of revolution, i.e., when coils follow almost along parallels. For the realization of this idea, substitute t for u and at for v into the equation of a surface of revolution, where the real number a is sufficiently large, for example, $a = 27$: $\vec{\mathbf{r}}(t) = [\, f(t)\cos(at), f(t)\sin(at), g(t)\,]$, or in cylindrical coordinates (see below) $\vec{\mathbf{r}}(t) = [\, f(t),\ at,\ g(t)\,]$.

For example, we take $f = R$ and $g = t$ for the circular cylinder, $f = t,\ g = bt$ for the cone, $f = R\cos(t),\ g = R\sin(t)$ for the sphere, $f = a + b\cos(t),\ g = b\sin(t)$ with $a > b$ for the torus of revolution, $f = a\sin(t),\ g = a\left(\cos(t) + \ln(\tan\frac{t}{2})\right)$ for the pseudosphere (the surface of revolution for the tractrix).

Example 8.2.1 Let us plot isometric and dimetric projections of a *helix on the catenoid* (the surface of revolution for the catenary); Figs. 8.4–8.5.

```
> with(plots): t1:=4: isoaxes:=polygonplot([[[0,0], [-t1,
  -t1/2]], [[0,0],[t1, - t1/2]], [[0,0], [0,t1]]]):# isometric
> f:=cosh(t): g:=4*t: # enter your data
> a:=20: X:=f*cos(a*t): Y:=f*sin(a*t): Z:=g:
> isom:=plot([Y-X, Z-(X+Y)/2, t=-3..3]):
> display([isom, isoaxes], axes=none, scaling=constrained);
```

```
> with(plots): t1:=4: diaxes:=polygonplot([[[0,0], [-t1/2,
  -t1/2]], [[0,0], [t1,0]], [[0,0], [0,t1]]]): # dimetric
> f:=cosh(t): g:=4*t: # enter your curve
> a:=20: X:=f*cos(a*t): Y:=f*sin(a*t): Z:=g:
> dimetr:=plot([Y-X/2, Z-X/2, t=-3..3]):
> display([dimetr, diaxes], axes=none, scaling=constrained);
```

These programs have a "teaching character", but the basic approach is also implemented in the computer algebra system DERIVE. For plotting images of space objects below, the corresponding commands in MAPLE are used.

8.2.2 Main Model of Projection

In addition to the space *coordinates of an object*, an important role is also played by the *coordinates of the viewer*. The option `orientation=[θ,ϕ]` (by assumption [45,45]) points out the direction (in spherical coordinates) from which we look at the object with respect to the coordinate system of the object.

Example 8.2.2 Let us write equations of the following space curves lying on a cylinder and a cone and then plot them:

1. The point P rotates uniformly about a straight line l (that is parallel to the axis OZ) and at distance R from l. At the same time, P moves with constant velocity parallel to l.

Hint. The *circular helix* $\vec{\mathbf{r}}(t) = [R\cos t,\ R\sin t,\ bt]$, Fig. 8.9. A generalization is a curve $\vec{\mathbf{r}}(t) = [a\cos t, b\sin t, ct]$ on an elliptic cylinder.

```
> R:=1: a:=1: spacecurve([R, t, a*t, t=0..16*Pi],
  coords=cylindrical, numpoints=250, orientation=[40, 70]);
```

A circular helix can be obtained as the straight line (oblique to ruling) on the development of the circular cylinder.

2. The straight line OL not orthogonal to OZ rotates uniformly around O with the angular velocity $\omega = 1$, and the point P moves along OL:

(a) with velocity proportional to the length OP, Fig. 8.10.

Hint. $\vec{\mathbf{r}}(t) = [Re^{at}\cos t,\ Re^{at}\sin t,\ be^{at}]$ is a *conic helix*.

```
> a:=.05: spacecurve([exp(a*t), t, exp(a*t), t=0..16*Pi],
  coords=cylindrical, numpoints=150, orientation=[40,70]);
```

(b) with constant velocity; Fig. 8.11.

Hint. $\vec{\mathbf{r}}(t) = [at\cos t, at\sin t, bt]$ is the *circular conic helix*. A generalization is the curve $\vec{\mathbf{r}}(t) = [at\cos t, bt\sin t, ct]$ on the elliptic cone.

```
> a:=.05: spacecurve([a*t, t, t, t=-14*Pi..14*Pi],
  coords=cylindrical, numpoints=350, orientation=[40,70]);
```

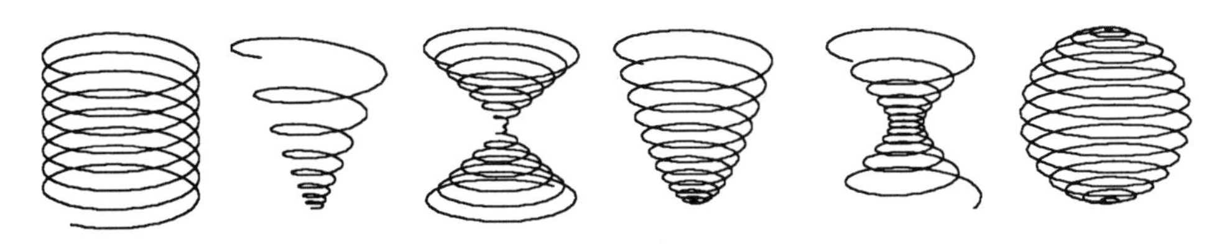

Figs. 8.9–8.14. Curves on algebraic surfaces of second order

Example 8.2.3 Springs on surfaces of revolution (see Section 20.2).

1. The curve $\vec{\mathbf{r}}(t) = [at\cos t,\ at\sin t,\ \frac{a^2t^2}{2p}]$ lies on the paraboloid of revolution, Fig. 8.12, and projects onto the plane XY as Archimedes' spiral.

```
> a:=1: spacecurve([a*t, t, t^2/2, t=0..24*Pi],
  coords=cylindrical, numpoints=500, orientation=[40,73]);
```

2. The curve $\vec{\mathbf{r}}(t) = [\cosh(at)\cos(t),\ \cosh(at)\sin(t),\ \sinh(at)]$ lies on a hyperboloid of revolution; Fig. 8.13.

```
> a:=.06:spacecurve([cosh(a*t),t,sinh(a*t),t=-12*Pi..12*Pi],
  coords=cylindrical, numpoints=300, orientation=[40,73]);
```

3. The curve (c) $\vec{\mathbf{r}}(t) = [a\cos t\ \cos(kt),\ b\cos t\ \sin(kt),\ c\sin t]$ lies on an ellipsoid (on a sphere if $a = b = c$, Fig. 8.14.)

```
> R:=1: a:=28: spacecurve([R, a*t, t, t=0..Pi],
  coords=spherical, numpoints=300, orientation=[20, 76]);
```

4. Let us plot several complicated curves on a sphere (similar to *macrame*).

```
> m:=7: n:=10: X:=cos(m*t)*cos(n*t):
  Y:=sin(m*t)*cos(n*t): Z:=sin(n*t):
> plots[tubeplot]([X, Y, Z, radius=.05], t=-Pi..Pi, color=
  [1,.8,.1],ambientlight=[.4,.4,.4],light=[75,50,1,.9,.5],
  orientation=[0,26],numpoints=450); # Fig. 8.16
```

For `m:=1: n:=17:` and `t=-Pi/2..Pi/2` we obtain the curve consisting of 17 meridians, Fig. 8.15.

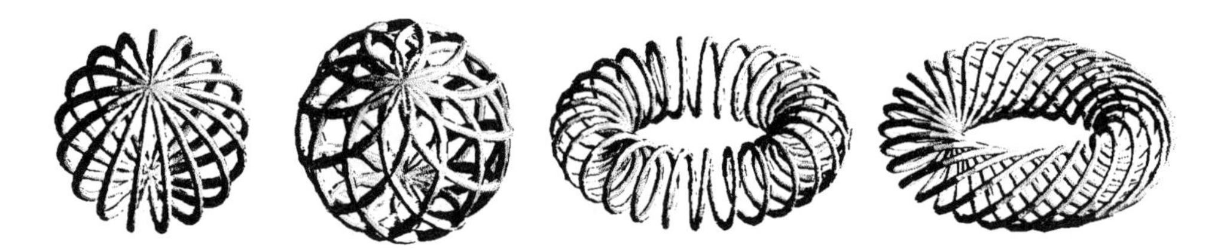

Figs. 8.15–8.18. Closed curves on a sphere and knots on a torus

5. We tie a *torus knot* of arbitrary type on the torus of revolution and also plot a part of the *irrational winding*. For the torus knot $\mathbf{K}_{8,27}$, we assume $R_1 = 3$, $R_2 = 1$ and $u = 27t$, $v = 8t$.

```
> m:=27: n:=8: X:=(3+cos(m*t))*cos(n*t):
  Y:=(3+cos(m*t))*sin(n*t): Z:=sin(m*t):
> plots[tubeplot]([X,Y,Z,radius=.1],t=-Pi..Pi,color=
  [1,.8,.1],ambientlight=[.4,.4,.4],light=[75,50,1,.9,.5],
  orientation=[0, 60], numpoints=450); # Fig. 8.18
```

8.3 Plotting Curves (Tubes) with Shadow

Analogous parameters as for the option `orientation=[θ, φ]` are used for the option `light=[φ,θ,R,G,B]` of *directed lighting* of the object (pay attention to the opposite order of angles), where R, G, and B are numbers between 0 and 1 corresponding to `red`, `green`, and `blue` components of the spectrum. If we do not fix the lighting direction, then `MAPLE` automatically uses one of its standard lighting schemes.

The additional option of *non-directed (diffuse) lighting* `ambientlight=[R,G,B]` gives the same effect as for auto-illumination of the surface of the object, similar when it has a mat surface.

The option `projection=`<*number* ∈ [0, 1]> allows one to plot an object in perspective (central projection). Moreover, the interval [0.5, 0.7] is usually most convenient; the extremal value 1 leads to the parallel projection.

We plot the shadow (as the plane curve) of the space curve (the circular helix) on the given plane using directed lighting.

Unfortunately, in `MAPLE` figures of 3-D objects have no shadows. Thus, we create the parallel projection (of the circular helix) ourselves using calculations with affine maps. Let the vector $\vec{\mathbf{d}} = [a, b, c]$ be nonparallel to the plane ω : $Ax + By + Cz = 0$ (with the normal vector $\vec{\mathbf{n}} = [A, B, C]$). Then the shadow of the point $P(x, y, z)$ on the plane ω under parallel lighting in the direction $\vec{\mathbf{d}}$ is the point $Q(x', y', z')$, where

$$\begin{aligned} x' &= \frac{(bB + cC)x - aBy - aCz}{aA + bB + cC}, \\ y' &= \frac{-bAx + (aA + cC)y - bCz}{aA + bB + cC}, \\ z' &= \frac{-cAx - cBy + (aA + bB)z}{aA + bB + cC}, \end{aligned} \iff \vec{\mathbf{r}}_Q = \vec{\mathbf{r}}_P - \frac{\vec{\mathbf{r}}_P \cdot \vec{\mathbf{n}}}{\vec{\mathbf{d}} \cdot \vec{\mathbf{n}}} \vec{\mathbf{d}}.$$

Example 8.3.1 Let us illustrate the following:

(a) The parallel projections of the circular helix onto the plane orthogonal to its axis, depending on the correlation of angles between the axis and the direction of projection with tangent lines to curves, are the usual curtate or prolate cycloids; Fig. 8.19.

(b) The central projection of the circular helix onto the plane orthogonal to its axis from a point on the helix is the cochleoid; Fig. 8.22.

Hint. (a) In the next program, `helix` denotes the circular helix (tube); `T` denotes vertices and faces of the tray (parallelepiped); `Tray` denotes the image of the tray; `shadow` denotes the shadow of the curve.

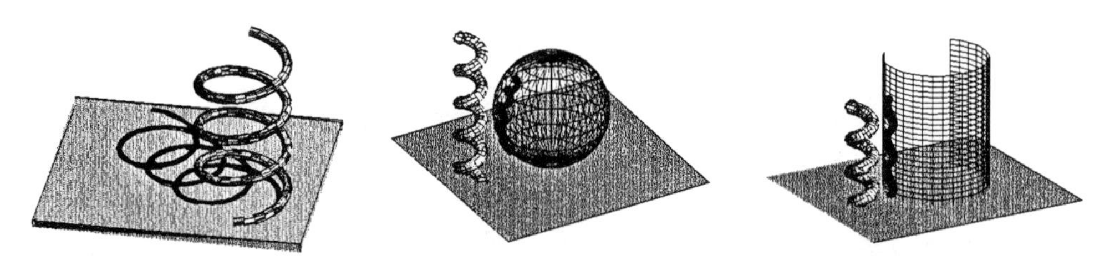

Figs. 8.19–8.21. Circular helix with a shadow on the plane, sphere, and cylinder

```
> A:=0: B:=0: C:=1: lv:=75: lh:=38: oh:=12: ov:=60:
> a:=evalf(sin((lv-(90-ov))/180*Pi)*cos((lh+oh)/180*Pi)):
> b:=evalf(sin((lv-(90-ov))/180*Pi)*sin((lh+oh)/180*Pi)):
> c:=evalf(cos((lv-(90-ov))/180*Pi)):
> R:=5: V:=.8: X:=R*cos(t): Y:=R*sin(t): Z:=V*t:
> with(plots): helix:=tubeplot([X, Y, Z, radius=0.5],
  t=0..7*Pi, numpoints=90, tubepoints=20,
  ambientlight=[.5, .5, .5], light=[lv, lh, .6, .8, .2]):
> Xs:=((b*B+c*C)*X-a*B*Y-a*C*Z)/(a*A+b*B+c*C):
> Ys:=(-b*A*X+(a*A+c*C)*Y-b*C*Z)/(a*A+b*B+c*C):
> Zs:=(-c*A*X-c*B*Y+(a*A+b*B)*Z)/(a*A+b*B+c*C):
> shadow:=tubeplot([Xs, Ys, Zs, radius=.5], t=0..7*Pi,
  numpoints=70, tubepoints=20, color=[.2, .3, .5]):
> T:=[[[8,-22,-.01], [-18,-22,-.01], [-18,8,-.01],
  [8,8,-.01]], [[8,-22,-1], [8,8,-1], [-18,8,-1],
  [-17,-22,-1]], [[8,8,-.01], [8,-22,-.01], [8,-22,-1],
  [8,8,-1]], [[8,8,-1], [8,8,-.01], [-18,8,-.01],
  [-18,8,-1]], [[-18,8,-.01], [-18,8,-1], [-18,-22,-1],
  [-18,-22,-.01]], [[-18,-22,-.01], [-18,-22,-1],
  [8,-22,-1], [8,-22,-.01]]]:
> tray:=polygonplot3d(T, color=COLOR(RGB, .4, .7, .2)):
> display3d([helix, shadow, tray], orientation=[oh, ov]);
```

(b) In the next program, the center of projection lies at the height `h`; `helix` denotes the circular helix (tube); T denotes vertices and faces of the tray (parallelepiped); `Tray` denotes the image of the tray; `shad1` denotes the shadow of the part of the curve below the center of projection; shad2 denotes the shadow of the part of the curve over the center; Fig. 8.22.

```
> lv:=75: lh:=38:  oh:=12: ov:=60:
> R:=.8: X:=R*(cos(t)-1): Y:=R*(sin(t)): Z:=.2*t: h:=.2*Pi*8:
> with(plots): helix:=tubeplot([X, Y, Z, radius=0.3],
  t=0..16*Pi, numpoints=90, tubepoints=15,
```

```
  ambientlight=[.5, .5, .5], light=[lv, lh, .6, .8, .2]):
> Xs:=h/(h-Z)*X: Ys:=h/(h-Z)*Y: Zs:=0:
> shad1:=tubeplot([Xs, Ys, Zs, radius=.3],
  t=0..7.95*Pi, numpoints=70, color=[.2, .3, .5]):
> shad2:=tubeplot([Xs,Ys,Zs,radius=.3], t=8.05*Pi..16*Pi,
  numpoints=70, color=[.2, .3, .5]):
> T:=[[[16,-22,-.01], [-18,-22,-.01], [-18,8,-.01],
  [16,8,-.01]], [[16,-22,-1], [16,8,-1], [-18,8,-1],
  [-18,-22,-1]], [[16,8,-.01], [16,-22,-.01], [16,-22,-1],
  [16,8,-1]], [[16,8,-1], [16,8,-.01], [-18,8,-.01],
  [-18,8,-1]], [[-18,8,-.01], [-18,8,-1],
  [-18,-22,-1], [-18,-22,-.01]], [[-18,-22,-.01],
  [-18,-22,-1], [16,-22,-1], [16,-22,-.01]]]:
> tray:=polygonplot3d(T, color=COLOR(RGB, .4, .7, .2)):
> display3d([helix, shad1, shad2, tray], orientation=[oh, ov]);
```

Example 8.3.2 For various impressions we plot projections of the (circular helix) curve onto the simple surfaces (a) cylinder and (b) sphere.

Hint. (a) We suppose that centers of projection lie on the axis of the cylinder; Fig. 8.21.

In the next program, `helix` denotes the circular helix (tube); `cyl` denotes the cylinder (frame) with the section; `Tray` denotes the surface of the tray (rectangular); `shad` denotes the shadow of the curve onto the cylinder.

```
> cyl:=plot3d([1, x, y], x=-Pi/4..5*Pi/4, y=0..2.5,
  coords=cylindrical, style=wireframe):
> R:=.2: X:=R*cos(t)+1*8*R: Y:=R*sin(t): Z:=R*t/2:
> with(plots): helix:=tubeplot([X, Y, Z, radius=R*.6],
  t=0..6*Pi, numpoints=99, light=[75, 40, .8, .7, .3],
  ambientlight=[.6, .6, .6]):
> Xp:=X/sqrt(X^2+Y^2): Yp:=Y/sqrt(X^2+Y^2): Zp:=Z:
> shad:=tubeplot([Xp, Yp, Zp, radius=R*.6/sqrt(X^2+Y^2)],
  t=0..6*Pi, numpoints=99, color=[.5, .2, .2]):
> tray:=polygonplot3d([[-2,-2,0], [-2,2,0], [2.5,2,0],
  [2.5,-2,0]], color=COLOR(RGB, .4, .7, .2)):
> display3d([shad, cyl, helix, tray], orientation=[60, 70]);
```

(b) We suppose that the center of the sphere is the center of projection; Fig. 8.20. In the next program, `helix` denotes the circular helix (tube); `sphere` denotes the sphere (frame); `Tray` denotes the tray (rectangular); `shad` denotes the shadow of the curve on the cylinder.

```
> sphere:=plot3d([1, x, y], x=-Pi/2..Pi/2, y=-Pi/2..3*Pi/2,
```

```
  coords=spherical, style=wireframe):
> R:=.2: X:=R*cos(t)+1*8*R: Y:=R*sin(t): Z:=R*t/2:
> with(plots): helix:=tubeplot([X, Y, Z, radius=R*.5],
  t=-2/R..6*Pi, numpoints=99, light=[75, 40, .8, .7, .3],
  ambientlight=[.5, .5, .5]):
> M:=sqrt(X^2+Y^2+Z^2): Xp:=X/M: Yp:=Y/M: Zp:=Z/M:
> shad:=tubeplot([Xp,Yp,Zp, radius=R*.5/M], t=-2/R..6*Pi,
  numpoints=99, color=[.5, .2, .2]):
> tray:=polygonplot3d([[-2,-2,-1], [-2,2,-1], [2,2,-1],
  [2,-2,-1]], color=COLOR(RGB, .4, .7, .2)):
> display3d([shad,sphere,helix,tray],orientation=[60,60]);
```

Example 8.3.3 Let us deduce explicit and parametric equations of the following curves, given as intersections of pairs of surfaces, and plot them.

(a) The axes (OX and OZ) of two circular cylinders of radii $R, r > 0$ intersect at a right angle, and at the intersection of the cylinders we obtain the curve *bicylinder*, which for $R \neq r$ consists of two closed curves, Fig. 8.25, and for $R = r$ breaks up into two intersecting ellipses that lie in the planes $x = y$ and $x = -y$.

Hint. The equations of the bicylinder are the following:

$$\begin{cases} x^2 + z^2 - r^2 = 0 \\ y^2 + z^2 - R^2 = 0 \end{cases} \iff \begin{cases} x = & r \cos t, \\ y = & \pm\sqrt{R^2 - r^2 \sin^2 t}, \\ z = & r \sin t, \end{cases} \quad t \in [0, 2\pi].$$

The option `style=wireframe` for the cylinder allows one to see the bicylinder in its entirety. In the next program we plot the family of the bicylinders under a change in the radius of one of the cylinders.

```
> with(plots): R:=2:m:=6: a:=array(1..m):b:=array(1..m):
> c:=plot3d([R, u, v], u=0..2*Pi, v=-2.5..2.5,
  coords=cylindrical, style=wireframe):
> for i from 1 to m do r:=R*((2*m+i)/(3*m)):
  a[i]:=spacecurve([R, t, sqrt(R^2-r^2*sin(t)^2),
  t=0..2*Pi], coords=cylindrical):
  b[i]:=spacecurve([R, t, -sqrt(R^2-r^2*sin(t)^2),
  t=0..2*Pi], coords=cylindrical) od:
> display(c, seq([a[i],b[i]], i=1..m),orientation=[30,80]);
```

(b) The sphere of radius R (with the center O) intersects the surface of the cylinder of the diameter R, one of whose rulings (parallel to the axis OZ)

contains the center of the sphere. The intersection is the *Viviani curve*; Fig. 8.23.

Hint. The equations of the curve are the following:

$$\begin{cases} x^2 + y^2 + z^2 = R^2 \\ x^2 + y^2 - Rx = 0 \end{cases} \iff \begin{cases} x = & R\cos^2 t, \\ y = & R\cos t\ \sin t, \\ z = & R\sin t, \end{cases}$$

where $t \in [0,\ 2\pi]$ is a longitude of points on a sphere. A point of self-intersection $(R, 0, 0)$ on a Viviani curve is divided into two loops. Projections of a Viviani curve onto coordinate planes YZ, XZ and XY look like a number 8, a parabola, and a circle, respectively; Fig. 8.24.

```
> f:=[cos(t)^2, cos(t)*sin(t), sin(t)]: V:=array(1..2,0..1):
> for i from 1 to 2 do for j from 0 to 1 do V[i,j]:=
  spacecurve(f, t=0..2*Pi, orientation=[90*j, 90*i]) od od:
> plots[display](V);   # Fig. 8.24
```

(c) The sphere of radius R (with the center O) intersects the ellipsoid with the axes $a > b > c$. Plot the intersection curves on the ellipsoid varying radius R of the sphere.

Hint. The best method for parametrizing these curves comes from mechanics, namely, from solutions of Euler's differential equations of the rotation of a rigid body about a fixed point in space

$$\begin{cases} l_1{}' + (\frac{1}{J_2} - \frac{1}{J_3})\, l_2 l_3 = 0 \\ l_2{}' + (\frac{1}{J_3} - \frac{1}{J_1})\, l_3 l_1 = 0 \\ l_3{}' + (\frac{1}{J_1} - \frac{1}{J_2})\, l_1 l_2 = 0\,. \end{cases}$$

Here the J_i are the moments of inertia of a rotating body about the principal axes at time t, ω_i are the angular velocities about these axes, and $l_i = J_i\omega_i$ are the moments of a motion. We suppose $J_1 < J_2 < J_3$ and $l^2 > 2EJ_2$. Euler's equations have two integrals, and the (periodical) trajectory of a motion lies at the intersection of the sphere and the ellipsoid

$$\begin{cases} l_1^2 + l_2^2 + l_3^2 = l^2 \\ \frac{l_1^2}{J_1} + \frac{l_2^2}{J_2} + \frac{l_3^2}{J_3} = 2E \end{cases} \iff \begin{cases} l_1 = & J_1\,\omega_1^0\,\mathrm{cn}(z,k) \\ l_2 = & J_2\,\omega_2^0\,\mathrm{sn}(z,k) \\ l_3 = & J_3\,\omega_3^0\,\mathrm{dn}(z,k), \end{cases}$$

where $\mathrm{cn}(z,k)$, $\mathrm{sn}(z,k)$ and $\mathrm{dn}(z,k)$ are the *Jacobi elliptical functions* (with period $4K$, where $K =$ `EllipticK(k)`, see Section 1.3); we call them `JacobiCN(z,k);` `JacobiSN(z,k);` and `JacobiDN(z,k);` in `MAPLE`,

and $\omega_1^0 = \sqrt{\frac{2EJ_3 - l^2}{J_1(J_3 - J_1)}}$, $\quad \omega_2^0 = \sqrt{\frac{2EJ_3 - l^2}{J_2(J_3 - J_2)}}$, $\quad \omega_3^0 = \sqrt{\frac{l^2 - 2EJ_1}{J_3(J_3 - J_1)}}$,

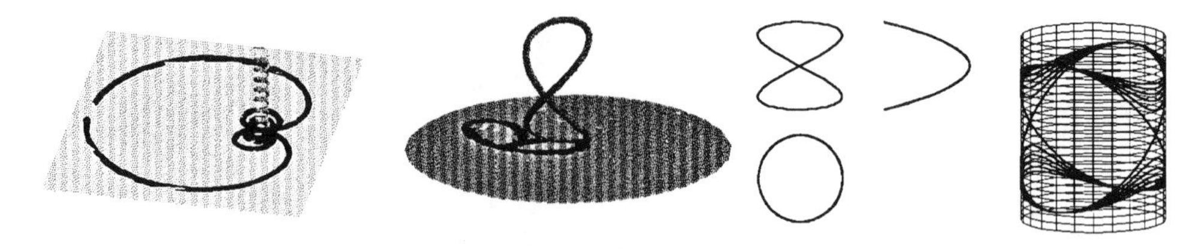

Figs. 8.22–8.25. Shadow of a circular helix. Viviani curve. Bicylinder

$$z = \sqrt{\frac{(J_3-J_2)(l^2-2EJ_1)}{J_1J_2J_3}}\, t, \qquad k = \sqrt{\frac{(J_2-J_1)(2EJ_3-l^2)}{(J_3-J_2)(l^2-2EJ_1)}}.$$

In the next program we plot the curve $\vec{\mathbf{r}}(t) = [X, Y, Z]$ (solution of Euler's equations with $Y(0) = 0$) with the corresponding intersecting sphere and ellipsoid.

```
> J_1:=1: J_2:=2: J_3:=16: L:=2.1: E:=1.1:      # enter your data
> L^2-2*E*J_2;        # check that the expression > 0
> omega_01:=sqrt((2*E*J_3-L^2)/(J_1*(J_3-J_1))):
  omega_02:=sqrt((2*E*J_3-L^2)/(J_2*(J_3-J_2))):
  omega_03:=sqrt((-2*E*J_1+L^2)/(J_3*(J_3-J_1))):
> z:=sqrt((J_3-J_2)*(L^2-2*E*J_1)/(J_1*J_2*J_3))*t:
> k:=sqrt(((J_2-J_1)*(2*E*J_3-L^2))/((J_3-J_2)*(L^2-2*E*J_1))):
  K:=EllipticK(k);
> X:=J_1*omega_01*JacobiCN(z,k):
  Y:=J_2*omega_02*JacobiSN(z,k):
  Z:=J_3*omega_03*JacobiDN(z,k):
> crv:=plots[spacecurve]([X,Y,Z], t=0..4*K+.25, thickness=3):
> sph:=plottools[sphere]([0,0,0],L):
> ell:=plottools[scale](sph, sqrt(2*E*J_1)/L, sqrt(2*E*J_2)/L,
  sqrt(2*E*J_3)/L):
> plots[display]([crv, sph, ell], scaling=constrained);
```

Example 8.3.4 Let us plot the Viviani curve with the shadow and the revolution about the axis OZ; Fig. 8.23.

Hints.

1. The vertical angles of orientation ov and lighting lv keep their values, but the horizontal angles depend on k. Assuming horizontal angles of orientation oh equal to $k \cdot 20$, we obtain the effect of rotation of the viewer in the opposite direction (each still corresponds to 20°). With 18 stills we have the whole circle (360°) around the object. To provide constant lighting with respect to the object (but not with respect to viewer), one must recalculate the values of the angle lh. We do not need this recalculation when using the traditional method of animation based on revolution of the object, but in this case the calculations

are more complicated. Namely, choose the initial lighting direction as follows: $lv = 55°$, $lh = 70°$; and using the correspondence $lh + oh = 70$, we obtain $lh = 70 - k \cdot 20$.

2. Details of the commands `ANIMATE` and `PLOT3D` can be found under the Maple help topic `plot,structure`.

In the next program, `tube,tube2` denote Viviani curves (tubes); `sphere` denotes a sphere (frame); `T` denotes a circular tray; `shad` denotes the parallel projection onto a tray; `op(tube)` denotes a Viviani curve (tube); and `op(shad(tube2)` denotes a shadow of a Viviani curve (tube). For different impressions, animation in the program is given in a form other than in previous examples.

```
> with(plots): with(plottools):
> A:=0: B:=0: C:=1: lv:=55: lh:=70-k*20: oh:=k*20: ov:=75:
> a:=evalf(sin((lv-(90-ov))/180*Pi)*cos((lh+oh)/180*Pi)):
> b:=evalf(sin((lv-(90-ov))/180*Pi)*sin((lh+oh)/180*Pi)):
> c:=evalf(cos((lv-(90-ov))/180*Pi)):
> shad:=transform((x,y,z,k)->[((b*B+c*C)*x-a*B*y-a*C*z)/
  (a*A+b*B+c*C), (-b*A*x+(a*A+c*C)*y-b*C*z)/(a*A+b*B+c*C),
  (-c*A*x-c*B*y+(a*A+b*B)*z)/(a*A+b*B+c*C)]):
> R:=2: f:=[R*cos(t)^2, R*cos(t)*sin(t), R*sin(t)+R]:
> tube:=tubeplot(f, t=0..2*Pi, radius=.1, color=[.8,.3,.5]):
> tube2:=tubeplot(f, t=0..2*Pi,radius=.1,color=[.2,.3,.3]):
> T:=plot3d([s,t,0], s=0..2*R,t=0..2*Pi, coords=cylindrical,
  color=[t/(2*Pi), .7, .2]):
> PLOT3D(ANIMATE(seq([op(T), op(tube), op(shad(tube2)),
  LIGHT(lv,lh, .8,.3,.5), PROJECTION(oh,ov,.5)], k=1..18)),
  AMBIENTLIGHT(.5, .5, .5), STYLE(PATCHNOGRID),
  SCALING(CONSTRAINED));
```

8.4 Trajectories of Vector Fields in Space

1. We plot three-dimensional vector fields using command `fieldplot3d` from the library `plots`.

```
> plots[fieldplot3d]([z*y, z*x, x*y], x=-1..1, y=-1..1,
  z=-1..1, grid=[5,5,5],arrows=SLIM,axes=boxed); # Fig. 8.26
```

An analogous result is obtained using the command `gradplot3d` from the library `plots`.

```
> plots[gradplot3d](x*y*z, x=-1..1, y=-1..1, z=-1..1,
```

```
grid=[5,5,5], arrows=SLIM, axes=framed);
```

2. Let us plot the three-dimensional curve that is the solution of a system of ODEs using the command `odeplot` from the library `plots`; Fig. 8.27.

```
> sys:=diff(y(x),x)=z(x), diff(z(x),x)=y(x):
  p:=dsolve({sys,y(0)=0,z(0)=1},{y(x),z(x)},type=numeric):
  plots[odeplot](p,[x,y(x),z(x)],-4..4,numpoints=25);
```

With several trajectories an analogous result (figure) is obtained using the command `DEplot3d` from the library `DEtools`.

```
> DEtools[DEplot3d]({D(x)(t)=y(t), D(y)(t)=-x(t)-y(t)},
  [x(t),y(t)],t=0..10, [[x(0)=0,y(0)=1], [x(0)=0,y(0)=.5]],
  stepsize=.1, tickmarks=[4,3,3]); # Fig. 8.28
```

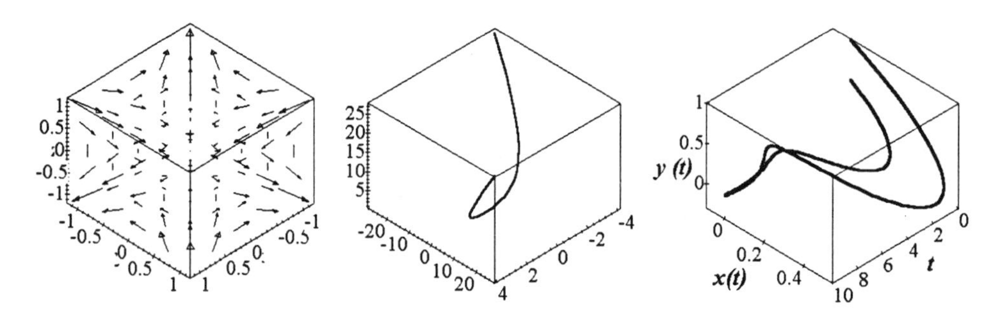

Figs. 8.26–8.28. Vector fields and trajectories in space

3. The command `textplot3d` from the library `plots` allows us to place text in three-dimensional plots.

```
> plots[textplot3d]([[1,2,3, `First solution`],
  [2,2,3, `Second solution`]]);
```

9

Tangent Lines to a Curve

In Section 9.1 we plot the (moving) tangent line and Frenet frame field for various types of equations of a curve; the equations of a tractrix are also derived. We then consider applications of tangent and normal lines to a curve.

In Section 9.2 curves appear as envelopes of families of straight lines, circles, and so on. In Section 9.3 we use `MAPLE` to realize some methods of mathematical embroidery from the literature. In Sections 9.4 and 9.5 we study evolutes, caustics and parallel curves (equidistants).

In Chapter 9 the reader will become acquainted with the command `minus`.

9.1 Tangent Lines

Definition 9.1.1 A *tangent line* to the curve γ at the point $P \in \gamma$ is the limit position of the secant PQ as the second point of intersection Q approaches P along the curve. *Semi-tangent lines from the right* and *from the left* are defined analogously.

Since the secant through the points at t and $t + \Delta t$ is parallel to the vector $\Delta\vec{\mathbf{r}} = \vec{\mathbf{r}}(t + \Delta t) - \vec{\mathbf{r}}(t)$ and by definition of the derivative $\vec{\mathbf{r}}'(t) = \lim\limits_{\Delta t \to 0} \frac{\Delta\vec{\mathbf{r}}}{\Delta t}$, the line through the point of the regular curve $\vec{\mathbf{r}}(t) \in \gamma$ in the direction of the velocity vector $\vec{\mathbf{r}}'(t) = [x'(t),\ y'(t),\ z'(t)]$ is the tangent line; Fig. 9.1.

The equation of the tangent line at the point t of the parametrized curve $\vec{\mathbf{r}}(t) \subset \mathbb{R}^3$ is the following:

$$\tilde{\mathbf{r}}(\tau) = \vec{\mathbf{r}}(t) + \tau\,\vec{\mathbf{r}}\,'(t) \quad (\tau \in \mathbb{R}) \iff \begin{cases} x = & x(t) + \tau\, x'(t), \\ y = & y(t) + \tau\, y'(t), \\ z = & z(t) + \tau\, z'(t). \end{cases}$$

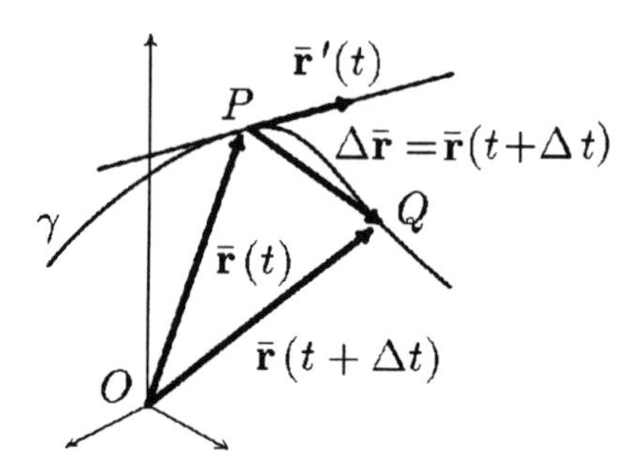

Fig. 9.1. Derivative of a vector-valued function and a tangent line

The *unit tangent and normal vectors of the plane curve* $\vec{\mathbf{r}}(t)$ are derived as follows: $\vec{\tau} = [\frac{x'}{\sqrt{x'^2+y'^2}}, \frac{y'}{\sqrt{x'^2+y'^2}}]$, $\vec{\nu} = [\frac{-y'}{\sqrt{x'^2+y'^2}}, \frac{x'}{\sqrt{x'^2+y'^2}}]$.

Equations of the tangent line to the plane curve (for various types of equations of the curve) are the following (below X, Y are coordinates of points on the tangent line, $\tau \in \mathbb{R}$):

graph $\quad Y = y + f'(x)(X - x),$

parametric $\quad X = x(t) + \tau\, x'(t),\ Y = y(t) + \tau\, y'(t),$

implicit $\quad \frac{\partial F}{\partial x}(x, y)(X - x) + \frac{\partial F}{\partial y}(x, y)(Y - y) = 0.$

Equations of the normal vector to the plane curve are the following:

graph $\quad X = x - \tau\, f\,'(x),\ Y = y + \tau,$

parametric $\quad X = x(t) - \tau\, y'(t),\ Y = y(t) + \tau\, x'(t),$

implicit $\quad X = x + \tau\, \frac{\partial F}{\partial x}(x, y),\ Y = y + \tau\, \frac{\partial F}{\partial y}(x, y).$

For an implicitly defined plane curve $F(x, y) = 0$, the *vector gradient* $\nabla F = [\frac{\partial F}{\partial x}, \frac{\partial F}{\partial y}]$ is parallel to its normal vector.

Example 9.1.1

1. Let us plot tangent lines to the graph of $y = f(x)$. Start with the parabola.

```
> f:=x^2/2: x1:=-2: x2:=2: m:=15: # enter your function
> fx:=diff(f, x): X:=i -> x1+i*(x2-x1)/m:
> p:=i -> subs(x=X(i), f+fx*(t-x)):
> Tp:=plot([seq(p(i), i=0..m)], t=x1..x2):
```

```
> Gf:=plot(f, x=x1..x2, thickness=2):
> plots[display]([Gf, Tp]); # Fig. 9.2
```

Continuation: We use two different methods to plot the moving normal vector along the graph, Fig. 9.3.

```
> N:=evalm([x, f]+s*[-fx, 1]/sqrt(1+fx^2)):
```
First method:

```
> Np:=plots[animate]([subs(x=X(i), N[1]), subs(x=X(i), N[2]),
  s=0..1], i=1..m):
> plots[display]([Np, Gf], scaling=constrained);
```

Second method (replace the line with Np):

```
> q:=i->plot([subs(x=X(i),N[1]),subs(x=X(i),N[2]),s=0..1]):
  Np:=plots[display]([seq(q(i), i=0..m)], insequence=true):
```

2. Let us plot the tangent lines and normal vectors to a parametrized plane curve; start with the cycloid:

```
> r:=[t-sin(t), 1-cos(t)]: m:=32: t1:=0: t2:=4*Pi:
> rt:=diff(r, t): Tp:=evalm(r+s*rt): # tangent line at a point t
> Np:=evalm(r+s*[-rt[2], rt[1]]):   # normal vector at a point t
> T:=i->t1+i*(t2-t1)/m:
  p:=i->plot([subs(t=T(i), Tp[1]), subs(t=T(i), Tp[2]),
  s=-2..2], color=blue):
  q:=i->plot([subs(t=T(i), Np[1]), subs(t=T(i), Np[2]),
  s=0..1], color=red):
> plots[display](seq([p(i), q(i)], i=0..m));
```

Continuation: We plot the *moving Frenet frame* along the curve.

```
> with(plots): B:=plot([r[1], r[2], t=t1..t2]):
> A1:=display(seq(p(i), i=0..m), insequence=true):
> A2:=display(seq(q(i), i=0..m), insequence=true):
> display([A1, A2, B], scaling=constrained); # Fig. 9.6
```

If the curve has no singular points (the ellipse `r:=[2*sin(t), cos(t)]` with
`t2:=2*Pi`) then we use its unit normal vector, Fig. 9.7.

```
> Np:=evalm(r+s*[-rt[2], rt[1]]/linalg[norm](rt, 2)):
```

3. *A dog runs along the axis OY starting from $O(0,0)$, and its owner (initially staying on the axis OX) follows the dog, pulling on the leash of length a. What is the trajectory of the owner ?*

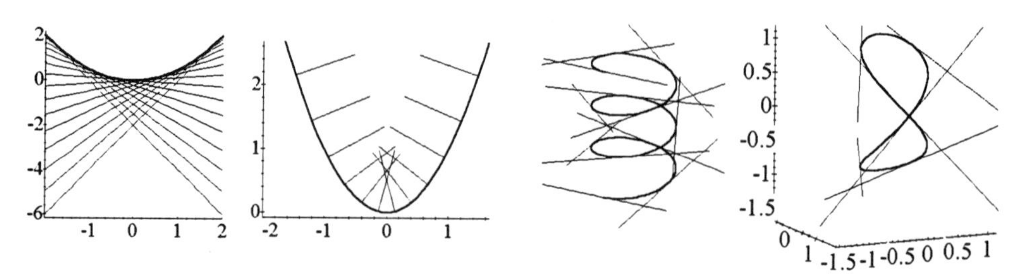

Figs. 9.2–9.5. Tangent lines and normals to a graph.
Tangent lines to a space curve

Derive its equation (the curve is called the *tractrix*, see Section 1.2) using the tangent line; Fig. 9.8. Find semi-tangent lines at singular points of the tractrix.

Hints: From a right triangle with the tangent line as the hypotenuse, derive $x = a\sin t$, where $t \in [0, \pi)$ is the angle between the tangent line and the axis OY. Thus $dx = a\cos t\,dt$. In view of the equality $dx = \tan t\,dy$ (by the geometrical sense of the tangent line), we obtain $dy = a\frac{\cos^2 t}{\sin t}\,dt$. After integration, deduce $\vec{\mathbf{r}}\,(t) = [a\sin t, a(\ln\tan(\frac{t}{2}) + \cos t)]$, which can be simplified to the form $y = a\ln(\frac{a-\sqrt{a^2-x^2}}{x}) - \sqrt{a^2-x^2}$; see Section 1.2.

We deduce the equation using symbolic manipulations in MAPLE and then plot the tractrix.

```
> with(plots): X:=t -> a*sin(t):
> eq1:=dx=diff(X(t), t)*dt; eq2:=dx=tan(t)*dy;
```

$$eq_1 := dx = a\cos(t)\,dt \qquad eq_2 := dx = \tan(t)\,dy$$

```
> eq3:=subs(dx=op(2,eq1),eq2); D1:=simplify(solve(eq3,dy)/dt);
```

$$eq_3 := a\cos(t)\,dt = \tan(t)\,dy \qquad D1 := \frac{a\cos(t)^2}{\sin(t)}$$

```
> simplify(int(D1, t)): Y:=unapply(%, t);
```

$$Y := t \to a\cos(t) + a\ln\left(\frac{\sin(t)}{\cos(t)+1}\right)$$

```
> q:=plot([X(t)/a, Y(t)/a, t=0.2..Pi-0.2], color=blue,
  scaling=constrained): %;
```

Continuation: Now we derive and plot the tangent line to the tractrix.

```
> simplify(Y(t)-X(t)*diff(Y(t), t)/diff(X(t), t)):
  y1:=unapply(%, t);
```

$$y_1 := t \to a\ln\frac{\sin(t)}{\cos(t)+1}$$

```
> p:=i -> plot([[X(Pi/2+i/9)/a, Y(Pi/2+i/9)/a],
```

```
  [0, y1(Pi/2+i/9)/a]]):
> display([seq(p(i), i=0..10), q], scaling=constrained);
```

Continuation: Finally we plot the moving tangent line to the tractrix.

```
> A:=animate([(1-s)*X(Pi/2+i/9)/a, (Y(Pi/2+i/9)+
  s*(y1(Pi/2+i/9)-Y(Pi/2+i/9)))/a, s=0..1], i=-10..10):
> display([q, A]);  # Fig. 9.8
```

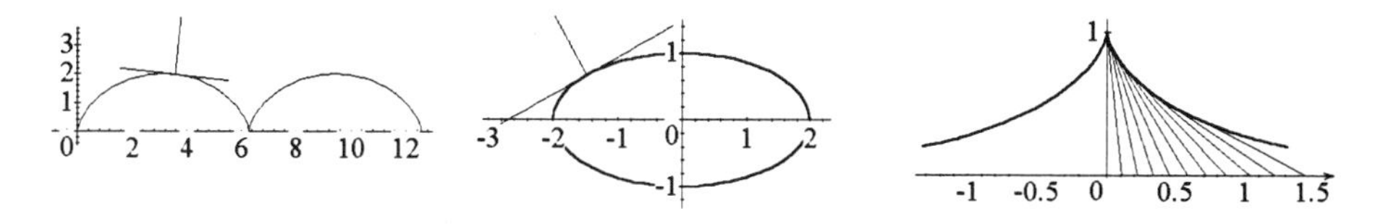

Figs. 9.6–9.8. Frenet frame field of cycloid, ellipse. Lines tangent to a tractrix

The following construction of the *tractrix of an arbitrary curve* generalizes the previous problem [Chapter 1, GH]: *A child goes along the curve* γ, *and his toy moves with him along the tractrix of this curve.*

For example, the *circular tractrix* in polar coordinates has the equation

$$\varphi := -\arcsin(\tfrac{\rho}{2a}) + \frac{\sqrt{4a^2-\rho^2}}{\rho} \quad (0 < \rho \le 2a).$$

4. Let us plot tangent lines to a space curve; start with the circular helix or the Viviani curve; Figs. 9.4–9.5.

```
> r:=[R*cos(t), R*sin(t), v*t]: t1:=0: t2:=6*Pi: v:=1:
># r:=[R*cos(t)^2,R*cos(t)*sin(t),R*sin(t)]:t1:=0:t2:=2*Pi:
> rt:=diff(r, t): m:=8: R:=1:
> with(plots):B:=spacecurve(r,t=t1..t2):Tp:=evalm(r+s*rt);
```

$$T_p := [\cos(t) - s\sin(t),\ \sin(t) + s\cos(t),\ t + s]$$

```
> T:=i -> t1+i*(t2-t1)/m: p:=i -> spacecurve([subs(t=T(i),
  Tp[1]),subs(t=T(i),Tp[2]),subs(t=T(i),Tp[3])],s=-2..2):
> display([seq(p(i), i=0..m), B], axes=boxed);
```

Continuation: We plot the moving tangent line along the curve.

```
> m:=32: A:=display(seq(p(i), i=0..m), insequence=true,
  orientation=[45, 60]): display([B, A]);
```

5. Find the equation of the parabola $y = x^2 + ax + b$ that is tangent to the circle $x^2 + y^2 = 2$ at the point $P(1, 1)$, and plot the curves.

```
> f:=y-(x^2+a*x+b): g:=x^2+y^2-2:
> with(linalg): df:=grad(f,[x,y]); dg:=grad(g,[x,y]);
```

```
> t:=Pi/4: P:=[sqrt(2)*cos(t),sqrt(2)*sin(t)]:  # enter your t
> a0:=solve(subs({x=P[1],y=P[2]}, det(matrix([df,dg]))),a);
  b0:=solve(subs({x=P[1], y=P[2], a=a0}, f));
  f0:=subs(subs({b=b0, a=a0}, f));
```

$$a_0 := -3 \qquad b_0 := 3 \qquad f_0 := y - x^2 + 3x - 3$$

```
> plots[implicitplot]({f0, g}, x=-2..3, y=-2..3);
```

9.2 Envelope Curve of a Family of Curves

In previous sections various curves appeared as sets of points satisfying certain geometrical conditions or equations. In this section the curves appear in a different guise, as envelopes of families of straight lines, circles, and so on.

Below we consider the smooth functions $\varphi(x, y, t)$.

Definition 9.2.1 The *envelope* of a family of curves $\{\gamma_t : \varphi(x, y, t) = 0\}$, where t is the parameter of the family, is the curve $\gamma : \{x(t), y(t)\}$ that is tangent at each of its points $\gamma(t)$ to the corresponding curve γ_t.

One derives the envelope of a family $\{\gamma_t : \varphi(x, y, t) = 0\}$ of curves from the system of equations (for instance, by elimination of t)

$$\varphi(x, y, t) = 0, \qquad \varphi'_t(x, y, t) = 0. \tag{9.1}$$

Remark 9.2.1 The system (9.1) defines the *discriminant set*, sometimes containing extraneous solutions (as points of inflection in the problem of deriving extrema of the function $f(x)$ using the condition $f'(x) = 0$).

Example 9.2.1

1. One can check that the discriminant $y = 0$ of the family $\varphi = y^3 - (x - t)^2$ consists of singular points of members of the family.

```
> p:=t -> plot([u^3+2*t, u^2/10, u=-1.1..1.1]):
> plots[display]([seq(p(t), t=1..7)]);
```

2. The following problem on the *parabola of safety* is useful in ballistics. *Prove that trajectories of points (shells), leaving the origin (a high-angle gun) with constant velocity v under different angles t $(0 < t < \pi)$ to the axis OX (a horizon), have a parabola in the role of envelope*; Fig. 9.12.

Hint: We have the equation of a trajectory (parabola): $x = v\tau\cos t$, $y = v\tau\sin t - \frac{1}{2}g\tau^2$, where τ is time. Eliminating τ, we derive the equation of the family $\varphi(x, y, t) = y - x\tan t + \frac{gx^2}{2v^2\cos^2 t}$. From the second equation

$\varphi_t'(x, y, t) = 0$, we obtain $\tan t = \frac{v^2}{gx}$. Then $\frac{1}{\cos^2 t} = 1 + \tan^2 t = 1 + \frac{v^4}{g^2x^2}$. Substituting in the first equation $\varphi(x, y, t) = 0$ with the aim of eliminating t, after simple transformations we obtain the equation $y = \frac{v^2}{2g} - \frac{g}{2v^2}x^2$ for the *parabola of safety*.

```
> with(plots): g:=9.8: v:=10:
> X:=v*t*cos(Pi/18*s): Y:=v*t*sin(Pi/18*s)-g*t^2/2:
> q:=plot(v^2/(2*g)-g/(2*v^2)*x^2, x=-v^2/g..v^2/g):
> T:=s -> 2*v*sin(Pi/18*s)/g: p:=s -> plot([X, Y, t=0..T(s)]):
> A:=display([seq(p(s), s=1..17)], insequence=true):
> display([q, A], scaling=constrained);  # Fig. 9.12
```

Without option `insequence=true` we obtain the stationary figure.

3. Find the envelope of the family of segments that cut out:

(a) triangles with area S from the coordinate angle XOY.
Hint: The equation of the family is $\gamma_t : \frac{x}{t} + \frac{yt}{2S} - 1 = 0$ or $\varphi = yt^2 - 2St + 2Sx = 0$. From $\varphi_t' = 2yt - 2S = 0$ we obtain $t = \frac{S}{y}$. Then substituting in the first equation we obtain $xy = \frac{S}{2}$: a *hyperbola*.

```
> with(plots): p:=n -> plot([t*n/10, (1-t)*10/n, t=0..1]):
  q:=n -> plot([t*n/10, -(1-t)*10/n, t=0..1]):
  display(seq([p(n),q(n)],n={$-10..10}minus{0})); # Fig. 9.9
```

(b) triangles with the hypotenuse a from the coordinate angle XY.
Hint: The equation of the family is $\gamma_t : \frac{x}{a\cos t} + \frac{y}{a\sin t} = 1$, the *astroid*. Recall that the line segment PQ between the points $P(x_1, y_1)$ and $Q(x_2, y_2)$ has the equations $x = tx_2 + (1-t)x_1,\ y = ty_2 + (1-t)y_1$, where $t \in [0, 1]$.

```
> p:=n->plot([t*cos(Pi/20*n), (1-t)*sin(Pi/20*n), t=0..1]):
  plots[display]([seq(p(n), n=-19..19)]);  # Fig. 9.10
```

(c) the segments with sum of lengths from the coordinate axes OX and OY equal to a. *Hint*: The equation of the family is $\gamma_t : \frac{x}{t} + \frac{y}{a-t} = 1$.

```
> p:=n -> plot([t*n/10, (1-t)*(1-abs(n)/10), t=0..1]):
  q:=n -> plot([t*n/10, -(1-t)*(1-abs(n)/10), t=0..1]):
  plots[display](seq([p(n), q(n)], n=-9..9)); # Fig. 9.11
```

The figures contain only families of curves, but one can see the complete illusion that such figures also contain envelopes.

Exercise 9.2.1 Write MAPLE programs to plot a parabola, an ellipse, and a hyperbola, using the algorithms below (from [Ped]).

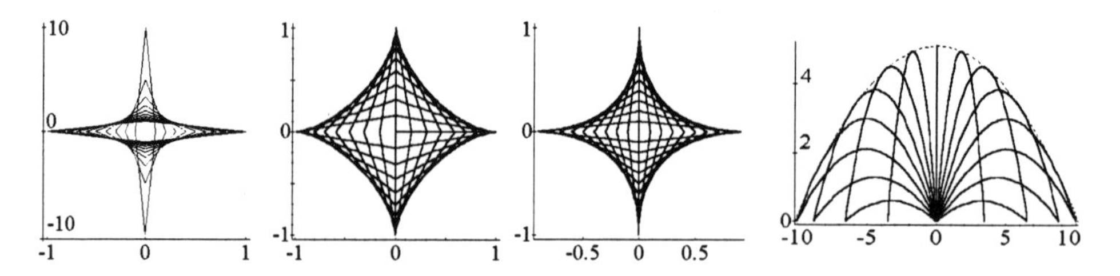

Figs. 9.9–9.12. Embroidery of the hyperbola $xy = 1$, astroid, and similarly, the curve $x^{3/2} + y^{3/2} = 1$. Parabola of safety

(a) Fix the straight line l and a point S that does not belong to this line; start to plot from points $P_t \in l$ straight lines l_t that are orthogonal to the segment P_tS. After plotting sufficiently many straight lines l_t, note that they envelop the curve γ, symmetric with respect to perpendiculars SA from points S onto the straight line l. The line l itself also is tangent to this curve (when the point P_t coincides with A). Prove that γ is a parabola.

(b) Fix a circle with center at the point O and a point S inside the circle. Join the point S by a straight line with any point P_t of the circle and plot the straight line orthogonal to the segment SP_t through the point P_t. Prove that all these straight lines are tangent to the ellipse.

(c) Fix a circle with center at the point C and a point S inside of the circle; join the point S with points P_t on the circle; and plot perpendiculars to segments SP_t from points P_t. Prove that the two branches of the envelope form a hyperbola.

9.3 Mathematical Embroidery

Mathematical embroidery is a method of plotting curves as envelopes of some families of straight lines (segments) or circles.

`MAPLE` can help us to realize on the display all the algorithms of *mathematical embroidery* [Ped] to find new and remarkable examples.

9.3.1 Mathematical Embroidery Using Line Segments

1. For embroidery of the *deltoid*, divide the circle of radius R with center O, starting with points $A(R, 0°)$, into arcs of $5°$ and number the points of partition counterclockwise. Then starting from points $A'(R, 180°)$, divide the circle into arcs of $10°$ and number the points of partition clockwise. Join by straight lines the points of the partition with matching numbers for A and A'. Moreover,

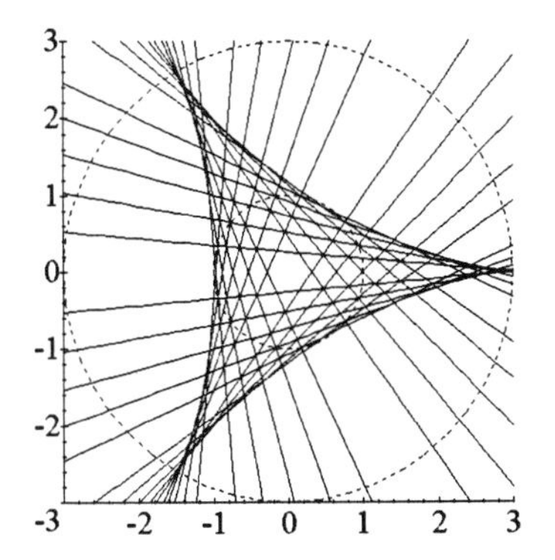

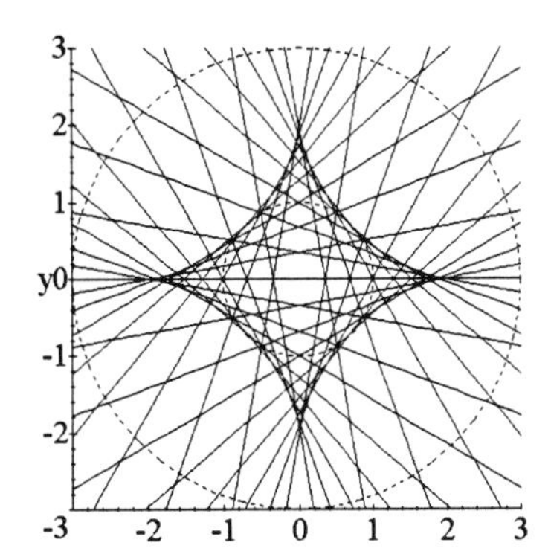

Figs. 9.13–9.14. Embroidery of an astroid and deltoid

continue the segments to their intersection with the larger circle of radius $3R$; Fig. 9.13.

2. Embroidery of the *astroid* is analogous to that of the deltoid. The only difference is that we fix the arcs from the points $A'(R, 180°)$ by 15°, and segments are continued till their intersection with the larger circle of radius $2R$; Fig 9.14.

For embroidery of the *astroid* and the *deltoid*, first we plot the base circle of radius $R = 1$ and the boundary circle of radius $R = 3$. Then we use integers $-36 \leq n \leq 35$ and plot straight lines from the family $y = \frac{y_2-y_1}{x_2-x_1}(x - x_1) + y_1$, where

$$x_1 = \cos(T),\ y_1 = \sin(T),\ x_2 = \cos(\pi - T \cdot m),\ y_2 = \sin(\pi - T \cdot m).$$

```
> with(plots): m:=3: # astroid,   for m = 2 deltoid
> T:=n->Pi/18*n: p:=n->plot((sin(Pi-T(n)*m)-sin(T(n)))/
  (cos(Pi-T(n)*m)-cos(T(n)))*(x-cos(T(n)))+sin(T(n)),
  x=-3..3, y=-3..3):
> q1:=plot([cos(t), sin(t), t=-Pi..Pi], linestyle=2):
  q2:=plot([3*cos(t), 3*sin(t), t=-Pi..Pi], linestyle=2):
>display([seq(p(i),i={$-36..35}minus{seq(18*i-27,i=0..3)}),
  q1, q2], scaling=constrained, axes=framed);   # astroid
> # display([seq(p(i),i={$-35..35}minus{seq(6*i,i=-5..5)}),
q1, q2], scaling=constrained, axes=framed);   # deltoid
```

3. Embroidery of the *cardioid* is analogous to that of the deltoid; the difference is that arcs from points $A(R, 0°)$ and $A'(R, 180°)$ are assumed equal, for example 10°, and are fixed in the same direction (for example, counterclockwise). Segments are chords of the base circle of radius R; they join points from A with points from A' having a *number twice as large*.

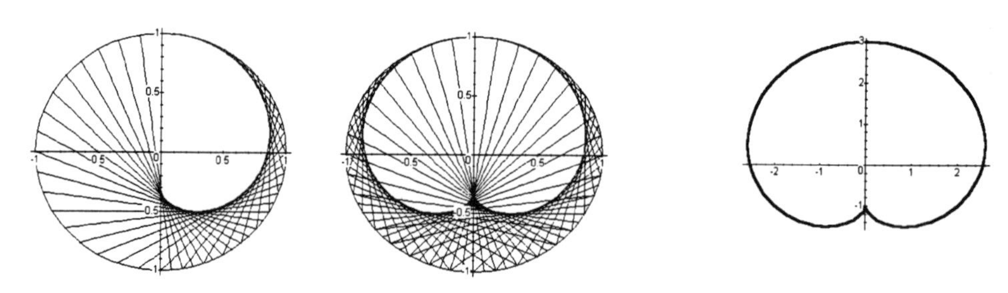

Figs. 9.15–9.16. Embroidery of the cardioid

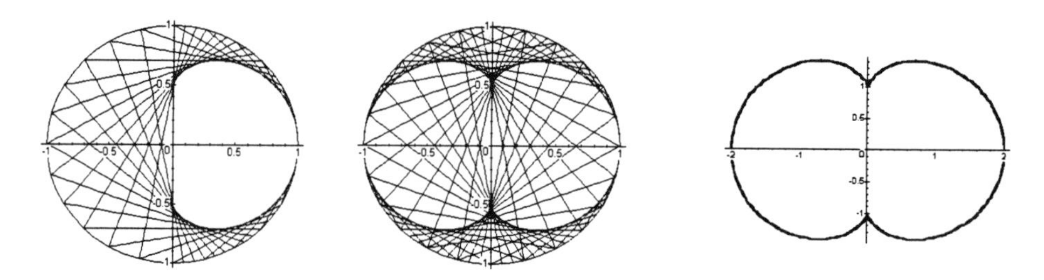

Figs. 9.17–9.18. Embroidery of the nephroid

4. Embroidery of the *nephroid* is analogous to that of the cardioid. The difference is that the chords of the base circle of radius R join points from A with points from A' having a number *three times as large.*

```
> with(plots): m:=2: # cardioid; m = 3 nephroid
>p:=n->plot([t*cos(Pi/36*(18+n*m))+(1-t)*cos(Pi/36*(-18+n)),
  t*sin(Pi/36*(18+n*m))+(1-t)*sin(Pi/36*(-18+n)), t=0..1]):
> q:=plot([cos(t), sin(t), t=-Pi..Pi]):
> display([seq(p(n), n=-35..35), q]);
  # Envelope: M = 1 – cardioid, M = 0.5 – nephroid.
> M:=1: plot([(1+M)*sin(M*t)-M*sin(t+M*t),
  -(1+M)*cos(M*t)+M*cos(t+M*t), t=0..2*Pi/M]);
```

9.3.2 *Mathematical Embroidery Using Circles*

5. Let us break the basic circle of radius R with center O, starting with the point $A(R, 0°)$, into arcs of $10°$. Then plot the circle of radius $R_i = |P_iA|$ around the center at each division point P_i. The heart-shaped envelope of all these circles is the *cardioid.* In other words, the union of circles through the point $A \in \omega$ whose centers lie on the given circle ω is the plane domain bounded by the cardioid; Fig. 9.19 b.

6. Embroidery of *Pascal's limaçon* by circles is analogous to that of the cardioid (case 5). The only difference is that the fixed point $A(kR, 0°)$ does not belong to the base circle. For $k > 1$, Pascal's limaçon has a loop, and for $0 < k < 1$, it does not; Fig. 9.19 a.

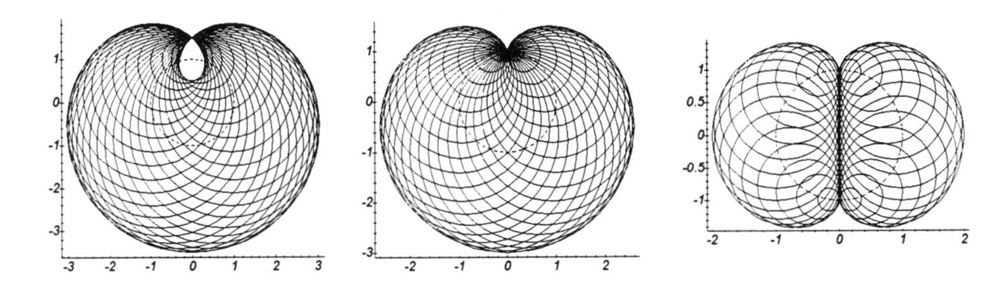

Figs. 9.19 a,b–9.20. Embroidery using circles: Pascal's limaçon, cardioid, and nephroid

7. Embroidery of the *nephroid* by circles is analogous to that of the cardioid. The only difference is that the radii of the circles R_i are assumed equal to the distances from points P_i to the diameter through the point A (the axis OX), i.e., $R_i = d(P_i, |OA|)$; Fig. 9.20.

```
> m:=1: T:=Pi/18*n: b:=n->sqrt(cos(T)^2+(m-sin(T))^2):
  p:=n->plot([cos(T)+b(n)*cos(t), sin(T)+b(n)*sin(t),
  t=0..2*Pi]): # cardioid, for m ≠ 1 the Pascal's limaçon
> c:=n->abs(cos(T)): p:=n->plot([cos(T)+c(n)*cos(t),
  sin(T)+c(n)*sin(t), t=0..2*Pi]):  # nephroid
> q:=plot([cos(t), sin(t), t=-Pi..Pi], linestyle=2):
> plots[display]([seq(p(n), n=-18..17), q]);
```

9.4 Evolute and Evolvent (Involute): Caustic

The normal vectors at points of the given curve can be chosen as the family (of straight lines) for plotting the envelope.

Definition 9.4.1 The envelope γ_2 of a family of normals to the given curve γ_1 is called the *evolute*. The curve γ_1 is called the *evolvent (involute)* of the curve γ_2 if γ_2 is the evolute of γ_1.

In fact, the evolute is related to the evolvent in the same way that differentiation is related to indefinite integration. The evolute of the plane curve $\vec{\mathbf{r}}(t)$ coincides with the *curve of the centers of curvature* (see Chapter 12). In particular, a point is the evolute of a circle. An evolute is defined by the equation $\vec{\mathbf{r}}_{ev}(t) = \vec{\mathbf{r}}\,(t) + \frac{1}{k(t)}\,\vec{\nu}(t)$, or in the coordinate form

$$\begin{cases} x_{ev}(t) = & x(t) - \frac{y'(t)}{k(t)\sqrt{x'^2+y'^2}} = x(t) - y'(t)\frac{x'^2+y'^2}{x'y''-x''y'}, \\ y_{ev}(t) = & y(t) + \frac{x'(t)}{k(t)\sqrt{x'^2+y'^2}} = y(t) + x'(t)\frac{x'^2+y'^2}{x'y''-x''y'}. \end{cases}$$

Various evolvents of the unit-speed plane curve $\vec{\mathbf{r}}(s)$ are given by the equation $\vec{\mathbf{r}}_{evv}(s) = \vec{\mathbf{r}}(s) - (s - s_0)\,\vec{\tau}(s)$, where $\vec{\mathbf{r}},\ \vec{\tau},\ s$ are related to the given curve, but the constant s_0 depends on the choice of the evolvent.

Example 9.4.1

1. We define the parabola $y = x^2$ by parametric equations $x = t,\ y = t^2$ and prove that its evolute is the *half-cubic parabola*

$$x_{ev} = t - \frac{1+(2t)^2}{1\cdot 2-(2t)\cdot 0}\,2t = -4t^3, \quad y_{ev} = t^2 - \frac{1+(2t)^2}{(2t)\cdot 0 - 1\cdot 2} = 3t^2 + \tfrac{1}{2}.$$

```
> x:=t: y:=t^2: t1:=-2: t2:=2: # enter your curve
> xt:=diff(x, t): yt:=diff(y, t):
  xtt:=diff(xt, t): ytt:=diff(yt, t):
> x1:=simplify(x-yt*(xt^2+yt^2)/(xt*ytt-xtt*yt));
  y1:=simplify(y+xt*(xt^2+yt^2)/(xt*ytt-xtt*yt));
```

$x_1 := -4t^3 \qquad y_1 := 3t^2 + 0.5$

Then we obtain Fig. 9.21 with normal movement:

```
> p1:=plot([x, y, t=t1..t2], color=blue):
  p2:=plot([x1, y1, t=t1+1..t2-1]):
> with(plots): display([p1, p2], scaling=constrained);
> N:=evalm([x,y] + s*[x1-x, y1-y]);
> m:=16: T:=i->t1+i*(t2-t1)/m: p:=i -> plot([subs(t=T(i),
  N[1]), subs(t=T(i), N[2]), s=0..1]):
> A:=display([seq(p(i), i=m/4..3*m/4)], insequence=true):
> display([A, p1, p2], scaling=constrained);
```

2. The evolute of the ellipse $\vec{\mathbf{r}}(t) = [a\cos(t),\ b\sin(t)]$ is the *prolate astroid*: $x_{ev} = \frac{a^2-b^2}{a}\cos^3(t), \quad y_{ev} = \frac{b^2-a^2}{b}\sin^3(t)$; Fig. 9.22.
In the above program we assume `a:=3: b:=2: t1:=0: t2:=2*Pi:`

3. The catenary is the evolute of the tractrix; Fig. 9.23. In the above program we assume `a:=1: t1:=.25: t2:=Pi-.25:`

4. One can check that the *evolute of the epicycloid* (or of *the hypocycloid*) is again the epicycloid (resp., the hypocycloid), similar to the curve whose coefficient is $\frac{1}{2m+1}$ and which rotates around it at an angle of $m\pi$.

5. The evolvents of the unit-speed circle $\vec{\mathbf{r}}(s) = [R\cos(\frac{s}{R}),\ R\sin(\frac{s}{R})]$ are the curves, Fig. 9.24:

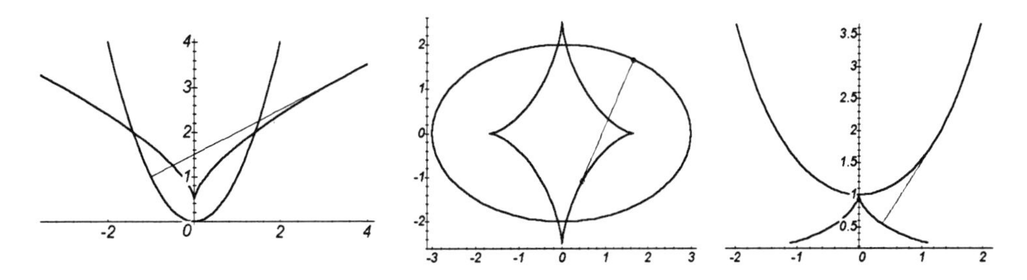

Figs. 9.21–9.23. Evolutes of the parabola, ellipse, and tractrix

$$\vec{\mathbf{r}}_{evv}(s) = [R\cos(\tfrac{s}{R}) + (s - s_0)\sin(\tfrac{s}{R}),\ R\sin(\tfrac{s}{R}) - (s - s_0)\cos(\tfrac{s}{R})].$$

Plot them by the following program.

```
> x:=R*cos(s/R)+(s-c)*sin(s/R): y:=R*sin(s/R)-(s-c)*cos(s/R):
> R:=1: p:=c -> plot([x, y, s=c..c+1.5*Pi]):
> q:=plot([R*cos(s/R), R*sin(s/R), s=0..2*Pi], color=blue):
> plots[display]([seq(p(c), c=0..5), q]);
```

The *evolvent of the circle* is also (analogous to the cycloid) the limited form of cycloidal curves. Namely, it is the trajectory of the point on the straight line rolling without gliding along the circle.

Plot the evolvents of the figure eight curve $\vec{\mathbf{r}}(t) = [\sin(t),\ \sin(2t)]$.

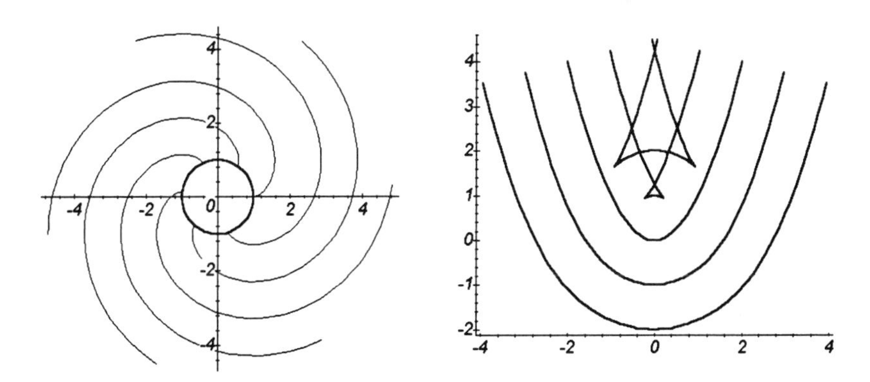

Figs. 9.24–9.25. Evolvents of a circle. Parallel curve of a parabola

Definition 9.4.2 The envelope of the pencil of rays emanating from a given point or direction and (a) reflected or (b) refracted by the given curve is called (a) the *catacaustic* or (b) the *diacaustic* (or *caustic*, for short) of this curve with respect to the given point.

Let γ be the graph $y = y(x)$ and consider a pencil of rays parallel to the unit vector $\vec{\mathbf{e}} = [a, b]$. The caustic $[X, Y]$ has the following equations:

$$X(x) = x - \frac{(b - ay')^2}{2y''}\, y' - \frac{(b - ay')(a + by')}{2y''}\,,$$
$$Y(x) = y + \frac{(b - ay')^2}{2y''} - \frac{(b - ay')(a + by')}{2y''}\, y'\,.$$

Let $\vec{\mathbf{r}} = \vec{\mathbf{r}}(s)$ be the natural parametrization of γ. If the rays are parallel to $\vec{\mathbf{e}}$, then the equation of the caustic $\tilde{\mathbf{r}}$ is

$$\tilde{\mathbf{r}} = \vec{\mathbf{r}} + \tfrac{(\vec{\mathbf{e}},\, \vec{\mathbf{n}})}{2k(s)}\, [(\vec{\mathbf{e}},\, \vec{\mathbf{n}})\, \vec{\mathbf{n}} - (\vec{\mathbf{e}},\, \vec{\tau})\, \vec{\tau}]\,.$$

If the rays run from the origin O of rectangular coordinates, then the equation of the caustic $\tilde{\mathbf{r}}$ is

$$\tilde{\mathbf{r}} = \vec{\mathbf{r}} + \tfrac{(\vec{\mathbf{r}},\, \vec{\mathbf{n}})}{2k(s)\vec{\mathbf{r}}^2 + (\vec{\mathbf{r}},\, \vec{\mathbf{n}})}\, [(\vec{\mathbf{r}},\, \vec{\mathbf{n}})\, \vec{\mathbf{n}} - (\vec{\mathbf{r}},\, \vec{\tau})\, \vec{\tau}]\,.$$

The catacaustic of the logarithmic curve $y = a \log \frac{x}{a}$ for rays parallel to OX is the catenary $x = a \cosh(\frac{y+a}{2})$; the catacaustic of a deltoid is an astroid.

Exercise 9.4.1 Write programs for the following problems.

(a) A pencil of parallel rays is reflected by the inner side of a half circle; plot the caustic of reflecting rays. *Answer.* A nephroid.

(b) A pencil of rays emanates from points $(a, 0)$ on the axis of the parabola $y^2 = 2px$ and reflects from the parabola; plot the caustic of reflecting rays. *Answer.* A sinusoidal spiral, see Example 6.1.1, with $m = -\frac{1}{3}$ (a *cubic of Tschirnhausen* or a *Catalan trisectrix*) when $a = \infty$.

9.5 Parallel Curves

Using the normal segments to the curve of constant length d (in one or the other side) we obtain *parallel curves (equidistants)*. Equations of equidistants γ_d of the curve $\gamma :\ x = x(t),\ y = y(t)$ are the following:

$$\vec{\mathbf{r}}_d(t) = \vec{\mathbf{r}}\,(t) + d \cdot \vec{\nu}(t) \iff \begin{cases} x_d(t) = x(t) - \frac{d \cdot y'}{\sqrt{x'^2 + y'^2}} \\ y_d(t) = y(t) + \frac{d \cdot x'}{\sqrt{x'^2 + y'^2}}\,. \end{cases}$$

If $|d|$ is less than the radius of curvature of the smooth curve γ, the equidistant γ_d is smooth; otherwise, γ_d may contain singular points. The difference between perimeters of two closed smooth parallel curves is equal to $2\pi d$.

A parallel curve of an ellipse (a *toroid*) appears as the boundary curve of a parallel projection of the torus of revolution at the same time the circular axis of the torus projects into the ellipse.

Example 9.5.1 We plot equidistants for several curves.

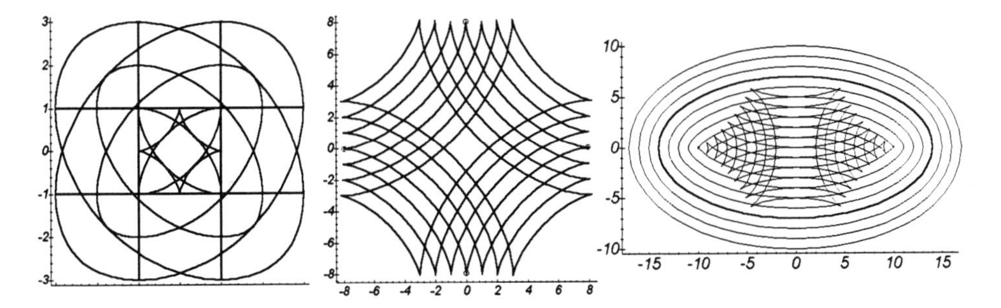

Figs. 9.26–9.28. Parallel curves of an astroid and ellipse

```
> a:=8: x:=a*cos(t)^3: y:=a*sin(t)^3: t1:=-Pi: t2:=Pi:   # Fig. 9.27
> a:=8: b:=7: x:=a*cos(t): y:=b*sin(t): t1:=0: t2:=2*Pi: # Fig. 9.28
> d:='d': x:=t: y:=t^2: t1:=-2: t2:=2:           # parabola, Fig. 9.25
> xt:=diff(x,t): yt:=diff(y,t):
  xtt:=diff(xt,t): ytt:=diff(yt,t):
> xd:=x-d*yt/sqrt(xt^2+yt^2): yd:=y+d*xt/sqrt(xt^2+yt^2):
> q:=plot([x, y, t=t1..t2], color=blue, thickness=2):
> p:=d -> plot([xd, yd, t=t1..t2], thickness=2):
> plots[display]([seq(p(d), d=-3..3), q]);
```

10
Singular Points on Curves

How does a curve look in the neighborhood of a singular point? Recall that a formal definition of a singular and regular point on a curve (see Section 5.1) depends on a class of parametrization. We present several types of singular points of plane curves, Fig. 10.1:

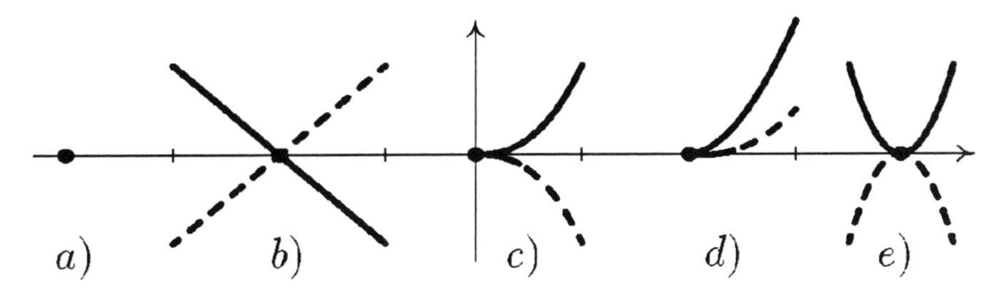

Fig. 10.1. Types of singular points on plane curves

(a) an *isolated* point, if a neighborhood of the singular point does not contain other points on a curve;

(b) a *double* point (or *node*), if two branches of the curve intersect at the point but are not tangent;

(c) a *cuspidal* point of the *first kind*, if two branches of the curve lie on different sides of the common semi-tangent line at the point;

(d) a *cuspidal* point of the *second kind*, if two branches of the curve lie on one side of the common semi-tangent line at the point;

(e) a point of *self-tangency*, if two branches of the curve are tangent at the singular point.

In Sections 10.1 and 10.2 we study singular points for different types of equations of a curve. Section 10.3 contains more advanced examples.

10.1 Singular Points on Parametrized Curves

There is a useful *sufficient* condition for the point t_0 on the plane curve $\vec{\mathbf{r}}(t)$ of the class C^∞ with the property that $\vec{\mathbf{r}}'(t_0) = 0$ is singular.

Theorem 10.1.1 *Let the curve γ of the class C^∞ be given by the equations $x = x(t)$ and $y = y(t)$. Then the point $(x_0, y_0) \in \gamma$ (for $t = t_0$) is*

(a) a regular point if the first nonzero derivative (in series) of the functions $x(t)$ and $y(t)$ at this point is odd,

(b) a singular (cuspidal) point if for $t = t_0$ the first nonzero derivative (in series) of the functions $x(t)$ and $y(t)$ is even.

Example 10.1.1

1. Let $\vec{\mathbf{r}}(t) = [t^2, t^3]$. Then (0, 0) is a cuspidal point of the first kind.

```
> plot([t^2, t^3, t=-1..1], scaling=constrained); # Fig. 10.2
```

2. Let $\vec{\mathbf{r}}(t) = [t^2, t^4 + t^5]$. Then (0, 0) is a cuspidal point of the second kind.

```
> plot([t^2, t^4+t^5, t=-1.2..0.8]);    # Fig. 10.3
```

3. The point $t = 0$ of the curve $x(t) = t^3$, $y(t) = t^5$ is regular with respect to parametrizations of the class C^1 (the curve is the graph of the function $y = x^{5/3}$ of the class C^1), but $t = 0$ is singular with respect to analytic parametrizations.

4. Let us plot the curve $\vec{\mathbf{r}}(t) = [2t - t^2, 3t - t^3]$ and *semi-tangent lines* to it at singular points; Fig. 10.4.

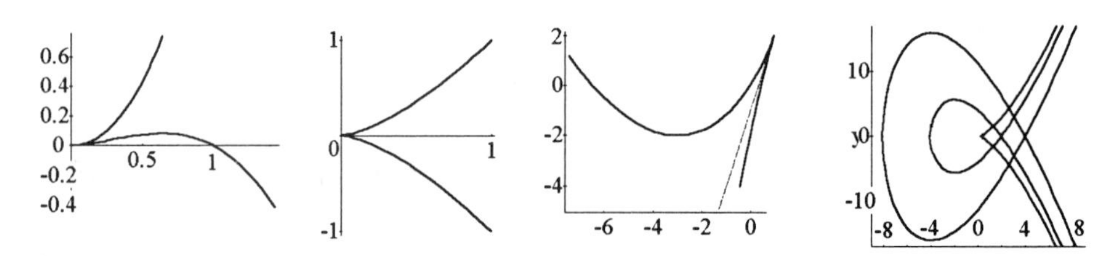

Figs. 10.2–10.5. Cuspidal points. Semi-tangent line. Family of curves

Hint. Show that $t_0 = 1$ is a singular point. We find the coefficient of inclination of a semi-tangent line by the formula $\lim\limits_{t \to -t_0} \frac{x'(t)}{y'(t)}$ (or $t \to t_0$).

```
> r:=array([2*t-t^2, 3*t-t^3]): rt:=map(diff, r, t);
```

$$r_t := [2 - 2t, 3 - 3t^2]$$

```
> p:=plot([r[1], r[2], t=-1.9..2.2], thickness=2):
> t1:=solve(rt[1]=0, rt[2]=0); pp:=subs(t1, [r[1],r[2]]);
  k:=limit(rt[2]/rt[1], t1);  Tp:=pp + [t, k*t];
```

$$t_1 := \{t = 1\} \qquad pp := [1, 2] \quad k := 3 \;\; T_p := [t+1, 3t+2]$$

```
> q:=plot([Tp[1], Tp[2], t=-2.3..0], linestyle=2):
> plots[display]([p, q], scaling=constrained); # Fig. 10.4
```

10.2 Singular Points on Implicitly Defined Plane Curves

What are the sufficient conditions for the point (x_0, y_0) to be singular on the implicitly defined plane curve $F(x, y) = 0$?

Denote by F_x, F_y, F_{xx}, F_{xy}, F_{yy}, F_{xxx}, F_{xxy}, F_{xyy}, F_{yyy} the partial derivatives at the given point. Assume that not all partial derivatives of second order at this point (x_0, y_0) are zero, and let $\Delta = F_{xx}F_{yy} - F_{xy}^2$.

Let $R = \mathrm{Res}(f, g)$ be the *resultant* of the following two polynomials: $f(t) = F_{xx}t^2 + F_{xy}t + F_{yy}$ and $g(t) = F_{xxx}t^3 + F_{xxy}t^2 + F_{xyy}t + F_{yyy}$, in other words, the determinant

$$R = \det \begin{vmatrix} F_{xx} & F_{xy} & F_{yy} & 0 & 0 \\ 0 & F_{xx} & F_{xy} & F_{yy} & 0 \\ 0 & 0 & F_{xx} & F_{xy} & F_{yy} \\ F_{xxx} & F_{xxy} & F_{xyy} & F_{yyy} & 0 \\ 0 & F_{xxx} & F_{xxy} & F_{xyy} & F_{yyy} \end{vmatrix}.$$

We also use the command `resultant` in programs of Section 20.1.

Theorem 10.2.1 *Three cases are possible for the point* (x_0, y_0) *on the curve* $F(x, y) = 0$ *with conditions* $F_x = 0,\ F_y = 0$:

$$\Delta \begin{cases} > 0, & \text{an isolated point of the level set } F = 0, \text{ type (a)}, \\ < 0, & \text{the intersection of two branches of the curve } F = 0, \text{ type (b)}, \\ = 0, & \begin{cases} R \neq 0, & \text{cuspidal point of the first kind, type (c)}, \\ R = 0, & \text{further investigation is required.} \end{cases} \end{cases}$$

If the Taylor series of the function $F(x, y)$ at the point (x_0, y_0) starts from some number $k > 2$, then we obtain by definition a *singular point of order* k. The case $k = 2$ was treated in Theorem 10.2.1.

Remark 10.2.1 The cochleoid (transcendental curve, Fig. 6.6) has an infinite number of branches through the pole and tangent to the polar axis; hence the pole is a singular point of *infinite order*.

Example 10.2.1
1. The equation $(x^2+y^2)(x-1)=0$ defines the straight line $x=1$ and the isolated point $(0,0)$ at which $F_{xx}=F_{yy}=-1$, $F_{xy}=0$ and $\Delta>0$ hold. Pascal's limaçon $(x^2+y^2-2ax)^2-b^2(x^2+y^2)=0$ for $b>2a>0$ has analogous properties, as does the curve $a^2y^2+x^4-y^4=0$.

2. The equation $(x^2+y^2)^2-2a^2(x^2-y^2)=0$ defines the lemniscate of Bernoulli, Fig. 6.18, with the node singular point $(0,0)$ at which $F_{xx}=-2a^2$, $F_{yy}=2a^2$, $F_{xy}=0$, and $\Delta<0$ hold. In a neighborhood of the point $(0,0)$ this curve consists of two elementary curves. An analogous situation appears for Pascal's limaçon for $0<b<2a$, Fig. 9.19, and for the folium of Descartes; Fig. 5.15.

3. The equation $y^2-x^3=0$ defines the half-cubic parabola, Fig. 9.21, with the cuspidal point of the first kind $(0,0)$ in which $F_{xx}=F_{xy}=0$, $F_{yy}=1$, and $\Delta=0$. An analogous situation appears for cycloid and cycloidal curves (astroid, deltoid, etc.)

4. The equation $y(y-x^2)=0$ defines the parabola $y=x^2$ and the straight line $y=0$ osculating at the singular point $(0,0)$ where $F_{xx}=F_{xy}=0$, $F_{yy}=1$, and $\Delta=0$ hold. The point $(0,0)$ on the curve $4x^2(x^2-a^2)+m^2y^4=0$ has analogous properties.

Exercise 10.2.1 1. Explain why the curve $x^6-2a^2x^3y-b^3y^3=0$ $(a,\ b\neq 0)$ has the singular point $(0,0)$ of order 3: Three branches of the curve are tangent to the straight line $y=0$. *Hint.* For $x\approx 0$, neglect the term x^6 and then divide by y: the curve is similar to the *half-cubic parabola* $2a^2x^3+b^3y^2=0$. For $y\approx 0$, neglect the term b^3y^3 and divide by x^3: the curve is similar to the *cubic parabola* $x^3-2a^2y=0$.

```
> f:=(a, b) ->x^6-2*a*2*x^3*y-b^3*y^3:
> plots[implicitplot](f(1, 1), x=-.5..0.5, y=-.5..0.5,
  grid=[70, 70], axes=framed, scaling=constrained); # Fig. 10.8
```

2. Find the connected curve whose singular points belong to k branches of the curve. *Hint.* Consider the roses $\rho=\sin(n\varphi)$ in rectangular coordinates, where $n=k$ for odd k, and $n=\frac{1}{2}k$ for even k.

3. What are the relations between a and b when the curve $y^2=x^3+ax+b$ has a double point? Plot some of these curves. *Answer*: $16a^3=-27b^2$.

```
> F:=y^2-x^3-a*x-b: Fx:=diff(F, x): Fy:=diff(F, y):
> s:=solve({F, Fx, Fy}, {y, a, b});
```

$$s := \{b = 2x^3,\ a = -3x^2,\ y = 0\}$$

```
> FF:=subs({a=-3*t^2, b=2*t^3}, F): FF;
```

$$y^2 - x^3 + 3t^2x - 2t^3$$

```
> Y:=t -> solve(FF, y): Y(t);
```

$$(2t + x)^{1/2}(-x + t), \ -(2t + x)^{1/2}(-x + t)$$

```
> plot([seq(Y(t),t={0,2,4})],x=-9..9,y=-17..17); # Fig. 10.5
```

10.3 Unusual Singular Points on Plane Curves

Some transcendental curves have more complicated types of singular points, as, for example, when the derivative of a function (in the equation of the curve) is not continuous.

Example 10.3.1

1. The right-hand derivative of the function $y = e^{1/x}$ has a discontinuity at the point $x = 0$. The graph of the curve is interrupted at the origin. Such points are called *stopping points*.

```
> p1:=plot(exp(1)^(1/x), x=-5..0, thickness=2):
> p2:=plot(exp(1)^(1/x), x=.5..5, thickness=2):
> h:=plot(1, x=-5..5, linestyle=2):
> plots[display]([p1,p2,h],scaling=constrained); # Fig. 10.6
```

2. The left-hand derivative of the function $y = \frac{x}{1+e^{1/x}}$ has a discontinuity at the point $x = 0$. The graph of the curve is broken at the origin: two branches of the curve form the angle $\varphi \in (0, \pi)$ at this point. Such points are called *angular points*.

```
> p:=plot(x/(1+exp(1)^(1/x)), x=-1..2, thickness=2):
> h:=plot(x,x=-1..0,linestyle=2):   g:=plot(0,x=0..2,linestyle=2):
> plots[display]([p,h,g],scaling=constrained); # Fig. 10.7
```

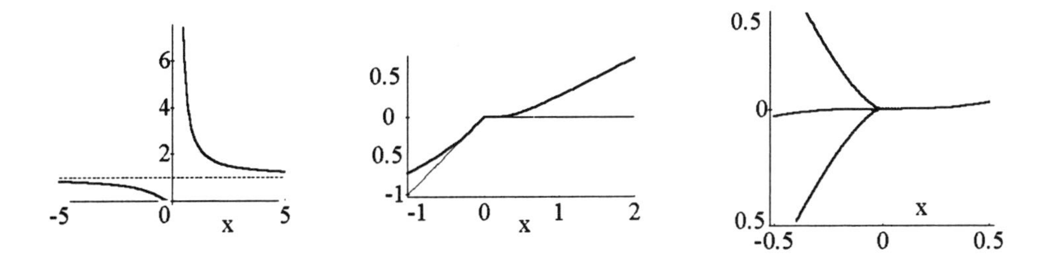

Figs. 10.6–10.8. Unusual singular points on plane curves

Exercise 10.3.1 Find singular points of the curve and explain their type:

(a) the cycloid $Rt - R\sin(t),\ y = R - R\cos(t)$,
(b) the cissoid $y^2 = \frac{x^3}{2a-x} \iff \vec{\mathbf{r}}(t) = [\frac{2a}{1+t^2}, \frac{2a}{t(1+t^2)}]$,
(c) the trisectrix of Maclaurin $x(x^2+y^2) = a(3x^2 - y^2)$,
(d) the astroid $x = R\cos^3(\frac{t}{4}),\ y = R\sin^3(\frac{t}{4})$, (e) $y^3 = ax^2 + x^3$,
(f) the trefoil or cloverleaf $\frac{27}{4}a^2(x^2+y^2-a^2)+ax^3-3axy^2-(x^2+y^2)^2 = 0$,
(g) the rosette, given by the polar equation $\rho = a\sin(\frac{\varphi}{2})$ or by the equation in rectangular coordinates $4(x^2+y^2)^3 - 4a^2(x^2+y^2)^2 + a^4y^2 = 0$.

Answer: The curve has a double point of self-tangency at the origin and two nodes at the points $(0, \pm\frac{a}{\sqrt{2}})$.

11
Length and Center of Mass of a Curve

Some basic formulas for calculating the length and center of mass of a curve in the plane and in space are given in Section 11.1. In programs presented in Section 11.2, we find characteristics for a regularly inscribed polygon (i.e., a polygonal approximation to the curve) and compare them with the results obtained by integrating velocity along the curve.

In Chapter 11 the reader will become acquainted with the commands

`restart, norm, pointplot3d.`

11.1 Basic Facts

Definition 11.1.1 A polygon σ with vertices $P_1,\ P_2, \dots, P_n$, is said to be *regularly inscribed in the curve* $\gamma = \vec{\mathbf{r}}\,(t)$ $(a \le t \le b)$ if there exists a partition $a = t_1 < \cdots < t_n = b$ of the segment $[a, b]$ such that $P_i = \vec{\mathbf{r}}\,(t_i)$.

The length of the polygon is equal to $l(\sigma) = \sum_{i=1}^{n-1} |P_{i+1}P_i|$. The length of the polygon $P_1(\rho_1, t_1), \dots, P_n(\rho_n, t_n)$ in polar coordinates is given by the formula $l(\sigma) = \sum_{i=1}^{n-1} \sqrt{\rho_{i+1}^2 + \rho_i^2 - 2\,\rho_{i+1}\,\rho_i \cos(t_{i+1} - t_i)}$, derived with the help of the cosine theorem.

Definition 11.1.2 The *length* $l(\gamma) < \infty$ *of the rectifiable curve* γ is the least upper bound of the lengths of all regularly inscribed polygon.

Examples of nonrectifiable curves ($l(\gamma) = \infty$) appear in Section 13.3. The length of the C^1 regular curve $\vec{\mathbf{r}}\,(t)$ is the integral of its velocity:

$$l(\gamma) = \int_a^b |\vec{\mathbf{r}}\,'(t)|\, dt.$$

Formulas for calculating the length of the curve γ:

1. $\gamma : \vec{\mathbf{r}}(t) = [x(t), y(t), z(t)] \subset \mathbb{R}^3 :$ $\quad l(\gamma) = \int_a^b \sqrt{x'^2 + y'^2 + z'^2}\, dt,$

2. $\gamma : y = f_1(x), z = f_2(x)$ in $\mathbb{R}^3 :$ $\quad l(\gamma) = \int_a^b \sqrt{1 + f_1'^2 + f_2'^2}\, dx,$

3. $\gamma : \vec{\mathbf{r}}(t) = [x(t), y(t)] \subset \mathbb{R}^2 :$ $\quad l(\gamma) = \int_a^b \sqrt{x'^2 + y'^2}\, dt,$

4. $\gamma : y = f(x)$ in $\mathbb{R}^2 :$ $\quad l(\gamma) = \int_a^b \sqrt{1 + f'^2}\, dx.$

5. $\gamma : \rho = \rho(\varphi)$ in polar coordinates: $\quad l(\gamma) = \int_a^b \sqrt{\rho^2 + \rho'^2}\, d\varphi.$

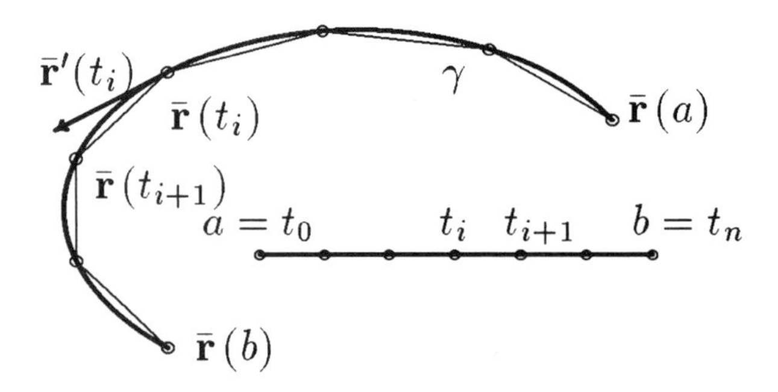

Fig. 11.1. Curve and its regularly inscribed polygon

The formula for the *mass of the curve* $\vec{\mathbf{r}}(t)$ constructed from thin wire with density $\mu(t)$ generalizes the formula $m(\gamma) = \int_a^b \mu(t)\, |\vec{\mathbf{r}}\,'(t)|\, dt$ for the length. The *center of mass* of this plane curve $\vec{\mathbf{r}}(t) = [x(t), y(t)]$ $(a \le t \le b)$ has the coordinates $x_c(\gamma) = M_y/m(\gamma)$, $\quad y_c(\gamma) = M_x/m(\gamma)$, where

$$M_x = \int_a^b \mu(t)\, y(t) \sqrt{x'^2 + y'^2}\, dt, \quad M_y = \int_a^b \mu(t)\, x(t) \sqrt{x'^2 + y'^2}\, dt$$

are *moments of the first order* of the curve with respect to OX and OY.

Definition 11.1.3 The *natural parametrization* of the regular curve $\vec{\mathbf{r}}\,(t)$ is the vector-valued function $\widetilde{\mathbf{r}}\,(s) = \vec{\mathbf{r}}\,(t(s))$, where $t(s)$ is the inverse function to $s(t) = \int_{t_0}^t |\vec{\mathbf{r}}'(t)| dt + s_0$.

The speed of this parametrization is unity: $|\widetilde{\mathbf{r}}\,'(s)| = 1$.

The *metrical form* of curvilinear coordinates $x = f_1(u, v)$, $y = f_2(u, v)$ in the plane is the quadratic form $ds^2 = E(u, v) du^2 + 2F(u, v) du dv + G(u, v) dv^2$, where

$$E = (\tfrac{\partial f_1}{\partial u})^2 + (\tfrac{\partial f_2}{\partial u})^2, \; F = \tfrac{\partial f_1}{\partial u}\tfrac{\partial f_1}{\partial v} + \tfrac{\partial f_2}{\partial u}\tfrac{\partial f_2}{\partial v}, \; G = (\tfrac{\partial f_1}{\partial v})^2 + (\tfrac{\partial f_2}{\partial v})^2.$$

For example, for *affine* coordinates $x = a_1 u + a_2 v + a_0$, $y = b_1 u + b_2 v + b_0$ the coefficients E, F, and G are constant.

The length of the curve $\gamma : u = u(t),\ v = v(t)\ (t_1 \le t \le t_2)$ in curvilinear coordinates in $\mathbb{R}^2$ is given by the formula

$$l(\gamma) = \int_{t_1}^{t_2} \sqrt{Eu'(t)^2 + 2Fu'(t)\,v'(t) + Gv'(t)^2}dt.$$

One can write down analogous formulas E, F, G, and $l(\gamma)$ for curvilinear coordinates $x = f_1(u, v, w),\ y = f_2(u, v, w),\ z = f_3(u, v, w)$ in space (see Section 19.3.2).

11.2 Calculation of Length and Center of Mass

Example 11.2.1 Using two methods (integrals and a regularly inscribed polygon), and then comparing the results, find the lengths and centers of mass (for $\mu \equiv 1$) of the following curves:
(a) parabola $y = x^2 \quad (-2 \le x \le 2)$,
(b) half-cubic parabola $y = x^{3/2} \quad (0 \le x \le 4)$,
(c) catenary $y = a \cosh \frac{x}{a} \quad (0 \le x \le x_0)$,
(d) arc of a cycloid $\vec{\mathbf{r}}\,(t) = [a(t - \sin(t)),\ a(1 - \cos(t))] \quad (0 \le t \le 2\pi)$,
(e) ellipse $\vec{\mathbf{r}}\,(t) = [a \cos(t),\ b \sin(t)] \quad (0 \le t \le 2\pi)$,
(f) astroid $\vec{\mathbf{r}}(t) = [a \cos^3(t), a \sin^3(t)] \quad (0 \le t \le \frac{\pi}{2})$,
(g) Archimedes' spiral $\rho(t) = a \cdot t \quad (0 \le t \le 2\pi)$,
(h) cardioid $\rho(t) = a(1 + \cos(t)) \quad (0 \le t \le 2\pi)$,
(i) circular helix $\vec{\mathbf{r}}(t) = [a \cos(t), a \sin(t), vt] \quad (0 \le t \le 4\pi)$,
(j) conic circular helix $\vec{\mathbf{r}}(t) = [t \cos(t), t \sin(t), vt] \quad (0 \le t \le 6\pi)$,
(k) Viviani curve $\vec{\mathbf{r}}(t) = [\cos^2(t), \cos(t) \sin(t), \sin(t)] \quad (0 \le t \le 2\pi)$.

Hint. We solve problems (a)–(k) with the help of the programs below using the following scheme:

(1) Enter the data for the curve and plot it.
(2) Derive the length and center of mass of the curve using an integral.
(3) Enter the number of edges n of the inscribed polygon and plot it.
(4) Derive the length and the center of mass of the polygon.
(5) Compare numeric results of steps (2) and (4) for some values of n.

Program 11.2.1 Curves: graphs $(a) - (c)$; Fig. 11.2.

```
> restart: a:=-2: b:=2: f:=x -> x^2:  # define your function
> p1:=plot(f, a..b, scaling=constrained): %;
> int(sqrt(1+ diff(f(x), x)^2), x=a..b); L1:=evalf(%);
```

$$2 \cdot 17^{1/2} - \frac{1}{2}\ln(-4 + 17^{1/2}) \qquad L_1 := 9.293567524$$

```
> My:=evalf(int(x*sqrt(1+diff(f(x), x)^2), x=a..b));
  Mx:=evalf(int(f(x)*sqrt(1+diff(f(x), x)^2), x=a..b));
  rc1:=[My/L1, Mx/L1];
```

$$M_y := 0 \qquad M_x := 16.94235094 \qquad rc_1 := [0,\ 1.823019082]$$

```
> n:=6: A:=i-> [a+i*(b-a)/n, f(a+i*(b-a)/n)]:
> p2:=plot([seq(A(i), i=0..n)], color=blue): %;
> plots[display]([p1, p2], scaling=constrained);
> L2:=evalf(sum(linalg[norm](A(i+1)-A(i), 2), i=0..n-1));
  d:=abs(L1-L2)/L1;
```

$$L_2 := 9.224027429 \qquad d := .007482605019$$

```
> rc2:=evalm(sum(0.5*(A(i+1)+A(i))*linalg[norm](
  A(i+1)-A(i), 2),i=0..n-1)/L2);  evalm(rc2-rc1);
```

$$rc_2 := [0,\ 1.850995700] \qquad [0,\ 0.027976618]$$

Program 11.2.2 Parametrized plane curves $(d) - (f)$; Fig. 11.3.

```
> a:=0: b:=2*Pi: X:=t->t-sin(t): Y:=t->1-cos(t): # cycloid
> p1:=plot([X, Y, a..b], scaling=constrained): %;
> int(sqrt(diff(X(t), t)^2+diff(Y(t), t)^2), t=a..b);
  L1:=evalf(%);
```

$$4 \cdot 4^{1/2} \qquad L_1 := 8.000000000$$

```
> Mx:=evalf(int(Y(t)*sqrt(diff(X(t),t)^2+diff(Y(t),t)^2),
  t=a..b));  My:=evalf(int(X(t)*sqrt(diff(X(t),t)^2+
  diff(Y(t),t)^2),t=a..b)); rc1:=[My/L1,Mx/L1];
```

$$M_x := 10.66666667 \qquad M_y := 25.13274123$$

$$rc_1 := [3.141592654,\ 1.333333334]$$

```
> A:=i-> [X(a+i*(b-a)/n), Y(a+i*(b-a)/n)]: n:=10:
> p2:=plot([seq(A(i), i=0..n)], scaling=constrained): %;
> plots[display]([p1, p2], scaling=constrained);
> L2:=evalf(sum(linalg[norm](A(i+1)-A(i), 2), i=0..n-1));
  d:=abs(L1-L2)/L1;
```

$L_2 := 7.968009175 \qquad d := 0.006167275250$

```
> rc2:=evalm(sum(0.5*(A(i+1) + A(i))*linalg[norm](
  A(i+1) - A(i), 2), i=0..n-1)/ L2);  evalm(rc1 - rc2);
```

$rc_2 := [3.141592655,\ 1.299518122] \qquad [-.110^{-8},\ 0.033815212]$

Program 11.2.3 Curves in polar coordinates: $(g) - (h)$; Fig. 11.4.

```
> a:=0: b:=2*Pi: rho:=t -> 0.2*t: n:=12: # Archimedes' spiral
> p1:=plots[polarplot](rho, a..b, scaling=constrained): %;
> L1:=evalf(int(sqrt(rho(t)^2+diff(rho(t),t)^2),t=a..b));
```

$L_1 := 4.251258829$

```
> T:=i -> a+i*(b-a)/n: p2:=plot([seq([rho(T(i)), T(i)],
  i=0..n)], coords=polar): %; plots[display]([p1, p2]);
> L2:=evalf(sum(sqrt(rho(T(i+1))^2+rho(T(i))^2-2*rho(T(i+1))
  *rho(T(i))*cos((b-a)/n)), i=0..n-1)); d:=(L2 - L1)/L1;
```

$L_2 := 4.189197254 \qquad d := -0.01459839956$

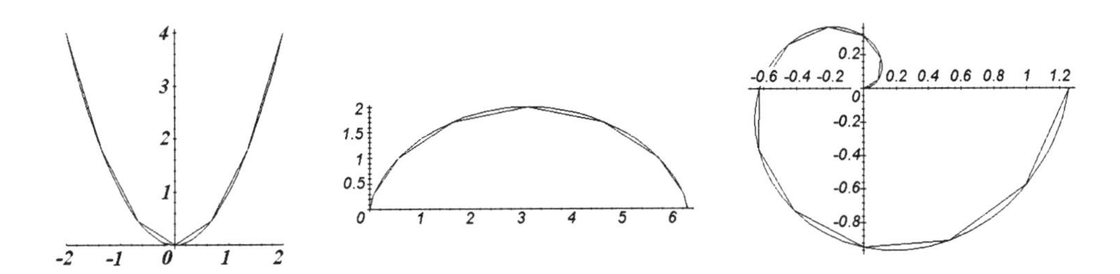

Figs. 11.2–11.4. Length of a plane curve

Program 11.2.4 Parametrized space curves $(i) - (k)$; Figs. 11.5–11.7.

```
> restart: a:=0: b:=2*Pi: X:=t -> cos(t)^2:
  Y:=t -> cos(t)*sin(t): Z:=t -> sin(t): # Viviani curve
> p1:=plots[spacecurve]([X, Y, Z, a..b], axes=boxed): %;
> int(sqrt(diff(X(t), t)^2 + diff(Y(t), t)^2 +
  diff(Z(t), t)^2), t=a..b);  L1:=evalf(%);
```

$4\ 2^{1/2}\ \mathrm{EllipticE}(\frac{1}{2}\ 2^{1/2}) \qquad L_1 := 7.640395576$

```
> Myz:=evalf(int(X(t)*sqrt(simplify(diff(X(t),t)^2+
  diff(Y(t), t)^2+diff(Z(t), t)^2, trig)), t=a..b));
  Mxz:=evalf(int(Y(t)*sqrt(simplify(diff(X(t), t)^2+
  diff(Y(t), t)^2+diff(Z(t), t)^2, trig)), t=a..b));
```

```
Mxy:=evalf(int(Z(t)*sqrt(simplify(diff(X(t), t)^2+
diff(Y(t), t)^2+diff(Z(t), t)^2, trig)), t=a..b));
```

$M_{yz} := 4.144318840 \qquad M_{xz} := 0 \qquad M_{xy} := 0$

```
> rc1:=[Myz, Mxz, Mxy]/L1;
```

$rc_1 := [0.5424220249,\ 0,\ 0]$

```
> T:=i->a+i*(b-a)/n; A:=i->[X(T(i)),Y(T(i)),Z(T(i))]:n:=8:
> p2:=plots[pointplot3d]([seq(A(i), i=0..n)], style=line): %;
> plots[display]([p1, p2], scaling=constrained);
> L2:=evalf(sum(linalg[norm](A(i+1) - A(i), 2), i=0..n-1));
  d:=(L2 - L1)/L1;
```

$L_2 := 7.544824170 \qquad d := -.01250869867$

```
> rc2:=evalm(sum(0.5*(A(i+1) + A(i))*linalg[norm]
  (A(i+1) - A(i), 2), i=0..n-1)/L2); evalm(rc1 - rc2);
```

$rc_2 := [0.5410404411,\ 0,\ -0.3313529838\ 10^{-10}]$

$[0.0013815838,\ 0,\ 0.3313529838\ 10^{-10}]$

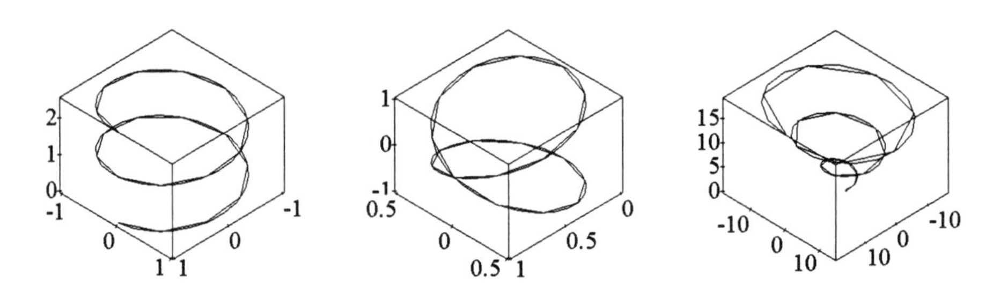

Figs. 11.5–11.7. Length of a space curve

Exercise 11.2.1

1. Find the center of mass of the arc of a circle with angle φ and density $\mu(x) = 1$.

2. The arclength of the ellipse $\vec{\mathbf{r}} = [a\cos(t), b\sin(t)]$ is expressed by the *elliptic integral of the second kind* $l(t) = a\int_0^t \sqrt{1-\varepsilon^2\sin^2(t)}\,dt = a\,E(\varepsilon, t)$ (see Section 1.3), where $\varepsilon = \sqrt{1-(b/a)^2}$ is its eccentricity. The integral $E(\varepsilon) = E(\varepsilon, \frac{\pi}{2})$ related to one-fourth of the length of the ellipse is called the *full elliptic integral*. Show that the length of the Viviani curve $\vec{\mathbf{r}} = [\cos^2(t), \cos(t)\sin(t), \sin(t)]$ is equal to $4\,E(\sqrt{2}/2)$.

3. Find the natural parametrization of

(a) circle, (b) circular helix, (c) catenary, (d) ellipse, (e) conic helix.

Hint. (a) For the circle $\vec{\mathbf{r}}\,(t) = [R\cos t,\, R\sin t]$ we have

$$s' = R \;\Rightarrow\; s = Rt \;\Rightarrow\; t = \tfrac{s}{R},$$

where t is the angle of the vector $\vec{\mathbf{r}}\,(t)$ with the axis OX. Its natural parametrization is $\tilde{\mathbf{r}}(s) = \vec{\mathbf{r}}\,(\frac{s}{R}) = [R\cos(\frac{s}{R}),\; R\sin(\frac{s}{R})]$.

4. Find the length of the torus knot $\mathbf{K}(m,n)$; see Section 8.2.2.

5. Derive the lengths of the edges of a triangle and its area through the coordinates of its vertices in affine coordinates.

6. Can one cut a hole in a sheet of paper (of size 10 cm × 10 cm) such that a man can get through it?

12

Curvature and Torsion of Curves

In this chapter we illustrate the use of some global theorems regarding the curvature of curves. The definition and basic calculating formulas for the curvature and the torsion of a curve are given in Section 12.1. In Sections 12.2 – 12.3 we calculate the geometrical characteristics of plane and space curves, plot the moving Frenet frame and an osculating circle, and present the the fundamental theorem of algebra as an example. Section 12.4 deals with the main theorem in the classical theory of curves.

In Chapter 12 the reader will become acquainted with the commands

`coeff, collect, randpoly, dotprod, crossprod.`

12.1 Basic Facts

Definition 12.1.1 The *curvature* of a curve γ at a point P is the real number $k(P) = \lim\limits_{P' \to P} \frac{\Delta\varphi}{\Delta s} = \lim\limits_{\Delta s \to 0} \frac{\Delta\varphi}{\Delta s}$ (if the limit exists), where Δs is the arclength of γ between points P and P', and $\Delta\varphi$ is the angle formed by tangent lines at these points. The *radius of a curvature* of γ at the point P is the real number $R(P) = \frac{1}{k(P)}$.

In other words, the *curvature* k is the rotating velocity of a tangent line along the curve with natural parametrization.

Theorem 12.1.1 *If* $\vec{\mathbf{r}}\,(s)$ *is a natural parametrization of a* C^2*-regular curve in* $\mathbb{R}^3$*, then the curvature is* $k(s) = |\vec{\mathbf{r}}\,''(s)|$ *(acceleration).*

For an arbitrary C^2–regular parametrization of this curve $\vec{\mathbf{r}}(t)$, the curvature is given by the formula $k(t) = \frac{|\vec{\mathbf{r}}'\times\vec{\mathbf{r}}''|}{|\vec{\mathbf{r}}'|^3}$.

Formulas for calculating the curvature of the curve γ:

1. $\gamma: \vec{\mathbf{r}} = [x(t), y(t), z(t)]$ in $\mathbb{R}^3$:

$$k(t) = \frac{\sqrt{(x'y'' - x''y')^2 + (x'z'' - x''z')^2 + (y'z'' - y''z')^2}}{((x')^2 + (y')^2 + (z')^2)^{3/2}}.$$

2. $\gamma: \vec{\mathbf{r}} = [x(t), y(t)]$ (i.e., $z(t) = 0$): $\quad k(t) = \frac{|x'y''-x''y'|}{((x')^2+(y')^2)^{3/2}}$.

3. $\gamma: y = y(x)$, a graph in $\mathbb{R}^2$: $\quad k(t) = \frac{|y''|}{(1+(y')^2)^{3/2}}$.

4. $\gamma: F(x, y) = 0$, a point $(x, y) \in \gamma$: $k(x, y) = \frac{|F_{xx}F_y^2-2F_{xy}F_xF_y+F_{yy}F_x^2|}{(F_x^2+F_y^2)^{3/2}}$.

5. $\gamma: \rho = \rho(\varphi)$, a graph in polar coordinates: $k(\varphi) = \frac{|\rho^2+2\rho'^2-\rho\rho''|}{(\rho^2+\rho'^2)^{3/2}}$.

The *osculating circle* at the point t of a curve $\vec{\mathbf{r}}(t)$ lies in the *osculating plane* (which by definition contains the vectors $\vec{\mathbf{r}}'(t)$ and $\vec{\mathbf{r}}''(t)$ at a fixed point t); its center coincides with the *curvature center* of the curve at the given point, i.e., the endpoint of the vector $\frac{1}{k(t)}\vec{\nu}(t)$ whose initial point is $\vec{\mathbf{r}}(t)$; its radius is equal to $R(t) = \frac{1}{k(t)}$. This circle is the "nearest" to the given curve among all circles through the point on the curve.

The set $\tilde{\gamma}$ of all curvature centers of the plane curve γ is its *evolute*; see Section 9.4.

The *torsion* $\varkappa$ of a C^3–regular curve in $\mathbb{R}^3$ is the rotating velocity of an osculating plane (or the binormal vector) along the curve with natural parametrization. Hence the torsion of a plane curve is identically zero.

Theorem 12.1.2 *If $\vec{\mathbf{r}}(s)$ is a C^3–regular curve in $\mathbb{R}^3$ with a natural parameter s, the torsion at points with $k \neq 0$ is given by* $\varkappa(s) = \frac{(\vec{\mathbf{r}}',\vec{\mathbf{r}}'',\vec{\mathbf{r}}''')}{k^2}$.

For arbitrary C^3–regular parametrization $\vec{\mathbf{r}}(t)$ of this curve, the torsion at points with $k \neq 0$ is given by the formula $\varkappa(t) = \frac{(\vec{\mathbf{r}}',\vec{\mathbf{r}}'',\vec{\mathbf{r}}''')}{(\vec{\mathbf{r}}'\times\vec{\mathbf{r}}'')^2}$.

A property of a curve (or of a geometrical object) is called *global* if it claims some fact about the object as a whole (for example, length, area, closing of the curve, connectness, boundness).

The property is called *local* if the corresponding fact can be checked in an arbitrary small part of the object or neighborhoods of its points (for example, the value of curvature or torsion, singular points).

In a *global theorem*, some global property is essential either in the conditions or in its claim. In a *local theorem*, all conditions and claims are local.

12.2 Curvature and Osculating Circle of a Curve in the Plane

Example 12.2.1 Let us plot the moving osculating circle of a plane curve (ellipse, logarithmic spiral). In the next program, `F.1` denotes the unit tangent vector, `F.3` the osculating circle, and `F.2` its radius as a line segment.

Program 12.2.1

```
> restart: with(plots): with(linalg): m:=12:
> a:=2: r:=[a*cos(t), sin(t)]: t1:=0: t2:=2*Pi: # ellipse
> rt:=diff(r, t): rt2:=diff(rt, t): rt3:=diff(rt2, t):
> tau:=evalm(rt/norm(rt,2)): n:=[-tau[2],tau[1]]:
> k:=simplify((rt[1]*rt2[2]-rt[2]*rt2[1])/norm(rt,2)^3):
> F.1:=evalm(r+s*tau): F.2:=evalm(r+s*n/k):
  F.3:=evalm(r+n/k+(n*cos(2*Pi*s)+tau*sin(2*Pi*s))/k):
> T:=i -> t1+i*(t2-t1)/m: for j from 1 to 3 do  p.j:=i ->
  plot([subs(t=T(i),F.j[1]),subs(t=T(i),F.j[2]),s=0..1]):
  A.j:=display([seq(p.j(i), i=0..m)], insequence=true) od:
> A.4:=plot([r[1], r[2], t=t1..t2], thickness=2):
> display([seq(A.i,i=1..4)],scaling=constrained); # Fig. 12.4
```

Example 12.2.2 Let us check that Pascal's limaçon has positive curvature $k(t)$ (Figs. 12.1–12.2) but is not a convex curve. The function $k'(t)$ has two extrema (Fig. 12.2), and hence the curve has two *vertices*; see the program. Also the curvature of a lemniscate of Bernoulli has only one maximum and one minimum.

This means that the convexity condition on the curve in the *theorem on the existence of at least four vertices on an oval* is not superfluous.

Exercise 12.2.1 Plot any oval (convex closed curve) with $2n > 4$ vertices. *Hint.* Such an oval intersects some circle at $2n$ points.

Example 12.2.3 For the closed curve $\vec{\mathbf{r}}(t)$ one defines the *coefficient of engagement with the point Q*, which shows how many times a point on $\vec{\mathbf{r}}(t)$ under monotone change of $t \in I$ rotates about Q. One can check that the coefficient of engagement keeps its (integer) value when the curve is continuously deformed in the plane without intersecting the point Q.

The *number of turnings m* of the closed curve $\vec{\mathbf{r}}(t)$ (i.e., the number of its loops taking into account their orientation $\pm$) is equal, by definition, to the coefficient of engagement of its unit tangent vector $\vec{\tau}(t)$ with respect to the origin O.

It is surprising that the *integral curvature* IC$=\int_\gamma k$ of the plane curve is equal to $2\pi\,(m+1)$, especially the fact that IC of a simple closed plane curve is equal to 2π. Let us check with MAPLE that the IC of an ellipse is equal to 2π; for Pascal's limaçon (it has one loop), IC$=4\pi$ holds.

```
> a:=1: b:=.2: r:=[(a*cos(t)-b)*cos(t),(a*cos(t)-b)*sin(t)]:
  t1:=0: t2:=2*Pi:
```

derive curvature of the limaçon using the program 12.2.1

```
> evalf(int(k, t=0..2*Pi)/(2*Pi));
```

2.041401700

```
> plot([k, diff(k,t)], t=0..2*Pi); # Figs. 12.1–12.2
```

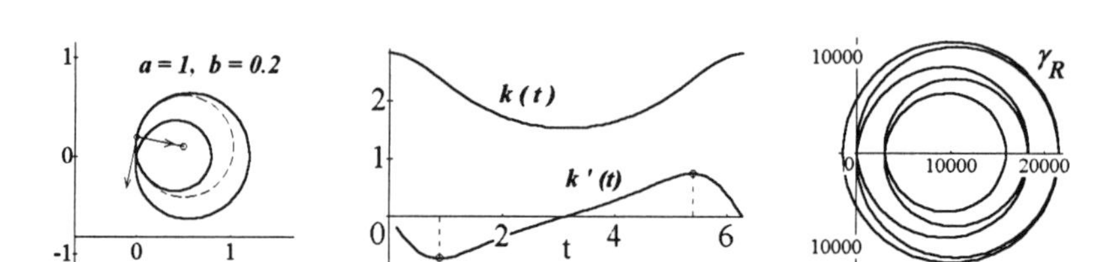

Figs. 12.1–12.3. Vertices of Pascal's limaçon. Existence of complex roots

Exercise 12.2.2 Calculate the IC for the lemniscate of Bernoulli and roses $\rho = \sin(5\varphi)$. Check that the IC of prolate epicycloids with modulus $m = \frac{a}{b}$, where a, b are relatively prime, is equal to $2\pi\, b$.

Example 12.2.4 The coefficient of engagement of a plane curve is used in the proof of the *Fundamental Theorem of Algebra*, which asserts the existence of a complex root of any polynomial $w = f_n(z)$ of degree $n > 0$. We do an experiment using the next program that illustrates this theorem; Fig. 12.3.

The idea of the program is as follows. Let $f_n(0) \neq 0$. The values of the polynomial along the circle $|z| = R$ are on the curve γ_R : $f_n(R(\cos(t) + i\sin(t)))$ $(t \in [0, 2\pi])$, which for "small" R lies near the point $f_n(0)$, i.e., *its coefficient of engagement with O is zero.* For "large" R, the curve γ_R almost coincides with the circle taken n times $\omega(R, n)$: $R^n(\cos(nt) + i\sin(nt))$, i.e., *its coefficient of engagement with O is equal to n.* (In the program the curve γ_R under increasing R will intersect the origin $n = 5$ times.) Hence there exists $R_0 > 0$ such that the curve γ_{R_0} contains the origin. In other words, the *polynomial has a root of the form* $z_0 = R_0(\cos(t_0) + i\sin(t_0))$; see [Pon, Chapters 7, 8].

```
> n:=5: f:=randpoly(z, terms=4, degree=n) + 10000;
```

$$f := -37z^5 - 35z^4 + 97z + 10050$$

```
> f1:=simplify(subs(z=R*(cos(t)+I*sin(t)), f)):
  f2:=collect(subs(I=y, f1), y):
> fy:=coeff(f2, y, 1); fx:=coeff(f2, y, 0);
```

$$f_y := -592R^5\cos(t)^4\sin(t) + 444R^5\cos(t)^2\sin(t) - 37R^5\sin(t)$$

$- 280\, R^4 \cos(t)^3 \sin(t) \;+\; 97\, R \sin(t) \;+\; 140\, R^4 \cos(t) \sin(t)$

$f_x \;:=\; -592\, R^5 \cos(t)^5 \;+\; 10050 \;+\; 740\, R^5 \cos(t)^3 \;+\; 97\, R \cos(t)$

$-185\, R^5 \cos(t) \;-\; 280\, R^4 \cos(t)^4 \;+\; 280\, R^4 \cos(t)^2 \;-\; 35\, R^4$

```
> plots[animate]([fx,fy, t=0..2*Pi], R=0..3, numpoints=200);
```

12.3 Curvature and Torsion of a Curve in Space

Exercise 12.3.1 Let us derive at the point t_0 of the curve $\vec{\mathbf{r}} = \vec{\mathbf{r}}(t) \subset \mathbb{R}^3$ the *tangent line, binormal vector, main normal vector; osculating plane, normal plane, rectifying plane; Frenet frame field; curvature and torsion.*

1. $\vec{\mathbf{r}}(t) = [e^t, e^{-t}, t\sqrt{2}],\ t_0 = 0,$
2. $\vec{\mathbf{r}}(t) = [\frac{t^2}{2}, \frac{2t^3}{3}, \frac{t^4}{2}],\ t_0 = 1,$
3. $\vec{\mathbf{r}}(t) = [t, t^2/2, t^3/6],\ t_0 = 1,$
4. $\vec{\mathbf{r}}(t) = [t, \frac{t^2}{3}, \frac{1}{2t}],\ t_0 = 1,$
5. $\vec{\mathbf{r}}(t) = [\sin(t), \cos(t), t^2],\ t_0 = \frac{\pi}{4},$
6. $\vec{\mathbf{r}}(t) = [2t, \ln t, t^2],\ t_0 = 1,$
7. $\vec{\mathbf{r}}(t) = [\cos(t), \sin(t), t^3 - 9t],\ t_0 = 0,$
8. $\vec{\mathbf{r}}(t) = [t, t^2, e^t],\ t_0 = 0,$
9. $\vec{\mathbf{r}}(t) = [t, t^3, t^2 + 4],\ t_0 = 1,$
10. $\vec{\mathbf{r}}(t) = [t, t^2 + 2, t^3 + 3],\ t_0 = 1.$

Hint. Use the next program for deriving characteristics of the curve in space at the given point (start with the circular helix).

Program 12.3.1

```
> restart: with(linalg): Digits:=2;
```

 # Data.

```
> x:=t -> R*cos(t): y:=t -> R*sin(t): z:=t-> a*t:
  t0:=1: x0:=evalf(limit(x(t),t=t0)); z0:=limit(z(t),t=t0);
```

$x_0 \;:=\; 0.54\, R \qquad z_0 \;:=\; a$

 # Calculations by formulas at the point t.

```
> assume( R>0, a>0, t>0, t<Pi/2): r:=[x(t), y(t), z(t)];
  rt:=diff(r, t); rt2:=diff(rt, t); rt3:=diff(rt2, t);
  tm:=[(x1-x(t))/rt[1], (y1-y(t))/rt[2], (z1-z(t))/rt[3]];
  p1:=simplify(det([[x1-x(t),y1-y(t),z1-z(t)],rt,rt2]))=0;
  tau:=evalm(rt/norm(rt, 2)); b1:=crossprod(rt, rt2);
  b:=evalm(b1/norm(b1, 2)); n:=crossprod(b, rt);
  p2:=simplify(dotprod(tau, [x1-x(t),y1-y(t),z1-z(t)],'orthogonal'))=0;
  p3:=simplify(dotprod(n, [x1-x(t), y1-y(t), z1-z(t)],'orthogonal'))=0;
  k:=norm(b1, 2)/norm(rt, 2)^3;
  kappa:=det([rt, rt2, rt3])/norm(b1, 2)^2;
```

$$k = \frac{R}{a^2+R^2}, \qquad \varkappa = \frac{a}{a^2+R^2}$$

 # Calculations at the given point t_0.

```
> t:=t0; tm0:=simplify(tm); p10:=simplify(p1);
  tau0:=evalm(tau); b0:=evalm(b); n0:=evalm(n);
  p20:=simplify(p2); p30:=simplify(p3);
  k0:=simplify(k); kappa0:=simplify(kappa);
```

For economy of space only the two results $k(t)$ and $\varkappa(t)$ are given.

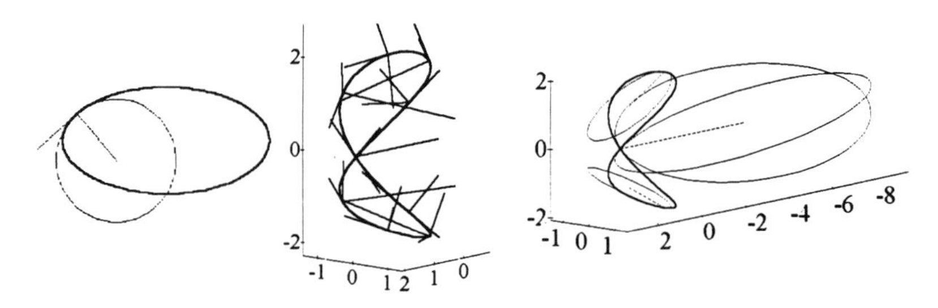

Figs. 12.4–12.6. Frenet frame field and osculating circles

Example 12.3.1 A program follows for plotting the moving Frenet frame and osculating circles of a curve in space, with the test example of the Viviani curve. In the program `F.1` denotes the tangent vector; `F.2` denotes the main normal vector; `F.3` denotes the binormal vector; `F.5` denotes the osculating circle; `F.4` denotes the segment of its position vector.

Program 12.3.2

```
> with(plots): with(linalg): R:=2: m:=32: t1:=0: t2:=2*Pi:
> r:=[R*cos(t)^2, R*cos(t)*sin(t), R*sin(t)]:
> rt:=diff(r, t): rt2:=diff(rt, t): rt3:=diff(rt2, t):
  tau:=evalm(rt/norm(rt, 2)): b1:=crossprod(rt, rt2):
  b:=evalm(b1/norm(b1, 2)): n:=crossprod(b, rt):
> k:=simplify(norm(b1,2)/norm(rt,2)^3): F.1:=evalm(r+s*tau):
  F.2:=evalm(r+s*n): F.3:=evalm(r+s*b): F.4:=evalm(r+s*n/k):
  F.5:=evalm(r+n/k + (n*cos(2*Pi*s)+tau*sin(2*Pi*s))/k):
> T:=i -> t1+i*(t2-t1)/m: for j from 1 to 5 do
  p.j:=i -> spacecurve(subs(t=T(i), evalm(F.j)), s=0..1):
  A.j:=display([seq(p.j(i), i=0..m)], insequence=true) od:
> B:=spacecurve(r, t=t1..t2, thickness=2):
> display([A.1, A.2, A.3, B], scaling=constrained,
  orientation=[40, 80], axes=framed);        # Fig. 12.5
> display([A.4, A.5, B], scaling=constrained);    # Fig. 12.6
```

Exercise 12.3.2 Calculate the *absolute integral curvature* AIC$=\int_\gamma |k|$ of the torus knot $\mathbf{K}(m,n)$ for small m,n (see Section 8.2.2):

$$\vec{\mathbf{r}}(t) = [(3+\cos(mt))\cos(nt),\ (3+\cos(mt))\sin(nt),\ \sin(mt)].$$

Remark 12.3.1 The AIC$=\int_\gamma |k|$ of a closed space curve γ is not less than 2π, and equality holds exactly for plane convex curves. If for a closed space curve AIC$< 4\pi$ holds, then the curve *is not engaged.*

12.4 Natural Equations of a Curve

The theorem on the existence and uniqueness up to congruence of naturally parametrized curves in space with given *continuous* curvature and torsion (i.e., rotating velocity of the binormal) completes the *classical theory of curves.* The condition on the continuity of the curvature and the torsion allows us to apply theorems of ODE to the Frenet formulas, but this a priori condition narrows the class of curves for which the theorem holds.

The following result [Bor] where the continuity of functions is replaced by the rather weak condition (b) removes this gap in the classical theory of curves and completes it.

Theorem 12.4.1 *Let $f_i : I \to \mathbb{R}$ $(i = 1,2)$ be two functions.*

1. If the following conditions are satisfied,

(a) $f_1(t) > 0,\ f_2(t) \geq 0$ on all of I,

(b) each of the functions f_i has a primitive function,

then there exists a unit-speed curve $\gamma : I \to \mathbb{R}^3$ with curvature $k = f_1$ and torsion $\varkappa = f_2$ that has a differentiable Frenet frame field.

2. For the existence of the unit-speed curve $\gamma : I \to \mathbb{R}^3$ with curvature $k = f_1$ and torsion $\varkappa = f_2$, up to congruence, it is necessary and sufficient that conditions (a), (b) be satisfied together with the following property:

(c) The set $\{t \in I : f_2(t) > 0\}$ is connected.

Example 12.4.1 Let us plot a unit-speed curve γ in $\mathbb{R}^2$ with curvature k that *has no primitive function.* In view of the theorem above, there *does not exist a naturally parametrized curve with curvature k.*

Hint. Define the function $\psi(t)$ by $\psi(0) = 0$ and $\psi(t) = t^2\sin(\frac{2\pi}{t^2})$ $(t \neq 0)$ and set $\vec{\mathbf{e}}_1(t) = \cos(\psi)\,\vec{\mathbf{i}} + \sin(\psi)\,\vec{\mathbf{j}}$. Obviously, there exists the derivative $\psi'(t)$, and $\vec{\mathbf{e}}_1' = \psi'\,\vec{\mathbf{e}}_2$ holds, where $\vec{\mathbf{e}}_2(t) = -\sin(\psi)\,\vec{\mathbf{i}} + \cos(\psi)\,\vec{\mathbf{j}}$. Deduce from this that there exists a curve $\vec{\mathbf{r}}(t) \subset \mathbb{R}^2$ such that $\vec{\mathbf{r}}' = \vec{\mathbf{e}}_1$ is naturally parametrized and $k(t) = |\psi'(t)|$ is its curvature.

Assume that $\varphi : \mathbb{R} \to \mathbb{R}$ is a primitive function of $k(t)$. Then φ does not decrease, but since the segments $I_n = [\frac{1}{\sqrt{n+.5}}, \frac{1}{\sqrt{n+.25}}]$ do not intersect, the sequence $a_n = \varphi(\frac{1}{\sqrt{n+.5}} - \frac{1}{\sqrt{n+.25}})$ converges. At the same time, $\varphi' \geq 0$ on each segment I_n (check this). Thus for I_n we have $\varphi' = k = |\psi'| = \psi$, and $a_n = \psi(\frac{1}{\sqrt{n+0.5}} - \frac{1}{\sqrt{n+0.25}}) = \frac{1}{n+0.25}$ holds, contrary to the convergence of the sequence $\{a_n\}$. Hence by Theorem 12.4.1 the curve $\vec{\mathbf{r}}(t)$ is not regular.

```
> f:=t^2*sin(2*Pi/t^2); ft:=diff(f, t);
> plot(f, t=0.2..2.5); plot(ft, t=0.45..2.5); # Figs. 12.7 a,b
```

Example 12.4.2 Let us plot a curve $\vec{\mathbf{r}}(t) \subset \mathbb{R}^3$ of the class C^3 that does not belong to the plane and whose curvature $k(t)$ vanishes at only one point, but with torsion $\varkappa(t) \equiv 0$, Fig. 12.8.

Hint. First we consider the plane graph of $y = x^4$. Obviously, $y'' = 12x^2$ and $k(0) = 0$. Then rotate the branch for $x \leq 0$ about the axes OY through the angle 90° and obtain the needed curve $\vec{\mathbf{r}}\,(t)$:

$$x(t) = t, \quad y(t) = \{0 \text{ if } t < 0 \text{ else } t^4\}, \quad z(t) = \{t^4 \text{ if } t < 0 \text{ else } 0\}.$$

Finally we check that the torsion $\varkappa \equiv 0$ (both branches are plane curves) and the curvature vanish only at the point $(0, 0, 0)$.

```
> readlib(piecewise):
> X:=t: Y:=piecewise(t<0,0,t^4): Z:=piecewise(t<0,t^4,0):
> plots[tubeplot]([X, Y, Z, radius=.1], t=-1.3..1.3,
  style=patchnogrid, scaling=constrained, axes=normal);
```

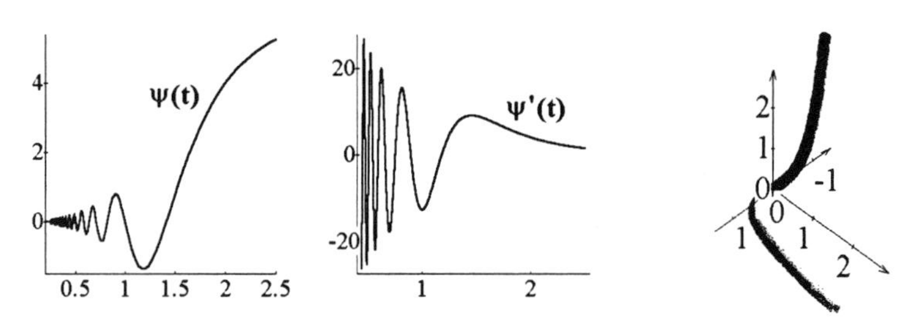

Figs. 12.7 a,b–12.8. No regular curve with $k = |\psi'|$. Curve with $\varkappa(t) \equiv 0$

Exercise 12.4.1

1. Write a program to solve the Frenet natural equations a) for plane curves, b) for curves in space. Then use it for drawing a curve with a given curvature and a torsion.

2. Write a program to show how a circle through three points on a curve tends to the osculating circle at P as three points approach P.

13
Fractal Curves and Dimension

In Chapter 13 we study geometrical properties of fractal curves and plot them. Programs are based on two methods for deriving and plotting self-similar (fractal) objects: (1) *symmetry and periodicity*, (2) *recursion.*

The programs in Section 13.1 for plotting Sierpiński curves, the Koch curve in Section 13.3, and the Menger curve in Section 13.5 are based on recursion. The programs for plotting Peano curves in the square and the triangle (method of D. Hilbert, 1891) in Section 13.2 are based on symmetry and periodicity. In the programs of Section 13.4 for plotting the dragon curve, we use points in logarithmic spiral and also symmetry properties.

(For plotting fractal curves using *turtle graphics*, see the file **turtle.mws** from the `MAPLE`, Release 4 subdirectory ..\\`share`\\`geometry`\\`turtle`. One can find the program for plotting the filled-in *Julia set* in [KK]).

In Chapter 13 the reader will become acquainted with the commands

`scale, rotate, translate, circle, cutout, hexahedron.`

13.1 Sierpiński's Curves

Recall that a function is *recursive* (i.e., is based on recursion) if it contains one or more calls to itself or to another function in which there are calls to the given function.

Example 13.1.1 We write a simple program for plotting with recursion.

Using the procedure **tree(L, lev)**, we first (for $s = 1$) plot an initial right angle `p[1]` with vertex (x_0, y_0) and edges making $\pm 45^\circ$ angles with the axis OY.

If the level of recursion is greater than 1, then in calling the given procedure we plot two smaller branches with vertices at endpoints of the angle p[0]. Finally, the sequence of branches is formed by seq(p[i], i=1..2^lev-1). It is returned by the procedure. The option remember is used for speeding up the calculations in recursive procedures.

```
> tree := proc(L:algebraic, lev:integer,
  x0:algebraic, y0:algebraic) local i; global s,p;
  options remember; s:=s+1;
  p[s]:=plot([[x0-L,y0+L],[x0,y0],[x0+L,y0+L]]);
  if lev > 1 then tree(L/2,lev-1,x0-L,y0+L);
  tree(L/2,lev-1,x0+L,y0+L) fi;
  RETURN(plots[display]([seq(p[i], i=1..2^lev-1)],
  scaling=constrained)) end:
```

```
> s:=0: tree(100, 3, 0, 0);
```

Let us replace the *angle* by a more complicated object; for example, write down in p[s] a *pair of leaves*, of a rose $\rho = \sin(2t)$ and obtain a *cactus*

```
p[s]:=plot([x0+L*sqrt(2)*sin(2*t)*cos(t),
y0+L*sqrt(2)*sin(2*t)*sin(t), t=-Pi/2..Pi/2],
color=green, style=point);
```

Figs. 13.1–13.2. Binary tree and *cactus* for $n = 3$

Problem 13.1.1 A given triangle (of zero rank) is broken by three internal line segments into four small triangles similar to a given triangle. The inner part of the middle triangle is removed, and for each of the other three closed triangles (of first rank) the process is repeated again, and so on. Denote by π_n the union of all 3^n triangles of nth rank. This set is connected and compact. Moreover, $\pi_{n+1} \subset \pi_n$ holds. The intersection of all π_n is called *Sierpiński's triangular curve*.

Write a program plotting this curve. Derive the common area of all the removed triangles.

Hint. Let the area of the given triangle be equal to 1. We calculate the area of all the removed triangles as the sum of the series

$$\frac{1}{4} + \frac{3}{4^2} + \frac{3^2}{4^3} + \cdots = \frac{1}{3}\sum_{i=1}^{\infty}\left(\frac{3}{4}\right)^i = \frac{1}{3}\left(\frac{1}{1-\frac{3}{4}} - 1\right) = 1,$$

which coincides with the area of the given triangle. Hence Sierpiński's triangular curve (the complement of the removed triangles) has zero area, which explains its status as a curve.

In the recursive procedure **serp1(L, lev, x0, y0)** we first (for $s = 0$) plot the given right triangle p[0] with legs of length 2*L (unfilled in view of the option style=line); then we plot a green triangle p[1] with vertices at the midpoints on the edges of the given triangle. If the level of recursion is greater than 1, then calling the procedure repeats the plotting of the green triangle inside three triangles of one-half the size, lying inside the angles p[0]. As a result, a sequence seq(p[i], i=1..(3^lev-1)/2) of green triangles is formed that is returned by the procedure.

```
> serp1 := proc(L::algebraic, lev::integer,
  x0::algebraic,y0::algebraic) global p,s; options remember;
  if s=0 then p[0]:=plots[polygonplot]([[x0,y0],[x0+2*L,y0],
  [x0,y0+2*L]], style=line) fi; s:=s+1;
  p[s]:=plots[polygonplot]([[x0,y0+L], [x0+L,y0+L],
  [x0+L,y0]], color=green);
  if lev>1 then serp1(L/2,lev-1,x0,y0);
  serp1(L/2,lev-1,x0+L,y0); serp1(L/2,lev-1,x0,y0+L) fi;
  RETURN(plots[display]([seq(p[i], i=0..(3^lev-1)/2)],
  scaling=constrained)) end:
```

```
> s:=0: serp1(100, 4, 0, 0);
```

Problem 13.1.2 A given square with edge a (the square of zero rank) is broken by four straight lines into nine equal squares with edge $\frac{1}{3}a$, and the interior of the middle square is removed. For each of the other eight closed squares (of first rank), forming the set C_1, the above process is repeated. We then obtain 64 squares of the second rank, whose union is C_2, and so on. Denote by C_n the union of all 8^n squares of nth rank with edges $\frac{a}{3^n}$. This set is connected and compact. Moreover, $C_{n+1} \subset C_n$ holds.

The intersection C of all the C_n is called the *Sierpiński carpet.*

Note that C is an *universal plane curve,* because if the curve γ can be embedded in the plane, then it can be embedded in the Sierpiński carpet, i.e., there exists a curve $\gamma_1 \subset C$ homeomorphic to γ.

Write down a program for plotting this curve. Derive the common area of all the removed squares.

Hint. Find the area of all removed squares as the following sum:

$$a^2(\tfrac{1}{3^2}+\tfrac{8}{3^4}+\tfrac{8^2}{3^6}+\ldots)=\tfrac{a^2}{8}\sum_{i=1}^{\infty}\left(\tfrac{8}{9}\right)^i=\tfrac{a^2}{8}\left(\tfrac{1}{1-\frac{8}{9}}-1\right)=\tfrac{a^2}{8}\cdot 8=a^2,$$

which coincides with the area of the given square. Hence the Sierpiński's carpet (the complement to all the removed squares) has zero area, which explains its status as a curve.

The recursive procedure **serp2(a, lev, x0,y0)** is structured in a similar way to the procedure **serp1(L, lev, x0,y0)**.

First for $s = 0$, we plot (unfilled in view of `style=line`) the given square `p[0]` with edge **a**; then we plot a green square `p[1]` homothetic to `p[0]` with respect to the center and having one-third the size. If the level of recursion is more than 1, then calling the procedure in the second level cycle, we repeat the plotting of the green square inside the eight squares of one-half the size lying inside the angles and along the edges of `p[0]`. As the result, a sequence `seq(p[i], i=1..(8^lev-1)/7)` of green squares is formed, which is returned by the procedure.

```
> serp2 := proc(a::algebraic, lev::integer, x0::algebraic,
  y0::algebraic) local i,j; global s,p; options remember;
  if s=0 then p[0]:=plots[polygonplot]([[x0,y0],[x0,y0+a],
  [x0+a,y0+a],[x0+a,y0]], style=line) fi; s:=s+1;
  p[s]:=plots[polygonplot]([[x0+a/3,y0+a/3],[x0+a/3,
  y0+2*a/3],[x0+2*a/3,y0+2*a/3],[x0+2*a/3,y0+a/3]],
  color=green);  if lev > 1 then for i from 0 to 2 do
  for j from 0 to 2 do if abs(i-1)+abs(j-1) > 0 then
  serp2(a/3,lev-1,x0+i*a/3,y0+j*a/3) fi od od fi;
  RETURN(plots[display]([seq(p[i], i=0..(8^lev-1)/7)],
  axes=none)) end:
```

```
> s:=0: serp2(100, 2, 0, 0);
```

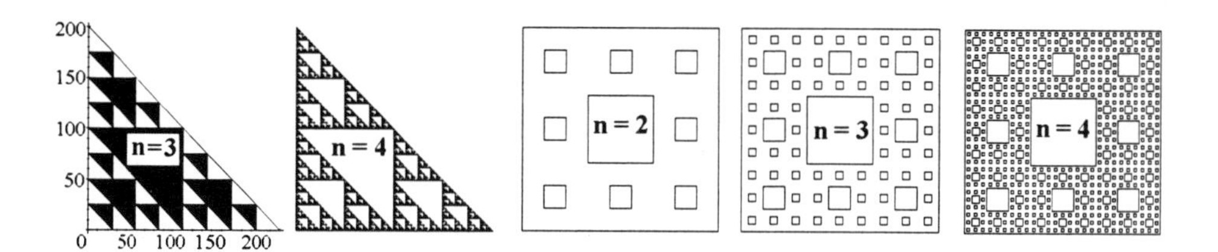

Figs. 13.3–13.4. Sierpiński's triangular curve and carpet

13.2 Peano Curves

If one starts with parametric equations of the curve in the form of a continuous vector-valued function $\vec{\mathbf{r}}(t)$, where t ranges over the segment $[a, b]$, but one considers only the image, that is, the set of points without taking account of their order, then one comes to the notion of a curve formed by C. Jordan in the nineteenth century. Moreover, such continuous images of a segment can fill the square, the cube, etc. On the other hand, there exist objects similar to curves that are not continuous images of the segment. For example, plot the union of graphs of the function $\sin\frac{1}{x}$ $(0 < x \leq 1)$ and the segment $\{(0, y) : -1 \leq y \leq 1\}$.

Problem 13.2.1 The *Peano curve* (studied by G. Peano in 1890) is the continuous image of a segment filling the interior of a square (or triangle). The Peano curve is related to the existence of simple curves in a space, whose projection onto the plane is in the form of filled areas, such as, for example, the curve $[f_1(t),\ f_2(t),\ t]$, where the first two functions define the Peano curve. Although this curve would serve nicely as a roof to keep out the rain, it cannot be identified with any continuous surface.

Write a program for plotting the Peano curve in a square and in a triangle. Plot a Pythagorean triangle using this program; Fig. 13.7 d.

Hint. Self-repeating curves can be generated by recursive functions. The first and second programs for plotting the Peano curve are based on the symmetry properties of a curve in a square or in a triangle. The third program realizes recursive plotting of the Peano curve using the Hilbert method.

(a) The Peano curve in the square is symmetric. It can be combined with itself under rotations about the center of a square through angles that are multiples of $\frac{\pi}{2}$. From this we obtain the following method.

Let us put the vertex of a square with edge `L` at the origin of its coordinates, and direct its edges along axes OX and OY.

We delete from the nth polygon P_n the segment AB nearest the origin of coordinates (plot it using `color=white`) and plot two segments parallel to the diagonal of the square until they intersect with the axes of the coordinates. Then we put in `p[0]` two times less image. Rotating `p[0]` about the origin on angles $\frac{\pi}{2}$, π, $\frac{3\pi}{2}$, we obtain `p[1]`,`p[2]`,`p[3]`. We make the union of the above four objects and parallel translate on the vector `[L,L]`, and we obtain the $(n+1)$th polygon P_{n+1}.

```
> peano1 := proc(L::algebraic, lev::integer)
  local A,s,p,i; global Q;
  Q:=plots[polygonplot]([[L,L/2],[L/2,L],[L,3*L/2],
```

```
[3*L/2,L]]); s:=1; while s <= lev do A:=L/2^(s-1);
p[4]:=plot([[A/2,0],[A/2,A/4]]);
p[5]:=plot([[0,A/2],[A/4,A/2]]);
p[6]:=plot([[A/2,A/4],[A/4,A/2]], color=white);
p[0]:=plots[display]([p[6],plottools[scale](Q,1/2,1/2),
p[4],p[5]]); for i from 1 to 3 do
p[i]:=plottools[rotate](p[0],i*Pi/2) od;
Q:=plottools[translate](plots[display](
[seq(p[i],i=0..3)]),L,L); s:=s+1 od; RETURN(Q) end:
```

```
> peano1(10, 2); # Fig. 13.5
```

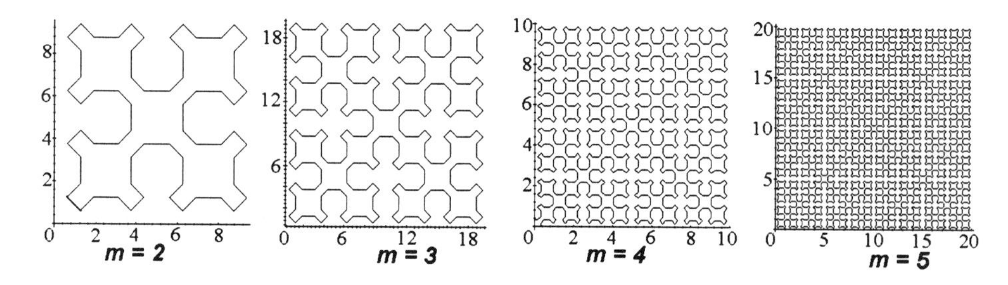

Figs. 13.5 a–d. Peano curve in the square

(b) We plot the triangular Peano curve by the program **peano2(L,lev)**.

The polygon P_{n+1} is obtained from P_n as the result of two symmetries with respect to the leg of the right triangle, and then its size is decreased by a factor of 2. But before making use of the first of these symmetries we must replace the last segment AC by the perpendicular AB to the axis of symmetry, Fig. 13.6 a.

For the sth step we take the point $A(a(1-\frac{k+2}{2^{s/2+1}}),\ a\frac{k+2}{2^{s/2+1}})$ as well as $C(a(1-\frac{k+2}{2^{s/2+1}}), 0)$. For deriving the points $B(x_b, y_b)$ using the point $A(x_a, y_a)$, find t such that the point $B(x_a+t, y_a+t)$ belongs to the straight line (the leg of the triangle) $x+y=a$. We have $t=(a-x_a-y_a)/2 = a\frac{1-k}{2^{s/2+1}}$.

Finally, $x_b = x_1 + t = a(1-\frac{3-k}{2^{s/2+1}}),\quad y_b = y_1 + t = a\frac{3-k}{2^{s/2+1}}$.

```
> peano2 := proc(L,k::algebraic, lev::integer)
local f,s,p; global Q;
p[0]:=plot([[-L*k/2,0], [-L*k/2,L/2], [L*k/2,L/2],
[L*k/2,0]],thickness=1);
p[1]:=plots[polygonplot]([[-L,0],[0,L],[L,0]],
linestyle=2); Q:=plots[display]([p[0],p[1]]);
f:=plottools[transform]((x,y) -> [-x,y]);
s:=2; while s < lev+2 do if (s mod 2=0) then
```

```
p[0]:=plot([[L*(1-(2-k)/2^(s/2)),L/2^(s/2)],[L*(1-(3-k)/
2^(s/2+1)),L*(3-k)/2^(s/2+1)]],thickness=1);
p[1]:=plot([[L*(1-(2-k)/2^(s/2)),L/2^(s/2)],
[L*(1-(2-k)/2^(s/2)),0]],thickness=1,color=white);
Q:=plots[display]([p[1],Q,p[0],]) fi;
p[2]:=plottools[scale](plottools[rotate](plottools[translate](
Q,0,-L),3*Pi/4),1/sqrt(2),1/sqrt(2));
Q:=plots[display]([p[2],f(p[2])]); s:=s+1 od;
RETURN(Q) end: # 0 < k < 1 defines the smoothness of a curve.
```

```
> peano2(10, 0.7, 4); # Fig. 13.6 d
```

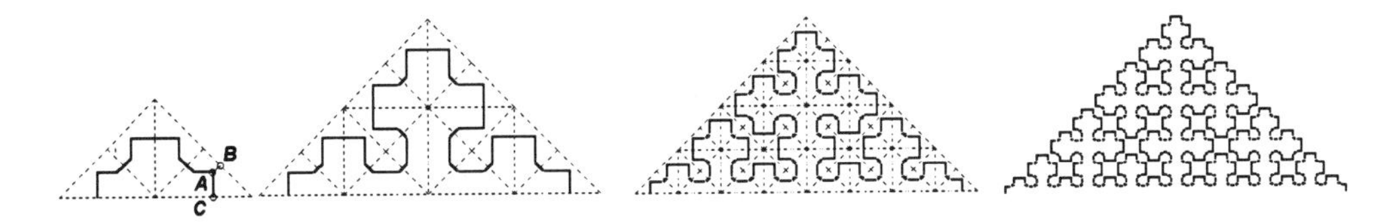

Figs. 13.6 a–d. Triangular Peano curve

(c) Recursion can be used for plotting the image known as Hilbert's curve (see [Amm]). This polygon H_1 is shaped like the letter Π, plotted in the form of three edges of a square. The polygon H_2 can be considered as a letter Π four parts of which are replaced by the same letter Π of one-third the size.

Applying this procedure to each of the four letters Π forming H_2, we obtain a polygon H_3 with elementary segments one-seventh the size of the edges of the given square.

The coefficient of similarity of the curve H_n is equal to $2^n - 1$. The points $A, B, C, D, E, F, G, H, I, J, K, M$ and segments (bunches) $dab = EM$, $dac = DF$ appearing in the procedure **peano3(a,b,c,dab,dac, lev)** are seen in Fig. 13.7 a.

Note that the angle $\angle CAB$ in the program is not assumed to be a right angle and that segments AB and AC are not necessarily equal; hence, instead of a square, the letter Π can be an arbitrary parallelogram.

```
> peano3 := proc(a,b,c,dab::array,dac::array, lev::integer)
  local d,e,f,g,h,i,j,k,m,p; global Q; options remember;
  d[1]:=(a[1]+c[1]-dac[1])/2; d[2]:=(a[2]+c[2]-dac[2])/2;
  f[1]:=d[1]+dac[1]; f[2]:=d[2]+dac[2];
  i[1]:=f[1]+b[1]-a[1]; i[2]:=f[2]+b[2]-a[2];
  e[1]:=(a[1]+b[1]-dab[1])/2; e[2]:=(a[2]+b[2]-dab[2])/2;
  g[1]:=f[1]+e[1]-a[1]; g[2]:=f[2]+e[2]-a[2];
  h[1]:=g[1]+dab[1]; h[2]:=g[2]+dab[2];
```

```
j[1]:=c[1]+h[1]-f[1]; j[2]:=c[2]+h[2]-f[2];
k[1]:=i[1]-dac[1]; k[2]:=i[2]-dac[2];
m[1]:=e[1]+dab[1]; m[2]:=e[2]+dab[2];
p[1]:=plot([[a[1],a[2]],[f[1],f[2]],[i[1],i[2]],[b[1],b[
2]]]); p[2]:=plot([[a[1],a[2]],[b[1],b[2]]],color=white);
Q:=plots[display]([p[2],Q,p[1]], scaling=constrained);
if lev > 0 then peano3(a,d,e,dac,dab,lev-1);
peano3(f,g,c,dab,dac,lev-1);peano3(h,i,j,dab,dac,lev-1);
dab[1]:=-dab[1]; dab[2]:=-dab[2];
peano3(b, k, m, dac, dab, lev-1);
dab[1]:=-dab[1]; dab[2]:=-dab[2] fi; RETURN(Q) end:
```

```
> a:=[1,1]: b:=[4,1]: c:=[1,4]: m:=2: N:=2^(m+1)-1:
  Q:=plot([[a[1],a[2]],[b[1],b[2]]], color=white):
  dac:=evalm((c-a)/N): dab:=evalm((b-a)/N):
> peano3(a,b,c,dab,dac,m); # Figs. 13 b–c
```

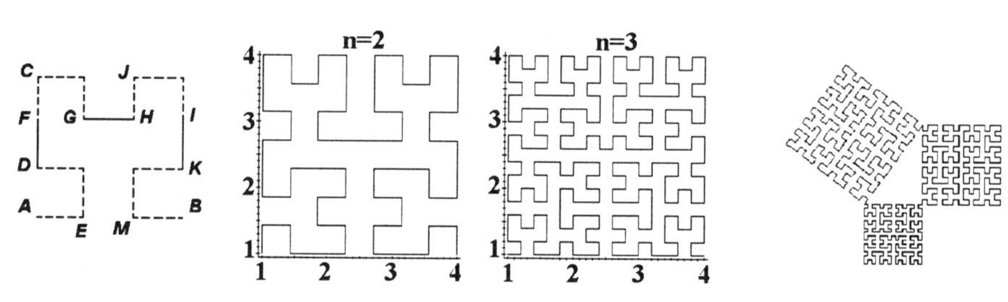

Figs. 13.7 a–d. Hilbert's curve

Program 13.2.1

```
> with(plottools): a:=[0,0]: b:=[0,3]: c:=[-3,0]:
  Q:=plot([[a[1],a[2]],[b[1],b[2]]], color=white):
> m:=2: N:=2^(m+1)-1:
  dac:=evalm((c-a)/N): dab:=evalm((b-a)/N):
> q[1]:=peano3(a,b,c,dab,dac,m):
  q[2]:=scale(rotate(translate(q[1],0,-3),Pi/2),4/3,4/3):
  q[3]:=translate(scale(rotate(q[1],Pi+arcsin(4/5)),5/3,
  5/3),0,3): plots[display]([q[1],q[2],q[3]]); # Fig. 13.7d
```

13.3 Koch Curves

Problem 13.3.1 A geometrical figure that can be broken into a finite number of equal figures similar to the given one is called a *self-similar figure.*

Simple examples are the segment, the triangle, the square, and the cube; Fig. 13.8. The self-similar figure in Fig. 13.9 looks more complicated, but it can be plotted very easily. Starting from a right triangle with edge a, repeat (indefinitely) the following process. Divide each segment connecting the vertices of the polygon into three parts and replace the middle part by two segments of length $\frac{a_1}{3}$, where a_1 is the length of the given segment. For the nth step, we obtain a polygon similar to a snowflake, which is called *Koch's snowflake* (in memory of its discoverer, a Swedish mathematician.)

Figs. 13.8 a–d. The simplest self-similar objects

There exist several definitions of *dimension* leading to essentially different mathematical results. The **first definition** is related to the minimal number of coordinates necessary for uniquely specifying the location of points in a figure. In this case the dimension is an integer.

In the **second definition** of the (topological) dimension of a figure, one takes one more than the dimension of the section dividing the figure into two separate parts. The set consisting of a finite (or countable) numbers of points is said to be null-dimensional. By this definition a smooth curve is one-dimensional, the plane (divided by a one-dimensional straight line) is two-dimensional, a ball in space is three-dimensional. Topological dimension is also an integer.

Consider the **third definition** of dimension, which is most closely related to our theme.

Definition 13.3.1 The *dimension of self-similarity* D is defined by the formula $D = \frac{\log N}{\log n}$, where N is the number of n-times smaller equal parts into which a self-similar figure can be broken.

We plot sections dividing the square into $N = 4$ squares with edge $n = 2$ times smaller than the initial edge, Fig. 13.8 c, since the dimension of self-similarity of a square is equal to $D = \frac{\log 4}{\log 2} = 2$.

Analogously, for a segment, $D = \frac{\log 2}{\log 2} = 1$, and for a cube (Fig. 13.8 d), $D = \frac{\log 8}{\log 2} = 3$ holds, as desired.

Calculate the dimension of the Koch curve, Sierpiński's triangular curve, Sierpiński's carpet, and the Peano curve. Try to measure the length of a Koch curve using a compass. Find the area of the triangle and square-shaped snowflakes and write programs for plotting them.

Hint. For the nth step of plotting, we obtain the polygon L_n consisting of $3 \cdot 4^n$ segments each of length $\frac{a}{3^n}$. Its total length is $L_n = 3(\frac{4}{3})^n a$. In order to measure the length of the Koch curve using a compass, fix the spread of the pair of compasses (i.e., *scale of measuring*) equal to λ, and moving it along the curve, calculate the number n of steps. The length of the curve is approximately equal to $L \approx \lambda n$. Fixing the scale $\lambda_n = \frac{a}{3^n}$, we find that the measured length of the Koch curve is equal to the length of the polygon corresponding to the nth stage of its plotting, $L_n = 3(\frac{4}{3})^n a$. For $n \to \infty$ we have $\lambda_n \to 0$, but the length of L_n runs to infinity. Meanwile, the area bounded by the Koch curve is finite. Let the area of the initial triangle be equal to S_0. Then the common area of the snowflake can be expressed by the geometrical series

$$S_0(1 + \tfrac{1}{3} + \tfrac{1}{3} \cdot \tfrac{4}{9} + \tfrac{1}{3}(\tfrac{4}{9})^2 + \cdots = S_0(1 + \tfrac{1}{3} \cdot \tfrac{1}{1-4/9}) = 1.6 S_0.$$

Note that the lengths L_n of the segment, the circle, and other smooth curves have a finite limit. Attempts at measuring the lengths of other self-similar curves lead to analogous results. When the scale of measuring is decreased, the length of the curve grows to infinity. This explains, for example, the difference of 20% in the length of the boundary between Portugal and Spain as perhaps given in reference books concerning these countries; different scales were probably used to measure the boundary. Write down the formula for the length of the Koch curve in the form

$$L = A\lambda^{-\beta}, \quad \text{where} \quad A = 3\,a^{\log 4/\log 3}, \ \beta = \frac{\log 4}{\log 3} - 1.$$

The coefficient β in the formula is related to the dimension of the object.

In deriving the dimensions of the polygon in Figs. 13.9, 13.3 and 13.5, we obtain that the dimension of each part (and hence the whole) of the Koch curve is equal to $\frac{\log 4}{\log 3} \approx 1.2618$; the Sierpiński triangular curve has dimension $\frac{\log 3}{\log 2} \approx 1.5849$; Sierpiński's carpet has dimension $\frac{\log 8}{\log 3} \approx 1.8727$. These *strange* curves *do not have integer dimension.* Moreover, the dimension of the Peano curve is equal to $\frac{\log 4}{\log 2} = 2$.

Using the definition of the dimension D, rewrite the formula for a length of the Koch curve in the form $L = 3a^D\lambda^{1-D}$, where λ is the scale of measuring (spread of the pair of compasses). Hence, a growth in the measured length of a self-similar curve with a decrease in the scale of measuring indicates that it does not have integer dimension.

A very simple program `triad1` with recursion plots the *snowflake* as the union of unfilled squares. The second program `triad1` plots the boundary of the square-shaped *snowflake*. The third program below plots the triangular *snowflake*.

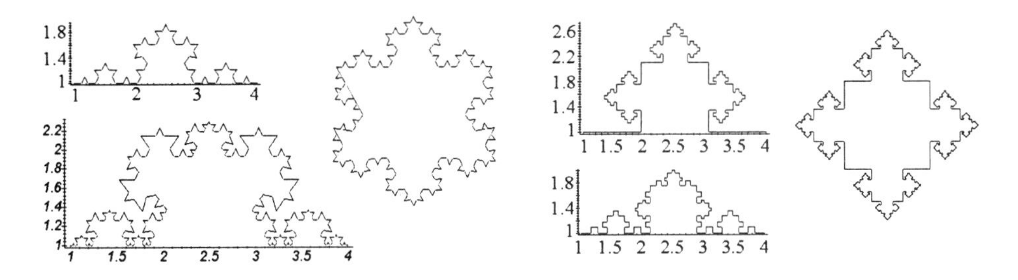

Fig. 13.9. Koch curves

The base objects in the second and third programs are square or triangular peaks placed on the segment; Fig.13.9.

```
> triad1 := proc(L::algebraic, lev::integer,
  x0::algebraic,y0::algebraic) global p,s; options remember;
  s:=s+1; p[s]:=plots[polygonplot]([[x0,y0],[x0,y0+L],
  [x0+L,y0+L],[x0+L,y0]], color=green);
  if lev>1 then triad1(L/3,lev-1,x0+L/3,y0-L/3);
  triad1(L/3,lev-1,x0-L/3,y0+L/3);
  triad1(L/3,lev-1,x0+L/3,y0+L);
  triad1(L/3,lev-1,x0+L,y0+L/3) fi;
  RETURN(plots[display]([seq(p[i],i=1..(4^lev-1)/3)])) end:

> s:=0: triad1(100, 3, 0, 0);

> triad2 := proc(a,b, lev::integer)
  local s,c,d,e,f; global p,Q; options remember; s:=.33;
  c[1]:=a[1]+(b[1]-a[1])*s; c[2]:=a[2]+(b[2]-a[2])*s;
  f[1]:=a[1]+(b[1]-a[1])*(1-s);
  f[2]:=a[2]+(b[2]-a[2])*(1-s);
  d[1]:=c[1]-(f[2]-c[2]); d[2]:=c[2]+(f[1]-c[1]);
  e[1]:=d[1]+f[1]-c[1]; e[2]:=d[2]+f[2]-c[2];
  p[1]:=plot([[c[1],c[2]],[d[1],d[2]],[e[1],e[2]],
  [f[1],f[2]]]);
  p[2]:=plot([[c[1],c[2]],[f[1],f[2]]], color=white);
  Q:=plots[display]([p[2], Q,p[1]], scaling=constrained);
  if lev>1 then triad2(c,d,lev-1); triad2(d,e,lev-1);
  triad2(e,f,lev-1) fi; RETURN(Q) end: # 0 < s < 0.5

> a:=[1,1]: b:=[4,1]: Q:=plot([[a[1],a[2]],[b[1],b[2]]]):

> triad2(a, b, 3);
```

Program 13.3.1

```
> with(plottools): setoptions(scaling=constrained,axes=none):
> f:= transform((x,y) -> [-x,y]):
> q.1:=translate(Q,-a[1],-a[2]): q.2:=rotate(f(q.1),Pi/2):
  q.3:=translate(rotate(q.1, -Pi/2), b[1]-a[1], 0):
  q.4:=translate(rotate(q.1, Pi), b[1]-a[1], a[1]-b[1]):
  plots[display]([seq(q.i, i=1..4)]);
```

```
> triad3 := proc(a,b,lev::integer) local s,k,c,d,f,p;
  global Q; options remember; s:=0.4; k:=2;
  c[1]:=a[1]+(b[1]-a[1])*s; c[2]:=a[2]+(b[2]-a[2])*s;
  f[1]:=a[1]+(b[1]-a[1])*(1-s);
  f[2]:=a[2]+(b[2]-a[2])*(1-s);
  d[1]:=(a[1]+b[1])/2-k*(f[2]-c[2]);
  d[2]:=(a[2]+b[2])/2+k*(f[1]-c[1]);
  p[1]:=plot([[c[1],c[2]],[d[1],d[2]],[f[1],f[2]]]);
  p[2]:=plot([[c[1],c[2]],[f[1],f[2]]], color=white);
  Q:=plots[display]([p[2],Q,p[1]],scaling=constrained);
  if lev>1 then triad3(a,c,lev-1); triad3(c,d,lev-1);
  triad3(d,f,lev-1); triad3(f,b,lev-1) fi; RETURN(Q) end:
> a:=[1,1]: b:=[4,1]: Q:=plot([[a[1],a[2]],[b[1],b[2]]]):
> triad3(a,b,3);
```

Program 13.3.2

```
> with(plottools): setoptions(scaling=constrained,axes=none):
> f:=transform((x,y) -> [-x,y]):
> q.1:=translate(Q,-a[1],-a[2]): q.2:=rotate(f(q.1),2*Pi/3):
  q.3:=translate(rotate(q.1, -2*Pi/3), b[1]-a[1], 0):
  plots[display]([seq(q.i, i=1..3)], axes=none);
```

Exercise 13.3.1 Modify programs 13.3.1 and 13.3.2 to get Koch curves that are inside the triangle and the square. For obtaining similar Koch curves, replace the equilateral triangle in the original Koch curve by an isoceles triangle or try to find your own "generators."

13.4 Dragon Curve (or Polygon)

Problem 13.4.1 Let us look at the fantastic trajectory plotted along the rectangular coordinate net in Fig. 13.10 b. The curve is reminiscent of a dragon with sharp claws, open jaws, and bent tail. Each 90° turn of the curve is smoothed to show visually the absence of points of self-intersection (i.e., the *polygon* does not go twice along any elementary segment).

The *dragon curve* can be constructed by repeatedly folding a long paper strip by some systematic method. Let us fold the paper strip once, then fold the strip in the middle again, and so on, n times. Arranging the strip so that its segments form angles of $90°$ near bends, we obtain the dragon curve of *rank* n. Several dragon curves can be joined together such that they cover the whole plane or form symmetric patterns.

Recall that the *logarithmic spiral* is the trajectory of the point P moving along the straight line OL with velocity proportional to the distance $|OP|$, while at the same time, the line OL uniformly rotates in the plane about the point O; see Section 6.5 for the equation of the curve in polar coordinates.

Show that points of the dragon curve that are ends of parts of length 2^n lie on the logarithmic spiral. Write a program for plotting the dragon.

Plot the clutch of four dragons of n th order, Fig. 13.10 c, and estimate the radius of the covered circle with center at their initial (common) point.

Hint. We consider two plotting methods.

The *first method* is based on the property of self-repetition of the dragon curve (starting at A and with endpoint at B): polygon `p[n+1]` is obtained by clutching `p[n]` with the same polygon, turned about the point `B[n]` at an angle of $90°$ clockwise. Assuming $A(0,0)$ and `B[0]=[x,y]`, we obtain `B[1]=[x-y,x+y]`. Hence

```
B[2]= [-2y,-2x],B[3]=[-2y-2x,2x-2y], B[4]= [-4x,-4y]=-4B[0],
B[5]=[-4x+4y, -4x-4y]=-4B[1],...
```

Let $\rho = \rho(\varphi)$ be the equation of the logarithmic spiral. From this follows the equality $\frac{\rho'(\varphi)}{\rho(\varphi)} = k$. Integrating both parts of this equality, we obtain $\ln\rho(\varphi) - \ln\rho(0) = k\cdot\varphi$. Finally, we deduce the equation

$$\rho = ae^{k\varphi} \qquad (\varphi \geq 0). \tag{13.1}$$

Let us show that points $B[n]$ $(n \geq 0)$ lie on the logarithmic spiral (13.1). For convenience we assume $x = 1$, $y = 0$, i.e., $B[0] = [1,0]$, $B[1] = [1,1]$, $B[2] = [0,-2]$, $B[3] = [-2,2]$, $B[4] = [-4,0] = -4B[0]$, $B[5] = [-4,-4] = -4B[1], \ldots$. The equalities $\rho[n+1] = \sqrt{2}\rho[n]$ and $\varphi[n] = n\frac{\pi}{2}$ for $n \geq 0$ show that the points belong to the logarithmic spiral. To find the parameters of the spiral, first we derive $a = \rho[0] = \sqrt{2}$. Since $e^{k\frac{\pi}{2}} = \sqrt{2}$ holds, then $k = \frac{2}{\pi}\ln 2$. Although M. Gardner [Gar 1] claims that points $B[n]$ are not used for plotting the dragon curve, we present a procedure `dragon1` that effectively uses these points.

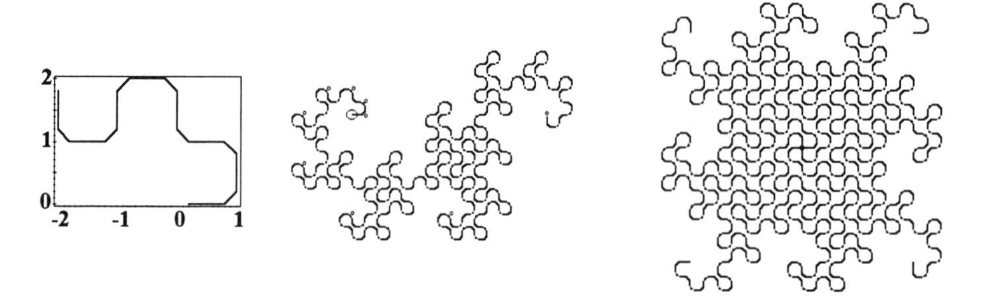

Figs. 13.10 a–c. Dragon curve, clutch of four dragon curves

The observation that the last elementary segment of the polygon p[n] is always directed along the axis OY allows us easily to carry out the smoothing (with parameter $0 < k < \frac{1}{2}$) of the dragon curve. Turning the dragon curve around the origin A at angles that are multiples of $\frac{\pi}{2}$ allows us to join four curves with intersection only at the point A, and for $n \to \infty$, the curve fills the whole plane.

Note that four dragon curves of rank 5 and 6 fill the circle of radius 3. Since the size of the dragon curve doubles through two iterations, one can suppose that the four dragon curves of ranks $2n - 1$ and $2n$, where $n \geq 3$, fill the circle of radius $3 \cdot 2^{n-3}$. The coefficient $k \in (0, \frac{1}{4})$ controls the smoothness of the curve during elementary turnings.

```
>with(plottools): with(plots):
  setoptions(scaling=constrained,axes=none):
>dragon1 := proc(m::integer, k::algebraic)
  local b,i; global c,p;
  p[1]:=plot([[0,0], [1-k,0], [1,k], [1,1]]);
  c[1]:=circle([1,0], 0.04); for i from 1 to m-1 do
  b:=[2^(i/2)*cos(i*Pi/4), 2^(i/2)*sin(i*Pi/4)];
  c[i+1]:=circle(b,0.05); p[i+1]:=plots[display](
  [plot([[b[1],b[2]-k], [b[1],b[2]], [b[1]-k,b[2]]],
  color=white), p[i], plot([[b[1],b[2]-k], [b[1]-k,b[2]]]),
  translate(rotate(translate(p[i],-b[1],-b[2]), -Pi/2),
  b[1], b[2])]) od;
  RETURN(plots[display]([p[m], seq(c[i], i=1..m)])) end:
```

```
> dragon1(7, 1/4);
> display([p[7], rotate(p[7],Pi/2, color=green),
 rotate(p[7],-Pi/2,color=blue),rotate(p[7],Pi, color=red)]);
```

The *second plotting method* is based on the law for turning the polygon at each of its nodes. Observing the motion of the curve from the first segment, assume the notation -1 for turning counterclockwise and 1 for turning clockwise. Denote by $t(i)$ the turning of the ith segment. From Fig. 13.20 a we see that $t(1) = -1$, $t(2) = -1$, $t(3) = 1$, $t(4) = -1$, $t(5) = -1$, $t(6) = 1$, $t(7) = 1$. Note that the sequence $t(i)$ can be defined recursively $t(i) = \begin{cases} t(i/2), & \text{if } i \text{ is even}, \\ (i \bmod 4) - 2, & \text{if } i \text{ is odd}. \end{cases}$

But the program can be written without using recursion. If $e(i) = [e_1(i), e_2(i)]$ is the direction of the ith segment, then its normal vector is defined by the well-known rule delta $(i) = [-e_2(i), e_1(i)]$. The direction of the $(i+1)$th segment is given by the rule `t(i)*delta(i)`. Again, the coefficient $k \in (0, \frac{1}{4})$ controls the smoothness of the curve on the turnings.

```
> dragon2 := proc(k::algebraic, N::integer)
  local t,i,q1,q2,q3,q4,delta; global p;
  q2:=[k,0]; q3:=[1-k,0]; delta:=evalm(q3-q2);
  p[0]:=plot([[q2[1],q2[2]],[q3[1],q3[2]]]);
  for i from 1 to N do if (i mod 2=0) then t[i]:=t[i/2]
  else t[i]:=(i mod 4)-2 fi; q4:=evalm(q3+k/(1-2*k)*delta);
  delta:=evalm(t[i]*[delta[2],-delta[1]]); q1:=evalm(q3);
  q2:=evalm(q4+k/(1-2*k)*delta); q3:=evalm(q2+delta);
  p[i]:=plot([[q1[1],q1[2]],[q2[1],q2[2]],[q3[1],q3[2]]])
  od; RETURN(plots[display]([seq(p[i], i=0..N)])) end:

> dragon2(0.2, 64);
```

13.5 The Menger Curve

Problem 13.5.1 An analogue of Sierpiński's carpet (see Section 13.1) in space is obtained as follows; Figs. 13.11 c,d. Let us divide a given cube (of zero rank) with edge a by six planes, parallel to its faces, into 27 equal cubes with edge $\frac{a}{3}$. Then we remove the central cube and all its neighbors along two-dimensional faces of this division. We obtain the set K_1 consisting of 20 cubes of the first rank. Doing the same with each of the other closed cubes of first rank, obtain the set K_2 consisting of 400 cubes of the second rank. Continuing this process indefinitely obtain the sequence of polyhedra $K_1 \supset K_2 \supset K_3 \supset \dots$, whose intersection M is called the *universal Menger curve*. Its universal property means that any space curve can be embedded into the set M.

a) Write a program for plotting the polyhedron K_n. Find the volume of all deleted cubes.

b) Write a program for plotting an analogue of Sierpiński triangular curve in a pyramid (see figure in [Cro, p. 16]).

Hint. We consider two plotting methods and recommend that the reader develop the programs.

First method. For the first step we plot the figure based on the cube of zero rank `hexahedron([x0, y0, z0], a0)`.

```
> display(cutout(hexahedron([x0, y0, z0], a0), 1/3));
```

This command (we use the library `plottools`) changes only plane faces of the cube, cutting from each a central square with edge $\frac{a}{3}$; we will agree to take this as the desired result. For the second step we apply the previous operation to each of the $20 = 3^3 - 7$ cubes (of the first rank), and so on.

Let us write a recursive procedure in which s plays the role of a counter. We save the data for each cube at the mth step in the array `p[s]` containing $S_m = \frac{20^m-1}{19}$ elements. In particular, `p[1]` contains cubes of the first rank. To plot the Menger curve at the mth step using the command `display(p[i], i=N_m)`, we save in the array `p[s]` only cubes of maximal rank. After the first step ($S_1 = 21$), their number is $N_1 = 20$; after the second step ($S_2 = 421$), the number of cubes of rank 2 is $N_2 = 400$; after the third step ($S_3 = 8842$), there would be $N_3 = 8400$ cubes of the third rank, etc. Elements of N_2 and N_3 can be obtained as follows:

```
> N2:={$ 3..421 } minus {seq(21*i+2, i=1..20)}: nops(N2);
  N3:={$ 3..8842} minus ({seq(421*j+2, j=0..20)} union
  {seq(seq(421*j+21*i+3, i=1..20), j=0..20)}): nops(N3);
```

400 8400

The following procedure helps us to plot Figs. 13.11 c,d.

```
> menger1 := proc(L::algebraic, lev::integer, x0,y0,z0)
  local i,j,k,f; global p,s; options remember;
  s:=s+1; f:=0.4; p[s]:=cutout(hexahedron([x0,y0,z0],L),f);
  if s=2 then p[1]:=p[2] fi; if lev > 1 then
  for i from -1 to 1 do for j from -1 to 1 do
  for k from -1 to 1 do if (k=0 and abs(i)+abs(j)>1) or
  (k<>0 and abs(i)+abs(j)>0) then menger1(L/3,lev-1,
  x0+2*i*L/3, y0+2*j*L/3, z0+2*k*L/3) fi od od od fi;
  RETURN(plots[display]([seq(p[i], i=1..(20^lev-1)/19)])) end:

> with(plottools): s:=0: menger1(100, 2, 0, 0, 0);
```

Using the *second method* we plot only the omitted cubes. Let us start plotting with the given cube `p[0]` in transparent view using the option `style=line`. In the first step we obtain 7 cubes (of first rank) in the form of a thickened system of rectangular coordinates; the central (seventh) cube is covered by the other 6 cubes, and we don't need to plot it. Hence the sequence p contains $1+6=7$ elements. The volume of the 7 cubes (of the first rank) is equal to $7(\frac{a}{3})^3$. In the second step, we obtain $140 = 7 \times 20$ cubes (of the second rank), in addition, and as before, the central cubes are covered so we don't need to plot these 20 cubes. Now the sequence p contains $1+6+120 = 127$ elements. The volume of the 140 cubes (of the second rank) is equal to $140(\frac{a}{9})^3$. In the third step we obtain $2800 = 7 \times 20 \times 20$ additional cubes (of the third rank); again the central cubes are covered, and we don't plot these $400 = 20 \times 20$ cubes. The sequence p contains $1+6+120+2400 = 2427$ elements. The volume of the 2800 cubes (of the third rank) is equal to $2800(\frac{a}{27})^3$.

The volume of the limit object is equal to the sum of the series

$$V = 7a^3(\tfrac{1}{3^3} + \tfrac{20}{3^6} + \tfrac{20^2}{3^9} + \ldots) = \tfrac{7a^3}{20} \sum_{i=1}^{\infty} \left(\tfrac{20}{27}\right)^i = \tfrac{7a^3}{20} \cdot \tfrac{20}{7} = a^3,$$

which coincides with the volume of the given cube. Hence the Menger curve (actually the complement to the object that we have plotted here) has zero volume.

```
> menger2 := proc(L::algebraic, lev::integer, x0,y0,z0)
  local i,j,k; global p,s; options remember; if s=0 then
  p[0]:=hexahedron([x0,y0,z0],L,style=line) fi;
  for i from -1 to 1 do for j from -1 to 1 do
  for k from -1 to 1 do if abs(i)+abs(j)+abs(k)=1 then
  s:=s+1; p[s]:=hexahedron([x0+2*i*L/3,y0+2*j*L/3,
  z0+2*k*L/3],L/3) fi od od od; if lev > 1 then
  for i from -1 to 1 do for j from -1 to 1 do
  for k from -1 to 1 do if abs(i)+abs(j)+abs(k)>1 then
  menger2(L/3,lev-1,x0+2*i*L/3,
  y0+2*j*L/3, z0+2*k*L/3) fi od od od fi; RETURN(plots
  [display]([seq(p[i],i=0..6*(20^lev-1)/19)])) end:
> with(plottools): s:=0: menger2(100, 3, 0, 0, 0);
```

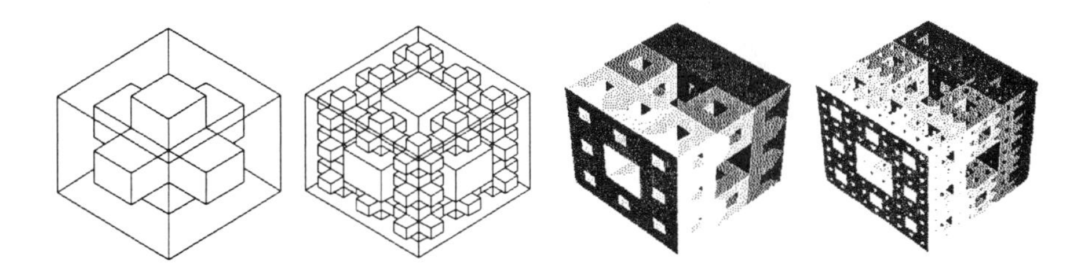

Figs. 13.11 a–d. The universal Menger curve

14

Spline Curves

The main problem of this chapter is as follows: given a sequence of *control points* $\mathbf{P} = \{P_0, P_1, \ldots, P_m\}$ arbitrarily placed in the plane or in space, construct a smooth curve passing near — through these points — and satisfying some additional conditions.

A polygon that joins neighboring points from the array $\mathbf{P}$ is called a *control polygon*; the points P_0 and P_m are called *boundary* points, and the points $P_1, \ldots, P_{m-1}$ are called *inner* points.

The equation of the curve will be written in the form $\vec{\mathbf{r}}(t) = \sum a_i(t) P_i$ $(t \in [a, b])$, where functions $a_i(t)$ must be derived.

Spline curves are obtained using a scheme similar to that in Section 3.2: we fix the net $a = t_1 < t_2 < \cdots < t_m = b$ (below, t_i are integers) and plot *segments*, i.e., vector-valued functions $\{\vec{\mathbf{r}}_i(t),\ t \in [t_i,\ t_{i+1}]\},\ i = 1, \ldots, m-2$ (as a rule, polynomials of small degree), gluing them at *nodes*: $\vec{\mathbf{r}}_i(t_{i+1}) = \vec{\mathbf{r}}_{i+1}(t_{i+1})$. For smooth splines the nodes do not need to coincide with the control points.

Five examples in Section 14.1 illustrate the importance of C^2-continuity (i.e., continuity of the osculating circle) of a parametrized curve. In Section 14.2 we complete our studies of Bezier curves begun in Section 4.3. In Sections 14.3 and 14.4 we calculate and plot B-spline curves and their C^2-continuous generalization of Beta-spline curves. Sections 14.5 and 14.6 are devoted to the Hermite and Catmull-Rom composed curves.

In this chapter the reader will become acquainted with the command

`assign`.

14.1 Preliminary Facts and Examples

Denote by $\vec{\mathbf{k}} = k\,\vec{\nu}$ the *vector of curvature*. For the C^2-continuous curve $\vec{\mathbf{r}}(t)$, this vector is derived by the formula (of the *double product*)

$$\vec{\mathbf{k}}(t) = \frac{(\vec{\mathbf{r}}\,' \times \vec{\mathbf{r}}\,'') \times \vec{\mathbf{r}}\,'}{|\vec{\mathbf{r}}\,'|^4}.$$

Definition 14.1.1 A curve is called *(geometrically) C^1-continuous* if its tangent line (i.e., the unit vector $\vec{\tau}$) changes continuously, *(geometrically) C^2-continuous* if its main normal vector and curvature (i.e., the vector $\vec{\mathbf{k}}$) change continuously.

Remark 14.1.1
1. C^2-continuity of a curve means the following geometrically visible condition: the osculating circle continuously varies in t.
2. C^2-continuity of a curve is a stronger condition than C^2-*regularity*, meaning C^1-continuity together with the continuity of $\vec{\mathbf{r}}''(t)$.

The next examples of composed plane curves illustrate the fact that the continuity of the vectors $\vec{\mathbf{r}}\,'$ and $\vec{\mathbf{r}}\,''$ (depending on the choice of parametrization of the curve) does not imply the continuity of the vectors $\vec{\tau}$ and $\vec{\mathbf{k}}$ (does not depend on the choice of parametrization of the curve) and hence does not always lead to a curve with nice geometry.

Example 14.1.1 Each curve in Examples 1–5 is composed of two segments $\vec{\mathbf{r}}_1(t)$ and $\vec{\mathbf{r}}_2(t)$ $(0 \le t \le 1)$. Moreover, at the nodes the *condition of continuity* $\vec{\mathbf{r}}_1(1) = \vec{\mathbf{r}}_2(0)$ holds. We do calculations to find characteristics of the curve at its node and then plot it.

1. $\vec{\mathbf{r}}_1(t) = [4t,\ 4t]$, $\vec{\mathbf{r}}_2(t) = [t+1,\ t+1]$ are the two segments; Fig. 14.1. The first derivatives at the node $(1, 1)$ are not equal: $\vec{\mathbf{r}}\,'_1(1) = [4, 4]$, $\vec{\mathbf{r}}\,'_2(0) = [1, 1]$, but the curve is C^2-continuous.

```
> r1:=[4*t,4*t]: r2:=[1+t,1+t]: t1:=0: t2:=1/4: t3:=0: t4:=1:
> p1:=plot([r1[1],r1[2],t=t1..t2]):
  p2:=plot([r2[1],r2[2],t=t3..t4]):
> c1:=plottools[circle](subs(t=t2, r1), 0.02):
> plots[display]([p1, p2, c1], scaling=constrained);
```

The programs for deriving and plotting the composed curves in Examples 2–5 differ only in their first lines; hence only this line is given below.

2. $\vec{\mathbf{r}}_1(t) = [-(1-t)^2,\ (1-t)^2]$, $\vec{\mathbf{r}}_2(t) = [t^2,\ t^2]$ are the two segments; Fig. 14.2. Although the first derivatives at the node $(0, 0)$ are equal, $\vec{\mathbf{r}}\,'_1(1) = \vec{\mathbf{r}}\,'_2(0) = [0, 0]$, and the curve $\vec{\mathbf{r}}(t)'$ is continuous, it is not C^1-continuous.

```
> r1:=[-(1-t)^2, (1-t)^2]: t1:=0: t2:=1:
  r2:=[t^2, t^2]: t3:=0: t4:=1:
```

3. $\vec{\mathbf{r}}_1(t) = [\sin(\frac{\pi}{2}t^2),\ \cos(\frac{\pi}{2}t^2)]$, $\vec{\mathbf{r}}_2(t) = [\cos(\frac{\pi}{2}t^2),\ -\sin(\frac{\pi}{2}t^2)]$ are two arcs of a circle; Fig. 14.3. The derivatives at the node $(1, 0)$ are not equal:

$$\vec{\mathbf{r}}_1'(1) = [0, -\pi],\ \vec{\mathbf{r}}_2'(0) = [0, 0],\quad \vec{\mathbf{r}}_1''(1) = [-\pi^2,\ -\pi],\ \vec{\mathbf{r}}_2''(0) = [0,\ -\pi].$$

However, the curvature vector $\vec{\mathbf{k}}$ is continuous, because the whole curve coincides with the half-circle. The curve is C^2-continuous.

```
> r1:=[sin(Pi/2*t^2),cos(Pi/2*t^2)]: t1:=0: t2:=1:
  r2:=[cos(Pi/2*t^2),-sin(Pi/2*t^2)]: t3:=0: t4:=1:
```

4. $\vec{\mathbf{r}}_1(t) = [\cos(\frac{\pi}{2}(1-t)^3),\ \sin(\frac{\pi}{2}(1-t)^3)]$, $\vec{\mathbf{r}}_2(t) = [2 - \cos(\frac{\pi}{2}t^3),\ -\sin(\frac{\pi}{2}t^3)]$ are arcs of two circles, Fig. 14.4. Although both derivatives at the node $(1, 0)$ are equal, $\vec{\mathbf{r}}_1'(1) = [0, 0] = \vec{\mathbf{r}}_2'(0)$, $\vec{\mathbf{r}}_1''(1) = [0, 0] = \vec{\mathbf{r}}_2''(0)$, i. e., the vectors $\vec{\mathbf{r}}'$ and $\vec{\mathbf{r}}''$ are continuous, the main normal vectors $\vec{\mathbf{n}}_1(1)$ and $\vec{\mathbf{n}}_2(0)$ at the common point of the two half-circles move in opposite directions. The curve is C^1-continuous but not C^2-continuous.

```
> r1:=[cos(Pi/2*(1-t)^3), sin(Pi/2*(1-t)^3)]:t1:=0:t2:=1:
  r2:=[2-cos(Pi/2*t^3), -sin(Pi/2*t^3)]: t3:=0: t4:=1:
```

5. $\vec{\mathbf{r}}_1(t) = [t^2 + \frac{1}{2}(t-3),\ -t^2 + 2t]$, $\vec{\mathbf{r}}_2(t) = [-t^2 + \frac{5}{2}t,\ -t^2 + 1]$, Fig. 14.5. The tangent lines at the node $(1, 0)$ coincide, i.e., the first derivatives are equal and nonzero: $\vec{\mathbf{r}}_1'(1) = [\frac{5}{2},\ 0]$, $\vec{\mathbf{r}}_2'(0) = [\frac{5}{2},\ 0]$. The second derivatives at the node are not equal, $\vec{\mathbf{r}}_1''(1) = [2,\ -2]$, $\vec{\mathbf{r}}_2''(0) = [-2,\ -2]$, but the curve is C^2-continuous because its curvature is continuous: $k_1(1) = k_2(0) = \frac{8}{25}$.

```
> r1:=[t^2+t/2-3/2, -t^2+2*t]: t1:=0: t2:=1:
  r2:=[-t^2+5*t/2, -t^2+1]: t3:=0: t4:=1:
```

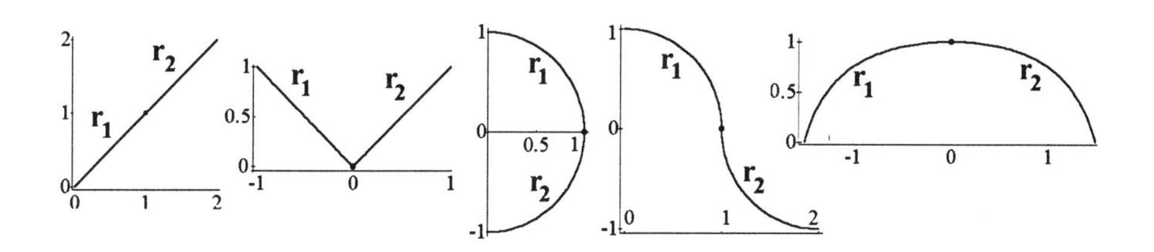

Figs. 14.1–14.5. Composed curves 1–5 from Example 14.1.1

14.2 Composed Bezier Curves

14.2.1 Elementary Cubic Bezier Curve

Definition 14.2.1 Given points P_0, P_1, P_2 and P_3, the *(elementary) cubic Bezier curve* (i.e., the Bezier curve with $m = 3$; see Section 4.2) is defined by the equation $\vec{\mathbf{r}}(t) = (((1-t)P_0 + 3tP_1)(1-t) + 3t^2P_2)(1-t) + t^3P_3$, where $0 \le t \le 1$.

Obviously, $\vec{\mathbf{r}}(0) = P_0$, $\vec{\mathbf{r}}(1) = P_1$, $\vec{\mathbf{r}}'(0) = 3(P_1 - P_0)$, $\vec{\mathbf{r}}'(1) = 3(P_3 - P_2)$.

Example 14.2.1 Using the program from Section 4.2 we plot three elementary cubic Bezier curves of different types:

```
1) xp:=[0,1,2,3]: yp:=[0,2,2,0]:      # Fig. 14.6 a
2) xp:=[0,1,2,3]: yp:=[0,2,0,2]:   # zigzag, Fig. 14.6 b
3) xp:=[2,4,1,3]: yp:=[0,2,2,0]:   # loop, Fig. 14.6 c
```

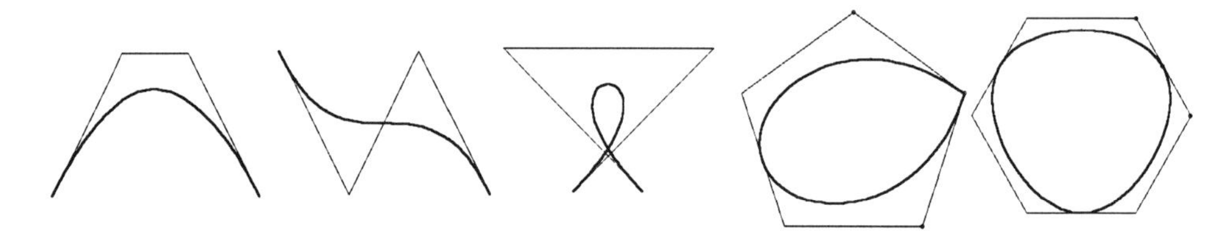

Figs. 14.6 a–e. Elementary and closed Bezier curves

Definition 14.2.2 The *composed cubic Bezier curve* defined by the array $P_0, \ldots, P_{3m}$ ($m \ge 1$) is a curve that can be represented as a union of elementary cubic Bezier curves $\gamma = \gamma^{(1)} \cup \cdots \cup \gamma^{(m)}$ ($0 \le t \le m-1$); the section $\gamma^{(i)}$ corresponds to points $P_{3i-3}, P_{3i-2}, P_{3i-1}, P_{3i}$ for $i - 1 \le t \le i$ and $i = 1, \ldots, m$.

One can check that a composed Bezier curve is C^1-*continuous* if the triples of vertices $P_{3i-1}, P_{3i}, P_{3i+1}$ are collinear (lie on the same straight line).

For C^2-*continuity* of a Bezier curve in $\mathbb{R}^3$, it is necessary that all five points $P_{3i-2}, P_{3i-1}, P_{3i}, P_{3i+1}, P_{3i+2}$ be coplanar (lie in the same plane).

For simplicity, assume the condition $P_{3i} = \frac{1}{2}(P_{3i-1} + P_{3i+1})$ and do not include points P_{3i} in the array $\mathbf{P}$. Thus in the procedure `bez_2d` (for deriving and plotting the plane curve), the array contains $2m+1$ control points, and the curve γ with $0 \le t \le m-1$ consists of m segments, defined as follows: $\gamma^{(1)}$ by points P_0, P_1, P_2, $\gamma^{(i)}$ by the points P_{2i-1}, P_{2i} for $2 \le i \le m-1$, and $\gamma^{(m)}$ by the points $P_{2m-2}, P_{2m-1}, P_{2m}$.

Note that the function `(1+signum(-t+i))*(1+signum(t-i+1))` is the characteristic function of the segment $[i-1, i]$ and that it coincides with the function `piecewise(t<i-1, 0, t<i, 1, t>i, 0)`; see also the programs in Sections 14.3–14.6.

```
> bez_2d := proc(xp, yp) local s,n,i,p,pp,m;
  n[0]:=t -> (1-t)^3; n[1]:=t -> 3*t*(1-t)^2;
  n[2]:=t -> 3*t^2*(1-t); n[3]:=t -> t^3; m:=nops(xp)/2;
  for s from 1 to 2*m do p[s-1]:=[xp[s], yp[s]] od;
  p[0]:=2*p[0]-p[1]; p[2*m-1]:=2*p[2*m-1]-p[2*m-2];
  pp:=n[0](t-i+1)*(p[2*i-2]+p[2*i-1])/2+n[1](t-i+1)*p[2*i-
  1]+n[2](t-i+1)*p[2*i]+n[3](t-i+1)*(p[2*i]+p[2*i+1])/2;
  RETURN(evalm(sum((1+signum(-t+i))*(1+signum(t-i+1))/4*
  pp, i=1..m-1))) end:
```

Example 14.2.2

1. Let us plot the plane composed Bezier curve, Fig. 14.7 a, using the command `bez_2d(xp, yp)`.

```
> xp:=[.056, .287, .655, .716, .228, .269, .666, .929]:
  yp:=[.820, .202, .202, .521, .521, .820, .820, .227]:
> polygon:=plot([seq([xp[j], yp[j]], j=1..nops(xp))]):
> bez:=bez_2d(xp, yp): curve:=plot([bez[1], bez[2], t=0.001..
  nops(xp)/2-1.001]): plots[display]([curve, polygon]);
```

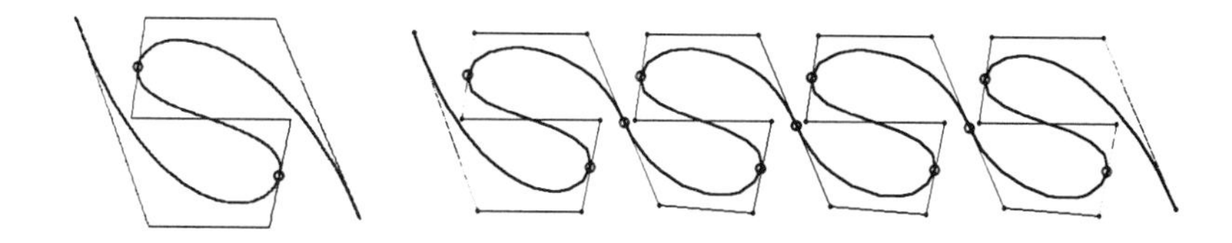

Fig. 14.7. Composed Bezier curves

Continuation. We plot the self-repeating composed Bezier curve, Fig. 14.7, where the circled points are derived by the program

```
> x1:=[op(xp),seq(seq(xp[i]+.61*j, i=3..nops(xp)), j=1..3)];
  y1:=[op(yp), seq(seq(yp[i], i=3..nops(yp)), j=1..3)];
> P1:=plot([seq([x1[j], y1[j]], j=1..nops(x1))]):
> curve1:=plot([bez1[1], bez1[2], t=0.001..nops(xl)/2-1.001]):
> plots[display]([curve1, P1]);
```

2. One can plot three elementary Bezier curves (see data in Example 14.2.1) using the procedure `bez_2d(xp, yp)`.

3. Let us plot closed two types of Bezier curves:

(a) with the loop having a singular point, $k \in \mathbb{N}$.

```
> k:=3: xp:=[seq(sin(2*Pi*i/(2*k-1)), i=1..2*k)]:
  yp:=[seq(cos(2*Pi*i/(2*k-1)), i=1..2*k)]:   # Fig. 14.6 d
```

(b) C^1-continuous curve, $k \in \mathbb{N}$.

```
> k:=3: xp:=[seq(cos(2*Pi*i/(2*k)), i=1..2*k+2)]:
  yp:=[seq(sin(2*Pi*i/(2*k)), i=1..2*k+2)]:
  xp[1]:=(xp[2]+xp[2*k+1])/2: yp[1]:=(yp[2]+yp[2*k+1])/2:
  xp[2*k+2]:=xp[1]: yp[2*k+2]:=yp[1]:   # Fig. 14.6 e
```

14.2.2 *Rational Bezier Curve*

A rational Bezier curve, see Section 4.3, with $m = 3$, is called an *elementary* Bezier curve.

Definition 14.2.3 A *composed rational Bezier curve* defined as $P_0, \ldots, P_{3m}$ $(m \geq 1)$ can be represented as a union of elementary rational Bezier curves $\gamma = \gamma^{(1)} \cup \cdots \cup \gamma^{(m)}$ $(0 \leq t \leq m-1)$; the section $\gamma^{(i)}$ corresponds to the points $P_{3i-3}, P_{3i-2}, P_{3i-1}, P_{3i}$ for $i - 1 \leq t \leq i$ and $i = 1, \ldots, m$.

Write a procedure for plotting composed rational Bezier curves analogous to the procedure `bez_2d(xp, yp)`.

Remark 14.2.1 By a special choice of weights on each segment, one obtains a C^2-continuous composed rational Bezier curve; see Exercise 2 below. This property allows one to insert an elementary rational Bezier curve in the gap between any two given C^2-continuous composed curves in such a way that the resulting composed curve is C^2-continuous.

Exercise 14.2.1

1. What are the conditions for a closed composed Bezier curve ?

Answer: In case of equality of the points $P_0 = P_{3m}$ and collinearity of the triples $P_{3m-1}P_{3m} = P_0P_1$, a closed composed Bezier curve is C^1-continuous (see the third task in Example 14.2.2).

2. Given the array $P_0,\ P_1,\ P_2,\ P_3$ and nonnegative real numbers k_0 and k_3, find the weights $w_0,\ w_1,\ w_2,\ w_3$ such that the values of the curvature of the *rational elementary Bezier curve* at the boundary control points coincide with the given real number (k_0 at the point P_0 and k_3 at P_3).

Answer: $w_0 = 1, \quad w_1 = \frac{4}{3}(\frac{{c_0}^2 c_3}{{k_0}^2 k_3})^{1/3}, \quad w_2 = \frac{4}{3}(\frac{{c_3}^2 c_0}{{k_3}^2 k_0})^{1/3}, \quad w_3 = 1,$

$$c_0 = \frac{|(P_1 - P_0) \times (P_2 - P_0)|}{2|P_1 - P_0|^3}, \quad c_3 = \frac{|(P_2 - P_1) \times (P_2 - P_0)|}{2|P_3 - P_2|^3}.$$

14.3 Composed B-Spline Curves

Definition 14.3.1 Given points $P_0,\ P_1,\ P_2$, and P_3, the *(elementary) cubic B-spline curve* is defined by the equation

$$\vec{\mathbf{r}}(t) = \tfrac{(1-t)^3}{6} P_0 + \tfrac{3t^3-6t^2+4}{6} P_1 + \tfrac{-3t^3+3t^2+3t+1}{6} P_2 + \tfrac{t^3}{6} P_3 \quad (0 \le t \le 1).$$

Its boundary nodes lie in the triangles $P_0,\ P_1,\ P_2$ and $P_1,\ P_2,\ P_3$. One can check using MAPLE that the functional weight multipliers

$$n_0(t) = \tfrac{(1-t)^3}{6},\ n_1(t) = \tfrac{3t^3-6t^2+4}{6},\ n_2(t) = \tfrac{-3t^3+3t^2+3t+1}{6},\ n_3(t) = \tfrac{t^3}{6}$$

are nonnegative for $0 \le t \le 1$ and their sum is equal to 1.

Definition 14.3.2 A *composed cubic B-spline curve* defined by the array P_0, $\ldots,\ P_m,\ (m \ge 3)$ is a curve that can be represented in the form of a union of elementary cubic B-spline curves $\gamma = \gamma^{(1)} \cup \cdots \cup \gamma^{(m-2)}$ $(0 \le t \le m-2)$; the section $\gamma^{(i)}$ corresponds to points $P_{i-1},\ P_i,\ P_{i+1},\ P_{i+2}$ for $i-1 \le t \le i$ and $i = 1, \ldots, m-2$.

The end nodes of the curve lie in triangles $P_0,\ P_1,\ P_2$ and $P_{m-2},\ P_{m-1},\ P_m$. In order for the B-spline to end smoothly at the control points P_0 and P_m, with tangency of the segments $[P_0,\ P_1]$ and $[P_{m-1},\ P_m]$, one must complete the array of vector-valued functions $\vec{\mathbf{r}}_i(t)$ so obtained with four additional segments (two at each side), for example, by the method of *multiple points*. The resulting curve is C^2-regular but only C^1-continuous.

Changing one point P_i in the array $\mathbf{P}$ leads to changing only part of the spline curve. We need to recalculate the equations of the *four segments* $\gamma^{(i-2)}$, $\gamma^{(i-1)}$, $\gamma^{(i)}$, $\gamma^{(i+1)}$.

Here is a procedure for deriving and plotting the data in the plane.

```
> bspl_2d := proc(xp, yp) local s,n,i,p,pp,m;
  n[0]:=t -> (1-t)^3/6; n[1]:=t -> (3*t^3-6*t^2+4)/6;
  n[2]:=t -> (-3*t^3+3*t^2+3*t+1)/6; n[3]:=t -> t^3/6;
  m:=nops(xp)-1; for s from 1 to m+1 do
  p[s-1]:=[xp[s], yp[s]] od;
  p[-1]:=p[0]; p[m+1]:=p[m]; p[-2]:=p[0]; p[m+2]:=p[m];
  pp:=n[0](t-i+1)*p[i-1]+n[1](t-i+1)*p[i]+n[2](t-i+1)*
  p[i+1]+n[3](t-i+1)*p[i+2]; RETURN(evalm(sum
  ((1+signum(-t+i))*(1+signum(t-i+1))/4*pp, i=-1..m))) end:
```

We use the procedure `bspl_2d(xp, yp)` to plot the plane curve.

```
> xp:=[.056, .287, .655, .716, .228, .269, .666, .929]:
```

```
  yp:=[.820, .202, .202, .521, .521, .820, .820, .227]:
> polygon:=plot([seq([xp[j], yp[j]], j=1..nops(xp))]):
> bs:=bspl_2d(xp, yp): curve:=plot([bs[1], bs[2], t=-1.999..
  nops(xp)-1.001]): %; plots[display]([curve, polygon]);
```

One can plot the self-repeating plane B-spline by a method similar to the Bezier curve; Fig. 14.7.

Here is an analogous procedure for deriving and plotting the B-spline in $\mathbb{R}^3$.

```
> bspl_3d:=proc(xp,yp,zp) local n,i,s,pp,p; global spl,m,t;
  n[0]:=t->(1-t)^3/6; n[1]:=t->(3*t^3-6*t^2+4)/6;
  n[2]:=t->(-3*t^3+3*t^2+3*t+1)/6; n[3]:=t->t^3/6;
  m:=nops(xp)-1; for s from 1 to m+1 do
  p[s-1]:=[xp[s], yp[s], zp[s]] od;
  p[-1]:=p[0]; p[m+1]:=p[m]; p[-2]:=p[0]; p[m+2]:=p[m];
  pp:=n[0](t-i+1)*p[i-1]+n[1](t-i+1)*p[i]+n[2](t-i+1)*
  p[i+1]+n[3](t-i+1)*p[i+2]; spl:=evalm(
  sum((1+signum(-t+i))*(1+signum(t-i+1))/4*pp, i=-1..m));
  RETURN(plots[spacecurve](spl, t=-1.999..m-0.001)) end:
```

We use the procedure `bspl_3d(xp, yp, zp)` to plot the space curve.

```
> xp:=[.056, .287, .655, .716, .228, .269, .666, .929]:
  yp:=[.820, .202, .202, .521, .521, .820, .820, .227]:
  zp:=[1/4, 1/4, 0, 0, 0, 0, -1/4, -1/4]:
> polygon:=plots[pointplot3d]([seq([xp[j], yp[j], zp[j]],
  j=1..nops(xp))]): bs:=bspl_3d(xp, yp, zp):
> plots[display]([bs,polygon],scaling=constrained); #Fig.14.10
```

Let us plot the knot *trefolium* using coordinates of vertices of the triangle $A(-2, 0)$, $B(0, 4)$, $C(2, 0)$ and midpoints of its edges (below, the arrays xp and yp contain the data).

```
> xp:=[2, 0, -1, 0, 1, 0, -2, -1, 1, 2, 2]:
  yp:=[0, 0, 2, 4, 2, 0, 0, 2, 2, 0, 0]:
  zp:=[0, 1, 2, 2, 0, 2, 2, 1, 2, 1, 0]:
> bs:=bspl_3d(xp, yp, zp):
>plots[tubeplot]([spl[1],spl[2],spl[3], radius=.2], t=-2..nops(xp)
  -1, tubepoints=50, style=patchnogrid, color=[1, .8, .1],
  light=[75,50,1, .9, .5], ambientlight=[.4,.4,.4]); # Fig. 14.12
```

Exercise 14.3.1

1. What are the conditions for a closed composed cubic B-spline curve? *Hints*. To plot a C^2-regular B-spline curve, we need to complete the array $P_0, \ldots, P_m$ with three points $P_{m+1} = P_0,\ P_{m+2} = P_1,\ P_{m+3} = P_2$.

2. Check that a composed B-spline curve lies in the union of convex hulls of the four points $P_{i-1}, P_i,\ P_{i+1},\ P_{i+2}$.

Definition 14.3.3 Given points $P_0,\ P_1,\ P_2$, and P_3, the *(elementary) rational B-spline curve* is defined by the equation

$$\vec{\mathbf{r}}(t) = \frac{\sum\limits_{i=0}^{3} w_i n_i(t) P_i}{\sum\limits_{i=0}^{3} w_i n_i(t)} \qquad (0 \le t \le 1).$$

3. Write a procedure for deriving a rational composed B-spline, taking procedures `bspl_2d(xp,yp,zp)` and `bspl_3d(xp, yp,zp)` as models.
Study the properties of this curve.

14.4 Beta-Spline Curves

14.4.1 Examples of Beta-Spline Curves

Definition 14.4.1 Given points $P_0,\ P_1,\ P_2$, and P_3, an *(elementary) cubic Beta-spline curve* is defined by the equation

$$\vec{\mathbf{r}}(t) = b_0(t)\, P_0 + b_1(t)\, P_1 + b_2(t)\, P_2 + b_3(t)\, P_3 \qquad (0 \le t \le 1),$$

where the four functional coefficients are defined by the formulas

$$\begin{aligned} b_0(t) =& \frac{2t^3}{\delta}, \\ b_1(t) =& \frac{1}{\delta}[2\beta_1^3 t(t^2 - 3t + 3) + 2\beta_1^2(t^3 - 3t^2 + 2) + 2\beta_1(t^3 - 3t + 2) \\ &+ \beta_2(2t^3 - 3t^2 + 1)], \\ b_2(t) =& \frac{1}{\delta}[2\beta_1^2 t^2(3 - t) + 2\beta_1 t(3 - t^2) + \beta_2 t^2(3 - 2t) + 2(1 - t^3)], \\ b_3(t) =& \frac{2\beta_1^3}{\delta}(1 - t)^3. \end{aligned}$$

Here $\beta_1 > 0$ and $\beta_2 \ge 0$ are called *parameters of the shape* (β_1 is the *parameter of slant* or *displacement*; β_2 is the *parameter of tension*) and the relation $\delta = 2\beta_1^3 + 4\beta_1^2 + 4\beta_1 + \beta_2 + 2$ holds.

In Section 14.4.2 we explain formulas for $b_i(t)$. One can check that

(1) For $\beta_1 = 1$ and $\beta_2 = 0$ one obtains an elementary cubic B-spline.

(2) The $b_i(t)$ are nonnegative for $0 \le t \le 1$, and their sum is equal to 1.

(3) The elementary cubic Beta-spline curve lies in the union of convex hulls of all four consecutive control points.

Definition 14.4.2 The *composed Beta-spline curve*, defined by the array $P_0, \ldots, P_m$ ($m \geq 3$), is the union of elementary Beta-spline curves $\gamma = \gamma^{(1)} \cup \cdots \cup \gamma^{(m-2)}$ ($0 \leq t \leq m-2$); the section $\gamma^{(i)}$ corresponds to points $P_{i-1}, P_i, P_{i+1}, P_{i+2}$ for $i-1 \leq t \leq i$ and $1 \leq i \leq m-2$.

The end nodes of the curve lie in the triangles P_0, P_1, P_2 and P_{m-2}, P_{m-1}, P_m. One can complete a curve with four segments $\gamma^{(-1)}, \gamma^{(0)}, \gamma^{(m-1)}$ and $\gamma^{(m)}$ (two from each side), for example, by the method of *multiple points*, and obtain a smooth Beta-spline curve $\gamma^{(-1)} \cup \cdots \cup \gamma^{(m)}$ ($-2 \leq t \leq m$) with endpoints P_0 and P_m and segments $\gamma^{(0)}, \gamma^{(m-1)}$ tangent to line segments $[P_0, P_1]$ and $[P_{m-1}, P_m]$.

Changing one point P_i in the array **P** leads to changing only part of the spline curve. We must recalculate the equations of the *four segments* $\gamma^{(i-2)}, \gamma^{(i-1)}, \gamma^{(i)}, \gamma^{(i+1)}$. For $\beta_1, \beta_2 \neq 0$, the composed curve is C^2-*continuous*. This is the main advantage of these curves over B-spline curves. Parameters of the form β_1 and β_2 can differ on different segments of the composed curve.

Let us write a procedure for deriving and plotting in $\mathbb{R}^2$.

```
> beta_2d :=proc(xp,yp,beta1,beta2)
  local s,i,pp,n; global d,b1,b2,p,m;
  b1:=beta1; b2:=beta2; d:=evalf(2*b1^3+4*b1^2+4*b1+b2+2);
  m:=nops(xp)-1; n[0]:=t -> 2*b1^3*(1-t)^3/d;
  n[1]:=t -> (2*b1^3*t*(t^2-3*t+3)+ 2*b1^2*(t^3-3*t^2+2)+
  2*b1*(t^3-3*t+2)+b2*(2*t^3-3*t^2+1))/d;
  n[2]:=t -> (2*b1^2*t^2*(-t+3)+2*b1*t*(-t^2+3)+
  b2*t^2*(-2*t+3)+2*(-t^3+1))/d; n[3]:=t -> 2*t^3/d;
  for s from 1 to m+1 do p[s-1]:=[xp[s], yp[s]] od;
  p[-1]:=p[0]; p[m+1]:=p[m]; p[-2]:=p[0]; p[m+2]:=p[m];
  pp:=n[0](t-i+1)*p[i-1]+n[1](t-i+1)*p[i]+n[2](t-i+1)*
  p[i+1]+n[3](t-i+1)*p[i+2]; RETURN(evalm(sum
  ((1+signum(-t+i))*(1+signum(t-i+1))/4*pp,i=-1..m))) end:
```

We use the procedure `beta_2d(xp,yp,zp)` to plot the curve, Fig. 14.8 b

```
> xp:=[.056, .287, .655, .716, .228, .269, .666, .929]:
  yp:=[.820, .202, .202, .521, .521, .820, .820, .227]:
> P:=plot([seq([xp[i], yp[i]], i=1..nops(xp))]):
> Bs:=beta_2d(xp, yp, 1, 7): curve:=plot([Bs[1], Bs[2], t=-2+
```

```
0.001..nops(xp)-1.001]): plots[display]([curve,P]);
```

Using the following program, which continues the previous one (it too can be applied to B-splines using bspl_2d), we derive the change in a Beta-spline when one of its control points (with number i_0) moves. One can rewrite this program as the procedure.

```
> i0:=4: delta:=[rand(0..10)()/40, rand(0..10)()/40]:
> char:=i -> (1+signum(-t+i))*(1+signum(t-i+1))/4:
> pnew:=p[i0]+delta: for i from i0-2 to i0+1 do
  pp:=n[0](t-i+1)*p[i-1]+n[1](t-i+1)*p[i]+n[2](t-i+1)*
  p[i+1]+n[3](t-i+1)*p[i+2]: old[i]:=pp:
  new[i]:=subs(p[i0]=evalm(pnew),pp) od: r_d:=sum(char(j)*
  (new[j]-old[j]), j=i0-2..i0+1): r_new:=evalm(Bs+r_d):
> new_curve:=plot([r_new[1], r_new[2], t=-1.99..m-0.01]):
> for i from -2 to m+2 do if i<> i0 then q[i]:=p[i]
  else q[i]:=pnew fi od:
> new_polygon:=plot([seq(q[i], i=0..m)]):
> poly2:=plot([[p[i0-1], pnew, p[i0+1]]]):
> display([new_polygon, poly2, new_curve, curve]); # Fig. 14.8 c
```

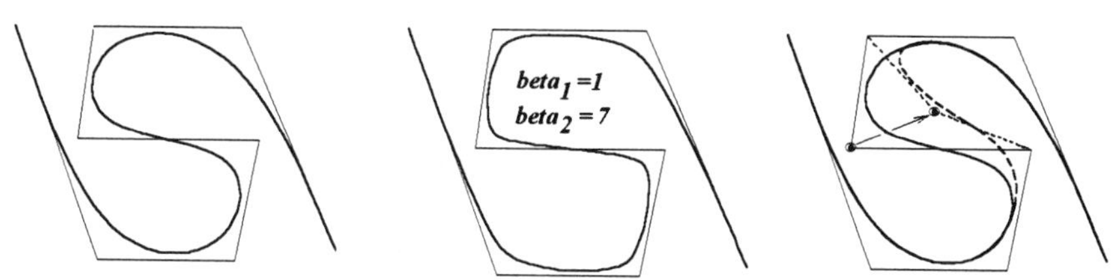

Figs. 14.8 a–c. B-spline. Beta-spline. Changing one point

One can play with beta_2d using the following data a) – b).

a) An example of the random generating of an array of points in the plane.

```
> mm:=7: xp:=[seq(rand(1..10)(), j=1..mm)]:
  yp:=[seq(rand(1..10)(), j=1..mm)]:
```

b) Data for car profile; Fig. 14.9.

```
> xp:=[226,348,356,400,410,452,470,452,438,496,416,412,430,
  428,394,344,232,164,58,38,36,98,100,148,158,228,226]:
> yp:=[21,17,5,4,18,24,55,62,82,89,83,79,79,63,68,94,87,
  62,53,34,21,16,2,1,13,21,21]:
```

Here is an analogous procedure for deriving and plotting in $\mathbb{R}^3$.

```
> beta_3d := proc(xp,yp,zp,beta1,beta2)
```

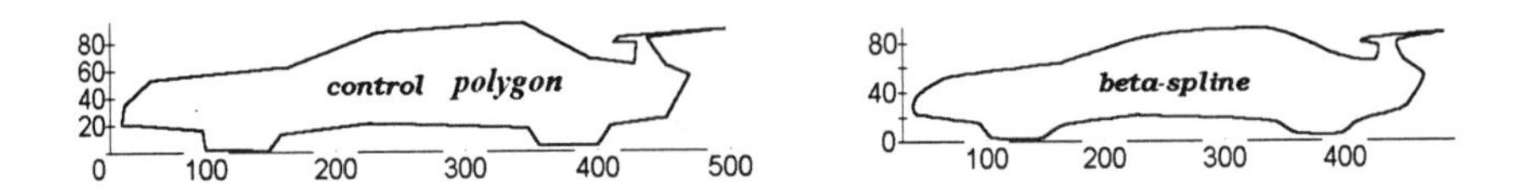

Fig. 14.9. Model of a car profile: $\beta_1 = 1,\ \beta_2 = 9$

```
local s,n,i,p,pp,m; global spl,d,b1,b2;
b1:=beta1; b2:=beta2; d:=evalf(2*b1^3+4*b1^2+4*b1+b2+2);
n[0]:=t -> 2*b1^3*(1-t)^3/d;
n[1]:=t -> (2*b1^3*t*(t^2-3*t+3)+ 2*b1^2*(t^3-3*t^2+2)+
2*b1*(t^3-3*t+2)+b2*(2*t^3-3*t^2+1))/d;
n[2]:=t ->(2*b1^2*t^2*(-t+3)+2*b1*t*(-t^2+3)+b2*t^2*
(-2*t+3)+2*(-t^3+1))/d; n[3]:=t -> 2*t^3/d;
m:=nops(xp)-1; for s from 1 to m+1 do
p[s-1]:=[xp[s], yp[s], zp[s]] od;
p[-1]:=p[0]; p[m+1]:=p[m]; p[-2]:=p[0]; p[m+2]:=p[m];
pp:=n[0](t-i+1)*p[i-1]+n[1](t-i+1)*p[i]+n[2](t-i+1)*
p[i+1]+n[3](t-i+1)*p[i+2]; spl:=evalm(
sum((1+signum(-t+i))*(1+signum(t-i+1))/4*pp, i=-1..m));
RETURN(plots[spacecurve](spl, t=-1.999..m-0.001)) end:
```

Let us use the procedure `beta_3d(xp, yp, zp)` to plot the space curve.

```
> xp:=[.056, .287, .655, .716, .228, .269, .666, .929]:
  yp:=[.820, .202, .202, .521, .521, .820, .820, .227]:
  zp:=[1/4,1/4,0,0,0,0,-1/4,-1/4]:
> polygon:=plots[pointplot3d]([seq([xp[j], yp[j], zp[j]],
  j=1..nops(xp))]): bs:=beta_3d(xp, yp, zp, 1, 1):
> plots[display]([bs,polygon],scaling=constrained) ; # Fig.14.11
```

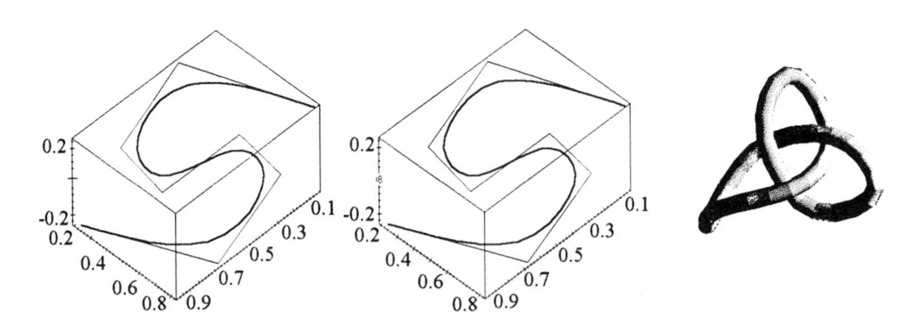

Figs. 14.10–14.12. B-spline and Beta-spline in space. Knot (trefoil)

Exercise 14.4.1

1. What are the conditions for a closed composed Beta-spline curve? *Hint.* They are similar to those for a B-spline curve; see Section 14.3.

2. Check that a composed Beta-spline curve lies in the union of the convex hulls of four consequtive control points $P_{i-1}, P_i, P_{i+1}, P_{i+2}$.

3. Do the following experiment with the test data given after procedure `beta_2d`:

(a) Change $\beta_1 > 0$ for fixed $\beta_2 \geq 0$ (for $\beta_1 = 0$ we obtain the control polygon).

(b) Change $\beta_2 > 0$ for fixed $\beta_1 > 0$ (for $\beta_2 \to \infty$ each node of the spline runs to the corresponding control point; Figs. 14.13 a–c).

(c) For a variety of impressions, consider $\beta_2 < 0$ for fixed $\beta_1 > 0$ (the convexity condition does not hold; Figs. 14.13 d–f). Check that if we don't allow nonnegativity of the parameters β_1 and β_2, then some *self-intersections* and *oscillations* of the curve can appear.

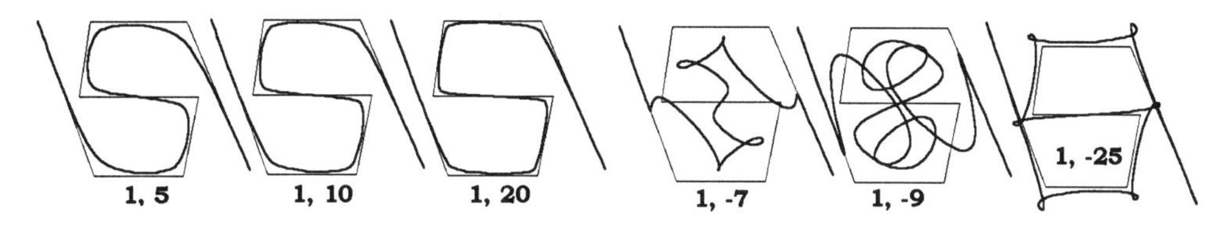

Fig. 14.13. Influence of β_1 and β_2 on the shape of a spline

Definition 14.4.3 Given points P_0, P_1, P_2, P_3 the *(elementary) rational Beta-spline curve* is defined by the equation

$$\vec{\mathbf{r}}(t) = \frac{\sum_{i=0}^{3} w_i b_i(t) P_i}{\sum_{i=0}^{3} w_i b_i(t)} \quad (0 \leq t \leq 1).$$

4. Write a procedure for deriving a rational composed Beta-spline similar to the following procedures:
`beta_2d(xp, yp, zp, beta1, beta2)` and
`beta_3d(xp, yp, zp, beta1, beta2)` (see the last task in Exercise 14.3.1).
Study the properties of this curve.

14.4.2 Theoretical Background to Beta-Spline Curves

The aim of this section is to motivate Definition 14.4.1.

Geometrical Continuity and Shape Parameters

Vectors $\vec{\tau}$ (tangent) and $\vec{\mathbf{k}}$ (curvature) are the main geometrical invariants used in Beta-spline curve construction (see [Bur]). We write continuity conditions for vectors $\vec{\tau}$ and $\vec{\mathbf{k}}$ of the composed curve at the common node of two segments $\vec{\mathbf{r}}_i(t)$ and $\vec{\mathbf{r}}_{i+1}(t)$ $(0 \le t \le 1)$ in the form of three vector equations, containing $\vec{\mathbf{r}}\,'$ and $\vec{\mathbf{r}}\,''$.

Continuity of a curve at a node is equivalent to the equation

$$\vec{\mathbf{r}}_{i+1}(0) = \vec{\mathbf{r}}_i(1). \tag{14.1}$$

Continuity of the vector $\vec{\tau}$ at a node of the composed curve is expressed in the equality $\vec{\tau}_{i+1}(0) = \frac{\vec{\mathbf{r}}\,'_{i+1}(0)}{|\vec{\mathbf{r}}\,'_{i+1}(0)|} = \frac{\vec{\mathbf{r}}\,'_i(1)}{|\vec{\mathbf{r}}\,'_i(1)|} = \vec{\tau}_i(1)$, from which follows

$$\exists\, \beta_1 \ne 0 : \quad \vec{\mathbf{r}}\,'_{i+1}(0) = \beta_1 \vec{\mathbf{r}}\,'_i(1). \tag{14.2}$$

Note that for the first curve in Example 14.1.1, we have $\beta_1 = \frac{1}{4}$.

The continuity condition for the main curvature vector $\vec{\mathbf{k}}$ at a node can be written as follows:

$$\begin{aligned} \vec{\mathbf{k}}_{i+1}(0) &= \frac{(\vec{\mathbf{r}}\,'_{i+1}(0) \times \vec{\mathbf{r}}\,''_{i+1}(0)) \times \vec{\mathbf{r}}\,'_{i+1}(0)}{|\vec{\mathbf{r}}\,'_{i+1}(0)|^4} \\ &= \frac{(\vec{\mathbf{r}}\,'_i(1) \times \vec{\mathbf{r}}\,''_i(1)) \times \vec{\mathbf{r}}\,'_i(1)}{|\vec{\mathbf{r}}\,'_i(1)|^4} = \vec{\mathbf{k}}_i(1). \end{aligned} \tag{14.3}$$

Lemma 14.4.1 *If, for vectors* $\bar{\mathbf{a}}, \bar{\mathbf{b}}, \bar{\mathbf{c}} \in \mathbb{R}^3$, *the equality* $(\bar{\mathbf{a}} \times \bar{\mathbf{b}}) \times \bar{\mathbf{a}} = (\bar{\mathbf{a}} \times \bar{\mathbf{c}}) \times \bar{\mathbf{a}}$ *holds, then* $\bar{\mathbf{b}} - \bar{\mathbf{c}} \,||\, \bar{\mathbf{a}}$.

Proof. From the condition it follows that $(\bar{\mathbf{a}} \times (\bar{\mathbf{b}} - \bar{\mathbf{c}})) \times \bar{\mathbf{a}} = 0$, which is true only if $\bar{\mathbf{a}} \times (\bar{\mathbf{b}} - \bar{\mathbf{c}}) \,||\, \bar{\mathbf{a}}$. But from the definition of the vector product, it follows that $\bar{\mathbf{a}} \times (\bar{\mathbf{b}} - \bar{\mathbf{c}}) \perp \bar{\mathbf{a}}$. Hence $\bar{\mathbf{a}} \times (\bar{\mathbf{b}} - \bar{\mathbf{c}})$ is zero, i.e., $\bar{\mathbf{b}} - \bar{\mathbf{c}} \,||\, \bar{\mathbf{a}}$. □

From (14.2) and (14.3) follows the equality
$((\vec{\mathbf{r}}\,'_i(1) \times \vec{\mathbf{r}}\,''_i(1)) \times \vec{\mathbf{r}}\,'_i(1) = (\vec{\mathbf{r}}\,'_i(1) \times \frac{1}{\beta_1^2}\vec{\mathbf{r}}\,''_{i+1}(0)) \times \vec{\mathbf{r}}\,'_i(1)$.

From this, in view of Lemma 14.4.1, it follows that $\frac{1}{\beta_1^2}\vec{\mathbf{r}}_{i+1}(0) = \vec{\mathbf{r}}\,''_i(1) + \alpha\vec{\mathbf{r}}\,'_i(1)$ for some α. Assume $\beta_2 = \alpha\beta_1^2$, and then obtain the vector equality

$$\vec{\mathbf{r}}\,''_{i+1}(0) = \beta_1^2 \vec{\mathbf{r}}\,''_i(1) + \beta_2 \vec{\mathbf{r}}\,'_i(1), \tag{14.4}$$

(where $\beta_1 \ne 0$ and $\beta_2 \in \mathbb{R}$). The equations (14.1), (14.2) and (14.4) are basic in beta-spline construction. That is, substituting vector polynomials of degree 3 in t, for $\vec{\mathbf{r}}$ we obtain several conditions on the coefficients of these polynomials.

Based on equations (14.2) and (14.4), we recognize particular cases that are related to the B-spline curve construction:

the equality $\beta_1 = 1$ *is equivalent to the continuity of* $\vec{\mathbf{r}}\,'$, *the system $\beta_1 = 1,\ \beta_2 = 0$ is equivalent to the continuity of $\vec{\mathbf{r}}\,'$ and $\vec{\mathbf{r}}\,''$.*

Height Functions of Beta-Spline Construction

Recall that various linear combinations of four points (vectors) $b_{-2}\mathbf{V}_{i-2} + b_{-1}\mathbf{V}_{i-1} + b_0\mathbf{V}_i + b_1\mathbf{V}_{i+1}$ (with $b_{-2} + b_{-1} + b_0 + b_1 = 1$), which do not lie on a plane, fill the space $\mathbb{R}^3$. (Similarly, with 2 or 3 points we can fill a plane or a straight line).

Let us fix an arbitrary set $\mathbf{V}_0, \dots, \mathbf{V}_n$ of $n+1$ points, called *control vertices*. Each *inner segment* $\vec{\mathbf{r}}_i(t)$ of a composed Beta-spline curve is defined by linear combinations of four neighboring vertices. Note that moving one of the consecutive vertices changes four neighboring segments and three neighboring nodes of the Beta-spline curve (a local property). The *boundary segments* of the spline curve are derived using, for example, the *multiple points method*; see Section 14.4.1. Thus we use linear combinations of four vectors (points)

$$\vec{\mathbf{r}}_i(t) = b_{-2}(\beta_1, \beta_2, t)\mathbf{V}_{i-2} + b_{-1}(\beta_1, \beta_2, t)\mathbf{V}_{i-1} + b_0(\beta_1, \beta_2, t)\mathbf{V}_i + b_1(\beta_1, \beta_2, t)\mathbf{V}_{i+1},$$

that can be written shortly using the $\sum$ notation

$$\vec{\mathbf{r}}_i(t) = \sum_{j=-2}^{1} b_j(\beta_1, \beta_2, t)\mathbf{V}_{i+j} \qquad (2 \le i \le n-1). \tag{14.5}$$

The coefficients $b_j(\beta_1, \beta_2, t)$ of linear combinations in (14.5) (called the *weight functions* and denoted for short by $b_j(t)$, $j = -2, -1, 0, 1$) depend in a certain way on variable t and the two parameters appearing in (14.2) and (14.4). Here we assume that the weight functions are polynomials of degree 3 in t with coefficients $c_{i,j}(\beta_1, \beta_2)$ depending on the parameters β_1 and β_2 (denoted for short by $c_{i,j}$), namely,

$$b_j(t) = c_{0,j} + c_{1,j}t + c_{2,j}t^2 + c_{3,j}t^3 = \sum_{k=0}^{3} c_{k,j}t^k. \tag{14.6}$$

We must take into account the additional condition $\sum_{j=-2}^{1} b_j(t) = 1$. In fact, we would like to obtain four weight functions $b_j(t)$ that do not depend on the choice of control vertices and guarantee continuity of the osculating circle of the Beta-spline curve. The derivation of such weight functions is based on the assumption that parametrizations of neighboring segments of a curve satisfy to geometrical continuity conditions (14.1), (14.2) and (14.4):

$$\begin{cases} \sum_{j=-2}^{1} b_j(0)\mathbf{V}_{i+j+1} = \sum_{j=-2}^{1} b_j(1)\mathbf{V}_{i+j}, \\ \sum_{j=-2}^{1} b'_j(0)\mathbf{V}_{i+j+1} = \beta_1 \sum_{j=-2}^{1} b'_j(1)\mathbf{V}_{i+j}, \\ \sum_{j=-2}^{1} b''_j(0)\mathbf{V}_{i+j+1} = \beta_1^2 \sum_{j=-2}^{1} b''_j(1)\mathbf{V}_{i+j} + \beta_2 \sum_{j=-2}^{1} b'_j(1)\mathbf{V}_{i+j}. \end{cases} \tag{14.7}$$

Since the vector equations (14.7) are satisfied for an arbitrary choice of the control vertices, RHS and LHS coefficients at the same control vertex coincide. We write (14.7 a) explicitly in the form

$$\begin{aligned} &b_{-2}(0)\mathbf{V}_{i-1} + b_{-1}(0)\mathbf{V}_i + b_0(0)\mathbf{V}_{i+1} + b_1(0)\mathbf{V}_{i+2} = \\ &b_{-2}(1)\mathbf{V}_{i-2} + b_{-1}(1)\mathbf{V}_{i-1} + b_0(1)\mathbf{V}_i + b_1(1)\mathbf{V}_{i+1}, \end{aligned} \tag{14.8}$$

and find five equalities for the 5 vertices

$$\begin{aligned} &0 = b_{-2}(1), \quad b_{-2}(0) = b_{-1}(1), \quad b_{-1}(0) = b_0(1), \\ &b_0(0) = b_1(1), \quad b_1(0) = 0. \end{aligned} \tag{14.9}$$

Analogously from (14.7 b) we deduce the following five equalities:

$$\begin{aligned} &0 = \beta_1 b'_{-2}(1), \quad b'_{-2}(0) = \beta_1 b'_{-1}(1), \\ &b'_{-1}(0) = \beta_1 b'_0(1), \quad b'_0(0) = \beta_1 b'_1(1), \quad b'_1(0) = 0. \end{aligned} \tag{14.10}$$

Finally, from (14.7 c) we obtain the following five equalities:

$$\begin{aligned} 0 &= \beta_1^2 8b''_{-2}(1) + \beta_2 b'_{-2}(1), \\ b''_{-2}(0) &= \beta_1^2 b''_{-1}(1) + \beta_2 b'_{-1}(1), \\ b''_{-1}(0) &= \beta_1^2 b''_0(1) + \beta_2 b'_0(1), \\ b''_0(0) &= \beta_1^2 b''_1(1) + \beta_2 b'_1(1), \quad b''_1(0) = 0. \end{aligned} \tag{14.11}$$

We must find the 16 coefficients $\{c_{i,j}\}$ of the four third-order vector polynomials $\{b_j(t)\}$ from the homogeneous linear system of 15 equations (14.9), (14.10) and (14.11).

Completing this with the sixteenth norming equation $\sum_{j=-2}^{1} b_j(0) = 1$, we obtain the following linear system (where $j = -1,\ 0,\ 1$):

$$\begin{cases} c_{3,-2} + c_{2,-2} + c_{1,-2} + c_{0,-2} = 0, \\ c_{0,j-1} = c_{3,j} + c_{2,j} + c_{1,j} + c_{0,j}, \\ c_{0,1} = 0, \\ \beta_1(3c_{3,-2} + 2c_{2,-2} + c_{1,-2}) = 0, \\ c_{1,j-1} = \beta_1(3c_{3,j} + 2c_{2,j} + c_{1,j}), \\ c_{1,1} = 0, \\ 3(2\beta_1^2 + \beta_2)c_{3,-2} + 2(\beta_1^2 + \beta_2)c_{2,-2} + \beta_2 2c_{1,-2} = 0, \\ 2c_{2,j-1} = 3(2\beta_1^2 + \beta_2)c_{3,j} + 2(\beta_1^2 + \beta_2)c_{2,j} + \beta_2 c_{1,j}, \\ c_{2,1} = 0, \\ c_{0,-2} + c_{0,-1} + c_{0,0} + c_{0,1} = 1. \end{cases} \tag{14.12}$$

The determinant of this non-homogeneous linear system (14.12) is equal to $\delta = 2\beta_1^3 + 4\beta_1^2 + 4\beta_1 + \beta_2 + 2$. Hence, if we assume the inequalities $\beta_1 > 0$ and $\beta_2 \geq 0$ (note that this natural assumption provides additional information about the shape of the curve, see Section 14.4.1), then we obtain the inequality $\delta > 2$. Hence, the system (14.12) has a unique solution:

$$\begin{aligned}
&c_{0-2} = \tfrac{\beta_1^3}{\delta}, \quad c_{1,-2} = -\tfrac{6\beta_1^3}{\delta}, \quad c_{2,-2} = -\tfrac{6\beta_1^3}{\delta}, \quad c_{3,-2} = -\tfrac{2\beta_1^3}{\delta},\\
&c_{0,-1} = \frac{4\beta_1^2+4\beta_1+\beta_2}{\delta}, \quad c_{1,-1} = \frac{6\beta_1(\beta_1^2-1)}{\delta},\\
&c_{2,-1} = \frac{3(-2\beta_1^3-2\beta_1^2-\beta_2)}{\delta}, \quad c_{3,-1} = \frac{2(\beta_1^3+\beta_1^2+\beta_1+\beta_2)}{\delta}\\
&c_{0,0} = \tfrac{2}{\delta}, \quad c_{1,0} = \tfrac{6\beta_1}{\delta}, \quad c_{2,0} = \frac{3(2\beta_1^2+\beta_2)}{\delta}, \quad c_{3,1} = \tfrac{2}{\delta},\\
&c_{3,0} = -\frac{2(\beta_1^2+\beta_1+\beta_2+1)}{\delta}, \quad c_{0,1} = 0, \quad c_{1,1} = 0, \quad c_{2,1} = 0.
\end{aligned} \tag{14.13}$$

Substituting (14.13) into (14.6) we finally find the four weight functions to be used in the Beta-spline construction; see Section 14.4.1.

We see that the sum of four weight functions is constant and equal to 1

$$\sum_{j=-2}^{1} b_j(t) \equiv 1 \qquad (0 \le t \le 1); \tag{14.14}$$

this is the reason for the additional sixteenth equation in (14.12) for $t = 0$.

14.5 Interpolation Using Cubic Hermite Curves

14.5.1 The Elementary Cubic Hermite Curve

Definition 14.5.1 Given points P_0 and P_1 and nonzero vectors Q_0 and Q_1, the *(elementary) cubic Hermite curve* is defined by the vector equation (for $0 \le t \le 1$)

$$\vec{\mathbf{r}}(t) = (1-3t^2+2t^3)P_0 + t^2(3-2t)P_1 + t(1-2t+t^2)Q_0 - t^2(1-t)Q_1.$$

Obviously, $\vec{\mathbf{r}}(0) = P_0$, $\vec{\mathbf{r}}(1) = P_1$, $\vec{\mathbf{r}}'(0) = Q_0$, $\vec{\mathbf{r}}'(1) = Q_1$.

Using elementary cubic Hermite curves, the composed curve is plotted by the array $P_0, \dots, P_m$ $(m \ge 1)$ and an array of nonzero vectors $Q_0, \dots, Q_m$. Each quadruple $P_{i-1}, Q_{i-1}, P_i, Q_i$ defines an elementary Hermite curve in the interval $t_{i-1} \le t \le t_i$. However, we obtain only a C^1-continuous composed curve. A C^2-continuous composed curve is obtained in this way in Section 14.5.2.

We write a procedure for deriving and plotting in $\mathbb{R}^2$.

```
> herm1_2d:=proc(xp,yp,xq,yq) local s,n,i,m,p,q,pp;
  global spl; n[0]:=t->2*t^3-3*t^2+1; n[1]:=t->t^2*(-2*t+3);
  n[2]:=t-> t*(t^2-2*t+1); n[3]:=t->t^2*(t-1);
```

```
m:=nops(xp)-1; for s from 1 to m+1 do
p[s-1]:=[xp[s], yp[s]]; q[s-1]:=[xq[s], yq[s]] od;
pp:=n[0](t-i+1)*p[i-1]+n[1](t-i+1)*p[i]+n[2](t-i+1)*
q[i-1]+n[3](t-i+1)*q[i]; RETURN(evalm(sum
((1+signum(-t+i))*(1+signum(t-i+1))/4*pp, i=1..m))) end:
```

Let us use the procedure herm1_2d(xp,yp,xq,yq) for plotting the plane curve.

```
> xp:=[.056, .287, .655, .716, .228, .269, .666, .929]:
  yp:=[.820, .202, .202, .521, .521, .820, .820, .227]:
> xq:=[.25, .25, -.25, -.25, .25, .25, -.25, -.25]:
  yq:=[-.25, .25, -.25, .25, -.25, .25, -.25, .25]:
> P:=plot([seq([xp[j], yp[j]], j=1..nops(xp))]):
> vec:=i->plottools[arrow]([xp[i-1], yp[i-1]], [xq[i-1],
  yq[i-1]], 0.01, .03, .1): V:=seq(vec(i+1), i=1..nops(xp)):
> H1:=herm1_2d(xp, yp, xq, yq):
  curve:=plot([H1[1], H1[2], t=.01..nops(xp)-1.01]):
  plots[display]([curve,P,V],scaling=constrained);# Fig. 14.4
```

14.5.2 Composed Cubic Hermite Curve

Definition 14.5.2 A *composed cubic Hermite curve*, determined by the array $P_0, \dots, P_m$ ($m \geq 1$) and a pair of nonzero vectors Q_0 and Q_1, is defined as a curve that can be represented as a union of elementary Hermite curves $\gamma = \gamma^{(1)} \cup \dots \cup \gamma^{(m)}$; the section $\gamma^{(i)}$ corresponds to the points $P_{i-1}, Q_{i-1}, P_i, Q_i$ for $i-1 \leq t \leq i$ and $i = 1, \dots, m$, and the vectors $Q_1, \dots, Q_{m-1}$ are derived from the matrix equation

$$
\begin{pmatrix}
1 & 4 & 1 & & & & \\
 & 1 & 4 & 1 & & & \\
 & & & \dots & & & \\
 & & & & 1 & 4 & 1 \\
 & & & & & 1 & 4 \; 1
\end{pmatrix}
\begin{pmatrix} Q_0 \\ Q_1 \\ \dots \\ Q_{m-1} \\ Q_m \end{pmatrix} =
$$

$$
\begin{pmatrix}
-3 & 0 & 3 & & & & \\
 & -3 & 0 & 3 & & & \\
 & & & \dots & & & \\
 & & & & -3 & 0 & 3 \\
 & & & & & -3 & 0 \; 3
\end{pmatrix}
\begin{pmatrix} P_0 \\ P_1 \\ \dots \\ P_{m-1} \\ P_m \end{pmatrix}.
$$

This composed curve is C^2-continuous (its curvature is continuous); it contains all control points and at the endpoints has tangent vectors Q_0 and Q_1.

Changing one point in the array **P**, adding one point, or changing one tangent vector at the endpoints leads to changing the whole Hermite curve.

We write a procedure for deriving and plotting in $\mathbb{R}^2$, solving the above system:

```
> herm2_2d := proc(xp, yp, xq, yq)
  local eq,a,b,ss,s,n,i,j,m,p,q,pp;
  n[0]:=t->2*t^3-3*t^2+1; n[1]:=t->t^2*(-2*t+3);
  n[2]:=t->t*(t^2-2*t+1); n[3]:=t->t^2*(t-1);
  m:=nops(xp)-1; p:=array(1..m+1); q:=array(1..m+1);
  a:=array(1..m-1, 1..m+1); b:=array(1..m-1, 1..m+1);
  for i from 1 to m-1 do for j from 1 to m+1 do
  if j=i then a[i,j]:=1; b[i,j]:=-3 elif j=i+1 then
  a[i,j]:=4; b[i,j]:=0 elif j=i+2 then a[i,j]:=1;
  b[i,j]:=-3 else a[i,j]:=0; b[i,j]:=0 fi od od;
  eq:=evalm(a&*q-b&*p);
  ss:=solve({seq(eq[i]=0, i=1..m-1)},{seq(q[i], i=2..m)});
  assign(ss); q[1]:=[xq[1],yq[1]]; q[m+1]:=[xq[2],yq[2]];
  for s from 1 to m+1 do p[s]:=[xp[s],yp[s]] od;
  i:='i'; pp:=n[0](t-i+1)*p[i]+n[1](t-i+1)*p[i+1]+
  n[2](t-i+1)*q[i]+n[3](t-i+1)*q[i+1]; RETURN(evalm(sum
  ((1+signum(-t+i))*(1+signum(t-i+1))/4*pp, i=1..m))) end:
```

Using the procedure `herm2_2d(xp,yp,xq,yq)` we plot the plane curve.

```
> xp:=[.056, .287, .655, .716, .228, .269, .666, .929]:
  yp:=[.820, .202, .202, .521, .521, .820, .820, .227]:
  xq:=[.1, -.1]: yq:=[.1, -.1]:
> vecs:=plots[display](plottools[arrow]([xp[1],yp[1]],
  [xq[1],yq[1]], 0.001, 0.02, 0.1), plottools[arrow]([xp[
  nops(xp)],yp[nops(yp)]], [xq[2],yq[2]],0.001,0.02,0.1)):
> polygon:=plot([seq([xp[j], yp[j]], j=1..nops(xp))]):
> H2:=herm2_2d(xp, yp, xq, yq):
  curve:=plot([H2[1],H2[2],t=0.01..nops(xp)-1.01]):
  plots[display]([vecs, curve, polygon]);      # Fig. 14.15
```

Exercise 14.5.1

What are the conditions for closed C^1-continuous (see Section 14.5.1) and C^2-continuous (see Section 14.5.2) composed Hermite curves?

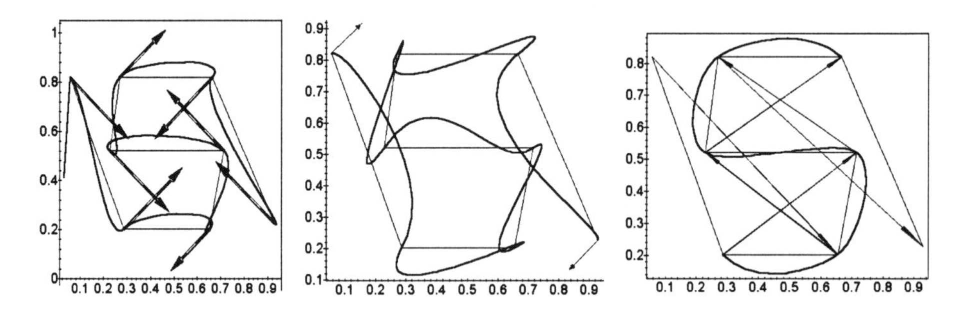

Figs. 14.14–14.16. Interpolated and composed Hermite curves.
Interpolated curve Catmull-Rom

14.6 Composed Catmull-Rom Spline Curves

Definition 14.6.1 Given points P_0, P_1, P_2, P_3, the *(elementary) Catmull-Rom spline curve* is defined by the vector equation where $0 \leq t \leq 1$

$$\vec{\mathbf{r}}(t) = \tfrac{1}{2}(-t(1-t)P_0+(2t-5t^2+3t^3)P_1+t(1+4t-3t^2)P_2-t^2(1-t)P_3).$$

The tangent line to an elementary Catmull-Rom curve at the endpoint $\vec{\mathbf{r}}(0) = P_1$ is parallel to the segment P_0P_2; at the endpoint $\vec{\mathbf{r}}(1) = P_2$, the tangent line is parallel to the segment P_1P_3; Fig. 14.16.

Definition 14.6.2 The *composed Catmull-Rom spline curve* defined by the array $P_0, \ldots, P_m$, $(m \geq 1)$ is a union of elementary Catmull-Rom curves $\gamma = \gamma^{(1)} \cup \cdots \cup \gamma^{(m-2)}$ $(0 \leq t \leq m-2)$; the section $\gamma^{(i)}$ corresponds to the points $P_{i-1}, P_i, P_{i+1}, P_{i+2}$, for $i-1 \leq t \leq i$ and $1 \leq i \leq m-2$.

The curve obtained is C^1-continuous, and it interpolates the points $\vec{\mathbf{r}}(0) = P_1, \ldots, \vec{\mathbf{r}}(m-2) = P_{m-1}$. The tangent vectors at these points are $\vec{\mathbf{r}}'(i) = \frac{1}{2}(P_{i+2} - P_i)$ for $0 \leq i \leq m-2$. Changing one point in the array $\mathbf{P}$ or adding one point changes only part of the Catmull-Rom curve. We need to recalculate the equations of the *four segments* $\gamma^{(i-2)}, \gamma^{(i-1)}, \gamma^{(i)}, \gamma^{(i+1)}$.

We write a procedure for deriving and plotting in $\mathbb{R}^2$.

```
> crom_2d := proc(xp, yp) local s,n,i,p,pp,m;
  n[0]:=t->-t*(1-t)^2/2; n[1]:=t->(3*t^3-5*t^2+2)/2;
  n[2]:=t->t*(-3*t^2+4*t+1)/2; n[3]:=t->t^2*(t-1)/2;
  m:=nops(xp)-1; for s from 1 to m+1 do
  p[s-1]:=[xp[s], yp[s]] od;
  pp:=n[0](t-i+1)*p[i-1]+n[1](t-i+1)*p[i]+n[2](t-i+1)*
  p[i+1]+n[3](t-i+1)*p[i+2]; RETURN(evalm(sum
  ((1+signum(-t+i))*(1+signum(t-i+1))/4*pp, i=1..m-2))) end:
```

Let us use the procedure `crom_2d(xp,yp)` for plotting a plane curve.

```
> xp:=[.056, .287, .655, .716, .228, .269, .666, .929]:
  yp:=[.820, .202, .202, .521, .521, .820, .820, .227]:
> P:=plot([seq([xp[j], yp[j]], j=1..nops(xp))]):
> for i from 1 to nops(xp)-2 do
  v[i-1]:=plots[display](plottools[arrow]([xp[i], yp[i]],
  [xp[i+2]-xp[i], yp[i+2]-yp[i]], 0.001, 0.01, 0.1)) od:
> vecs:=seq(v[i-1],i=1..nops(xp)-2): cr:=crom_2d(xp, yp):
> curve:=plot([cr[1],cr[2], t=0.01..nops(xp)-3.01]):
> plots[display]([curve,P,vecs],scaling=constrained);
```

15
Non-Euclidean Geometry in the Half-Plane

Non-Euclidean geometry has great historical, developing and methodological importance. The Cayley-Klein (disk) and Poincaré (half-plane) models of hyperbolic geometry are studied in the foundations of geometry and have various applications.

In Section 15.1 we recall basic facts about the Poincaré model. Then in Section 15.2 we do some calculations using the Poincaré metric and display the resulting figures (straight lines, θ-transversals, circles, equidistants, horocycles, fifth lines, etc.) in the half-plane.

In Chapter 15 the reader will become acquainted with the commands

`arccos, point.`

15.1 Preliminary Facts

The geometry induced on the half-plane $\mathbb{R}^2_+ = \{(x, y) \in \mathbb{R}^2 : y > 0\}$ by the *Poincaré metric* $ds^2 = \frac{k^2}{y^2}(dx^2 + dy^2)$ is the *hyperbolic geometry* of Lobachevsky.

Since the Gaussian curvature of the metric $ds^2 = f(x, y)(dx^2 + dy^2)$ is given by the formula $K = -\frac{1}{2f}\,\Delta \ln f$, the curvature of the Poincaré metric is a negative constant $K = -k^2$.

Lemma 15.1.1 *Each of the following maps ψ is an isometry of $\mathbb{R}^2_+$ onto itself (motion of the metric).*

(1) $\psi(x, y) = (x + a,\ y)$ is a parallel translation along the axis OX;

(2) $\psi(x, y) = (\lambda x, \lambda y)$ *is a homothety with center at* O*;*

(3) $\psi(x, y) = (-x, y)$ *is a symmetry with respect to the axis* OY*;*

(4) $\psi(x, y) = (\frac{x}{x^2+y^2}, \frac{y}{x^2+y^2})$ *is an inversion with respect to the unit circle with center at* O*.*

Proof. We check that the scalar product $(\vec{\mathbf{a}}, \vec{\mathbf{b}})_L := \frac{k^2}{y^2}(\vec{\mathbf{a}}, \vec{\mathbf{b}})$ holds. Then

(1), (3): $(d\psi(\vec{\mathbf{a}}), d\psi(\vec{\mathbf{b}}))_L = \frac{k^2}{y^2}(d\psi(\vec{\mathbf{a}}), d\psi(\vec{\mathbf{b}})) = \frac{k^2}{y^2}(\vec{\mathbf{a}}, \vec{\mathbf{b}}) = (\vec{\mathbf{a}}, \vec{\mathbf{b}})_L$.

(2): $(d\psi(\vec{\mathbf{a}}), d\psi(\vec{\mathbf{b}}))_L = \frac{k^2}{(\lambda y)^2}(d\psi(\vec{\mathbf{a}}), d\psi(\vec{\mathbf{b}})) = \frac{k^2}{\lambda^2 y^2}(\lambda\vec{\mathbf{a}}, \lambda\vec{\mathbf{b}}) = (\vec{\mathbf{a}}, \vec{\mathbf{b}})_L$.

(4) for basic vectors $\vec{\mathbf{i}}, \vec{\mathbf{j}}$ at the point (x, y),

$d\psi(\vec{\mathbf{i}}) = \left(\frac{y^2-x^2}{(x^2+y^2)^2}, \frac{-2xy}{(x^2+y^2)^2}\right)$, $d\psi(\vec{\mathbf{j}}) = \left(\frac{-2xy}{(x^2+y^2)^2}, \frac{x^2-y^2}{(x^2+y^2)^2}\right)$.

Thus $(d\psi(\vec{\mathbf{i}}), d\psi(\vec{\mathbf{i}}))_L = \frac{k^2}{(x^2+y^2)^2} / \left(\frac{y}{x^2+y^2}\right)^2 = \frac{k^2}{y^2} = (\vec{\mathbf{i}}, \vec{\mathbf{i}})_L$, and analogously for $\vec{\mathbf{j}}$. Moreover, $(d\psi(\vec{\mathbf{i}}), d\psi(\vec{\mathbf{j}}))_L = 0 = (\vec{\mathbf{i}}, \vec{\mathbf{j}})_L$ holds. □

Let G be the group generated by all motions of the form (1)–(4). Assume $k = 1$.

Theorem 15.1.1 *The geodesics (straight lines) of the Poincaré metric on the half-plane* $\mathbb{R}^2_+$ *are the following: vertical rays and half-circles with centers on the axis* OX*, with corresponding parametrization.*

Proof (First method). One can easily see that any of the curves in the theorem coincides with a set of fixed points of some transformation from G. But the curve of fixed points of an isometry (in a Riemannian metric) is always a geodesic [Tho].

Proof (Second method). Let us calculate the Christoffel symbols of the Poincaré metric $\Gamma^1_{12} = -\Gamma^2_{11} = \Gamma^2_{22} = \frac{1}{y}$; others are zero. Hence the differential equations of geodesics are the following:

$$x'' - \frac{2}{y}x'y' = 0, \qquad y'' + \frac{1}{y}(x'^2 - y'^2) = 0.$$

It is easy to calculate $y' = y'_x x' \Rightarrow y'' = y''_{xx}x'^2 + y'_x x'' = y''_{xx}x'^2 + \frac{y'}{x'}x''$. If $x' = 0$, then we obtain a vertical ray $\{x = \text{const}, \ y > 0\}$; otherwise we have $\frac{d^2y}{d^2x} = \frac{y''x' - x''y'}{x'^3} = -\frac{1}{y}(\frac{y'^2}{x'^2} + 1) = -\frac{1}{y}((\frac{dy}{dx})^2 + 1)$. From this follows the ODE

$$y\,y'' + y'^2 = -1 \Rightarrow (y\,y')' = -1 \Rightarrow y\,y' = -x + c_1.$$

We see that the solution geodesic $y^2 = -x^2 + 2c_1x + c_2$ is the circle $(x - c_1)^2 + y^2 = c_1^2 + c_2$ whose center is on the axis OX. □

15.2 Examples of Visualization

Example 15.2.1 Let us plot the segment (the straight line) through two given points.

```
> segment:=proc(A, B)  local x0,R,t1,t2,X,Y;
  if A[1]=B[1] then
  plot([[A,B], [[A[1],0], [A[1],2*max(A[2],B[2])]]],
  scaling=constrained,thickness=[2,1],linestyle=[1,2])
  else x0:=solve((B[1]-x)^2+B[2]^2=(A[1]-x)^2+A[2]^2,x);
  R:=sqrt((B[1]-x0)^2+B[2]^2); X:=x0+R*cos(t); Y:=R*sin(t);
  t1:=arccos((B[1]-x0)/R); t2:=arccos((A[1]-x0)/R);
  plot([[X,Y,t=t1..t2], [X,Y,t=0..Pi]],scaling=constrained,
  thickness=[2,1], linestyle=[1,2]) fi end:
```

```
> segment([2,3], [3,2]); # segment([2,3], [2,2]);
```

```
> plots[display]([seq(segment([0,1], [i/4,0]),i=-4..4)]);
```

The procedure `segment` is used below.

Example 15.2.2 One can derive the distance between points $A_1(x_1, y_1)$ and $A_2(x_2, y_2)$ (integrating the metric along the segment)

$$d(A_1, A_2) = \begin{cases} k \cdot |\ln \frac{y_2}{y_1}|, & \text{if } x_2 = x_1, \\ k \cdot |\ln \frac{\tan \frac{t_1}{2}}{\tan \frac{t_2}{2}}|, & \text{if } x_2 \neq x_1. \end{cases} \tag{15.1}$$

Here t_i are the angles between vectors $O'A_i$ and the axis OX, and $O' \in OX$ is the center of the Euclidean circle through the points A_1 and A_2, which represents the straight line A_1A_2. If the points are given in the form of complex numbers $z = x + \mathbf{i}y$, then the formula for distance takes the form $d(z_1, z_2) = k \ln \frac{1+\frac{z_1-z_2}{z_1-\bar{z}_2}}{1-\frac{z_1-z_2}{z_1-\bar{z}_2}}$.

The following procedure `distance` based on the formula (15.1) is used below.

```
> distance:=proc(A, B)  local x0,R,t1,t2,F;
  if A[1]=B[1] then RETURN(abs(log(B[2]/A[2]))) else
  x0:=solve((B[1]-x)^2+B[2]^2=(A[1]-x)^2+A[2]^2,x);
  R:=sqrt((B[1]-x0)^2+B[2]^2); t1:=arccos((B[1]-x0)/R);
  t2:=arccos((A[1]-x0)/R); F:=t -> ln(tan(t/2));
  RETURN(evalf(abs(F(t2)-F(t1)))) fi end:
```

```
> distance([0,20], [0,.05]);
```

5.99146

With the help of distance one can plot a) the center point of a segment, b) the doubling of a given segment, c) a given segment on a given ray.

Example 15.2.3 We plot a triangle with given vertices using the procedure triangle.

```
> triangle:=proc(A, B, C)  local text;
  text:=plots[textplot]({[A[1],A[2],`a`], [B[1],B[2],`b`],
  [C[1],C[2],`c`]}, align=ABOVE); plots[display]
  ([segment(A,B), segment(B,C), segment(A,C)], text) end:

> triangle([1,1], [5,2], [4,4]);
```

We derive its perimeter using the procedure perimeter.

```
> perimeter:=proc(A,B,C)
  evalf(distance(A,B)+distance(C,B)+distance(A,C)) end:

> perimeter([1,1], [2,1], [5,4]);
  1.21036
```

One can plot a triangle with 1, 2 or 3 infinite vertices. Note that a triangle with one infinite vertex is analogous to the Euclidean half-strip between two parallel lines; a triangle with two infinite vertices is analogous to the Euclidean angle between two rays, and a triangle with three infinite vertices has no analogous Euclidean figure. One can plot a triangle symmetric to another triangle with respect to (1) an axis or (2) a point.

Example 15.2.4
We calculate angles of a triangle using the procedure angle 1.

```
> angle1:=proc(B, A, C)  local x_AB,x_AC,R_AB,R_AC;
  if C[1]<>A[1] then
  x_AC:=solve((C[1]-x)^2+C[2]^2=(A[1]-x)^2+A[2]^2,x);
  R_AC:=sqrt((A[1]-x_AC)^2+A[2]^2) fi;
  if B[1]<>A[1] then
  x_AB:=solve((B[1]-x)^2+B[2]^2=(A[1]-x)^2+A[2]^2,x);
  R_AB:=sqrt((A[1]-x_AB)^2+A[2]^2) fi;
  if C[1]=A[1] and B[1]=A[1] then RETURN(0)
  elif B[1]=A[1] and C[1]<>A[1] then
  RETURN(evalf(arccos((x_AC-A[1])/R_AC))) elif
  C[1]=A[1] and B[1]<>A[1] then
  RETURN(evalf(arccos((x_AB-A[1])/R_AB))) else
  RETURN(evalf(arccos(
  (R_AB^2+R_AC^2-(x_AB-x_AC)^2)/(2*R_AB*R_AC))) fi end:
```

```
> angle([1,1], [1,2], [2,3]);
```

0.588003

We can calculate the sum of its angles and the area using the procedure `sum_angles`.

```
> sum_angles:=proc(A,B,C)
  evalf(angle1(B,A,C)+angle1(A,B,A)+angle1(B,C,A)) end:
```

```
> sum_angles([1,1],[3,1],[2,1]);
```

1.24905

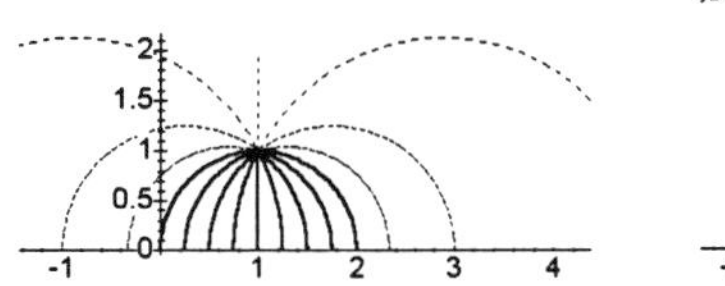

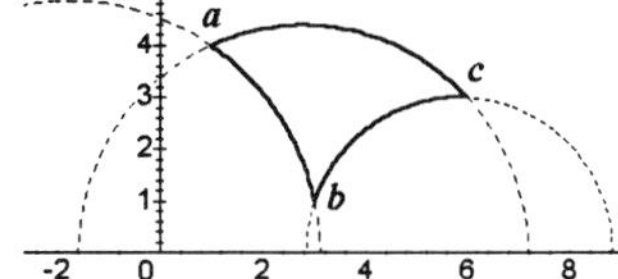

Figs. 15.1–15.2. Sheaf of straight lines. Triangle

Exercise 15.2.1 Using the above programs, check that for "small" triangles the sum of the angles is close to π. Plot the bisector of the angle and double the given angle.

The following facts from hyperbolic trigonometry help us solve more complicated problems with triangles.

Theorem 15.2.1 *Let T be a triangle with angles α, 0, $\frac{\pi}{2}$. Then*

(1) $\sinh(b)\tan(\alpha) = 1$, *(2)* $\cosh(b)\sin(\alpha) = 1$, *(3)* $\tanh(b)\sec(\alpha) = 1$.

Let T be a triangle with angles α, β, 0. Then

(4) $\cosh(c) = \frac{1+\cos(\alpha)\cos(\beta)}{\sin(\alpha)\sin(\beta)}$, *(5)* $\sinh(c) = \frac{\cos(\alpha)+\cos(\beta)}{\sin(\alpha)\sin(\beta)}$.

Theorem 15.2.2 (Pythagorean) *Let T be a triangle with angles α, β, $\frac{\pi}{2}$. Then* $\cosh(c) = \cosh(a)\cosh(b)$.

Theorem 15.2.3 (Sine and Cosine) $\frac{\sinh(a)}{\sin(\alpha)} = \frac{\sinh(b)}{\sin(\beta)} = \frac{\sinh(c)}{\sin(\gamma)}$.
$\cosh(c) = \cosh(a)\cosh(b) - \sinh(a)\sinh(b)\cos(\gamma)$.

Example 15.2.5 (*Polygons*). Let us plot a *square* (right quadrangle) with edge a. For the *Lambert quadrangle* with angles $\frac{\pi}{2}$, $\frac{\pi}{2}$, $\frac{\pi}{2}$, φ and the edges a_1, a_2 between right angles, we have $\sinh(a_1)\sinh(a_2) = \cos(\varphi)$.

If we reflect this quadrangle with respect to an edge that connects the right angles, then we obtain the *Sacceri quadrangle* with angles $\frac{\pi}{2}$, $\frac{\pi}{2}$, φ, φ.

We plot these quadrangles using the following procedure.

```
> lambert:=proc(a1,a2)  local t1,p,A,B,C,F,R1,x1,phi,text;
  if evalf(abs(sinh(a1)*sinh(a2)))>=1 then  no_solutions
  else A:=[0,1]; B:=[0, exp(a2)]; t1:=2*arctan(exp(-a1));
  C:=[cos(t1), sin(t1)]; R1:=tan(t1); x1:=1/cos(t1);
  F[1]:=(exp(2*a2)-R1^2+x1^2)/(2*x1);
  F[2]:=sqrt(exp(2*a2)-F[1]^2); F:=[F[1], F[2]]; phi:=angle1(B,F,C);
  p[1]:=segment(A, B); p[2]:=segment(A, C);
  p[3]:=segment(B, F); p[4]:=segment(C, F);
  text:=plots[textplot]({[A[1],A[2],`a`],[B[1],B[2],`b`],
  [C[1],C[2],`c`], [F[1],F[2],`f`]}, align=LEFT);
  plots[display]({seq(p[i], i=1..4), text}, axes=framed,
  title=convert(angle_f=evalf(phi*180/Pi), string)) fi end:
```

```
> lambert(0.7, 1); # Lambert
```

```
> plots[display](lambert(0.7,1), lambert(-0.7,1)); # Sacceri
```

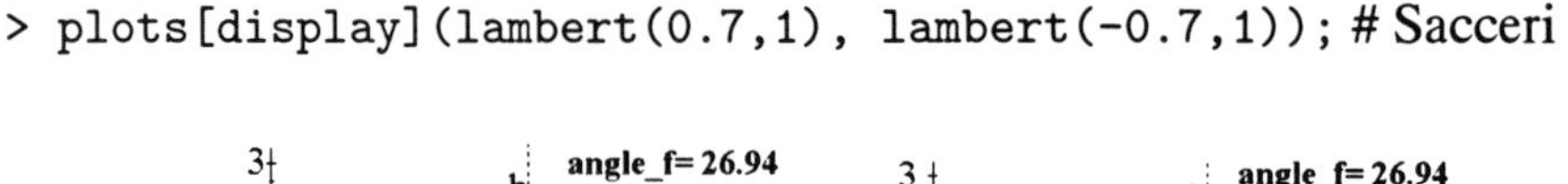

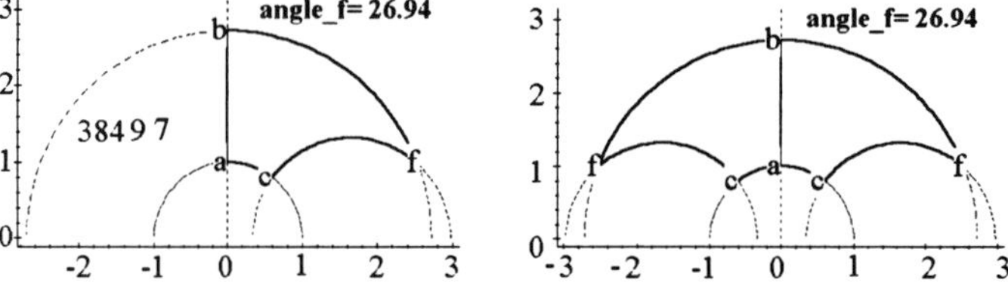

Figs. 15.3–15.4. Quadrangles of Lambert and Sacceri

Exercise 15.2.2 Calculate the edge and then plot the *regular 2n-polygon* with the sum of the angles equal to 2π (by gluing pairs of its edges, we have a model for a surface of constant negative curvature).

Example 15.2.6 Let us deduce the formula for the *angle of parallelism* (i.e., the angle between two rays with a common vertex that is *left* and *right parallel to a given straight line*), using Fig. 15.5.

```
> F:=t->k*ln(tan(t/2)): alpha=solve(F(Pi/2)-F(t)=d, t);
```

$$\alpha = 2 \arctan\left(\exp\left(-\frac{d}{k}\right)\right)$$

```
> T:=Pi/6: p[1]:=segment([0,1], [cos(T),sin(T)]):
  p[2]:=segment([cos(T),sin(T)], [cos(T),sin(T)+1]):
  p[3]:=plot([[0,0],[cos(T), sin(T)]]):
  p[4]:=segment([0,0], [0,2]):
```

```
p[5]:=plots[textplot]({[cos(T)-.1,sin(T)+.3,`t`],
[0.4,.1,`t`],[.3,1.1,`d`]}, align=RIGHT):
plots[display](seq(p[i], i=1..5)); # Fig. 15.6
```

Let us plot the right and left parallels to a given straight line through a given point using the procedure parallel. We also calculate the angle between them and the angle of parallelism.

```
> parallel:=proc(A,P,Q)  local x1,R1,g,AB1,AB2,B1,B2,phi,T;
  if P[1]=Q[1] then B1:=[P[1],0]; B2:=[A[1],A[2]+1] else
  x1:=solve((P[1]-x)^2+P[2]^2=(Q[1]-x)^2+Q[2]^2,x);
  R1:=sqrt((P[1]-x1)^2+P[2]^2);
  B1:=[x1-R1,0];B2:=[x1+R1,0] fi;
  AB1:=segment(A,B1); AB2:=segment(A,B2);
  T:=plots[textplot]({[A[1],A[2],`a`], [P[1],P[2],`p`],
  [Q[1],Q[2],`q`]}, align=ABOVE); g:=segment(P,Q);
  phi:=angle1(B1,A,B2)/2; plots[display]([AB1,AB2,g,T],
  title=convert(` angle of parallelism` =phi,string)) end:
```

```
> parallel([3,5], [5,2], [2,4]);
```

Using the program one can check that the limit angle, as the point approaches the given straight line, is equal to $\frac{\pi}{2}$, and the limit angle, as the point moves away from the given line, decreases to zero.

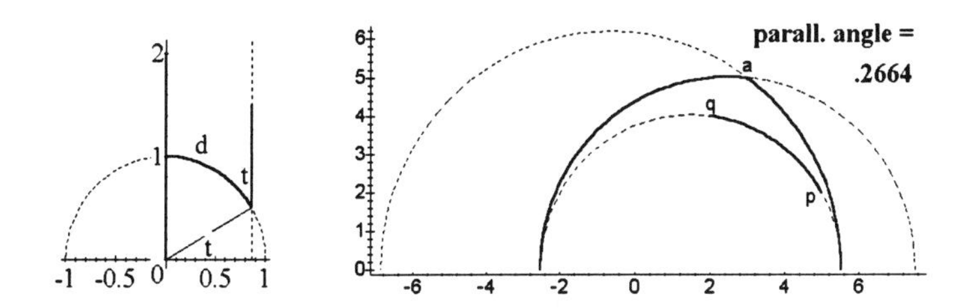

Figs. 15.5–15.6. The angle of parallelism. Left and right parallel lines

Example 15.2.7 Let us plot the perpendicular to a straight line through a given point and derive its length (i.e., the distance from the point to the line) using the procedure perp.

```
> perp:=proc(A,P,Q)  local x1,x2,R1,R2,PQ,AB,B,d,T;
  if P[1]=Q[1] then B:=[P[1], sqrt(A[1]^2+A[2]^2)] else
  x1:=solve((P[1]-x)^2+P[2]^2=(Q[1]-x)^2+Q[2]^2,x);
  R1:=sqrt((P[1]-x1)^2+P[2]^2);
  x2:=solve((A[1]-x)^2+A[2]^2=(x-x1)^2-R1^2,x);
  R2:=sqrt((A[1]-x2)^2+A[2]^2);
```

```
B[1]:=solve(R1^2-(x1-x)^2=R2^2-(x2-x)^2,x);
B[2]:=sqrt(R1^2-(x1-B[1])^2); B:=[B[1],B[2]] fi;
T:=plots[textplot]({[A[1],A[2],`a`], [B[1],B[2],`b`],
[P[1],P[2],`p`], [Q[1],Q[2],`q`]}, align=ABOVE);
PQ:=segment(P,Q); AB:=segment(A,B); d:=distance(A,B);
plots[display]([AB, PQ, T], title=convert(distance=d,
string)) end:
```

```
> perp([-7,1], [5,2], [2,4]);
```

Exercise 15.2.3 Plot the perpendicular to a segment through its midpoint using the procedure perp.

Example 15.2.8 Let us plot the common perpendicular to "*antiparallel*" (nonparallel and nonintersected) lines. Then calculate its length (i.e., the distance between these straight lines).

Idea of the solution. We plot the common tangent line EF to the Euclidean circles $\omega(AB)$ and $\omega(CD)$; then we find its midpoint and plot the perpendicular from this point to the axis OX (i.e., obtain the radical axis of two Euclidean circles). Through the above point $(x_3, 0)$ as the center, plot the Euclidean circle with radius equal to the length of the tangent line segment to both circles. The points H and K are the feet of the common perpendicular.

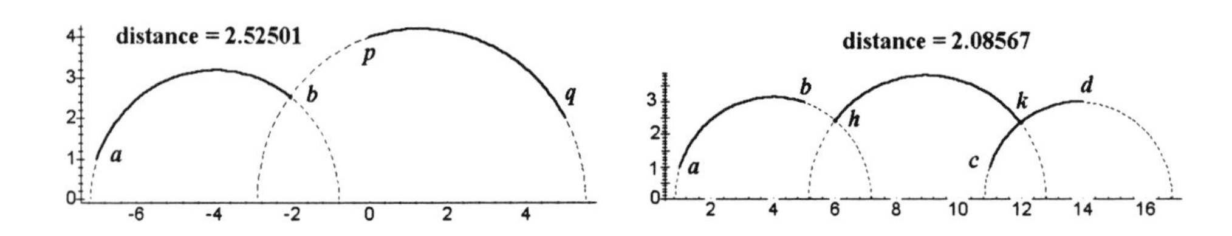

Figs. 15.7–15.8. Perpendicular to a line. The common perpendicular

In the following program we assume that the straight lines AB and CD are "antiparallel".

```
> biortho:=proc(A,B,C,D)
  local x0,x1,R1,x2,R2,x3,R3,g,x_E,x_F,H,K,d,T;
  if A[1]<>B[1] then
  x1:=solve((A[1]-x)^2+A[2]^2=(B[1]-x)^2+B[2]^2,x);
  R1:=sqrt((A[1]-x1)^2+A[2]^2) fi;
  if C[1]<>D[1] then
  x2:=solve((C[1]-x)^2+C[2]^2=(D[1]-x)^2+D[2]^2,x);
  R2:=sqrt((C[1]-x2)^2+C[2]^2) fi;
  if A[1]=B[1] then
```

```
x3:=A[1]; R3:=sqrt((x3-x2)^2-R2^2); K[1]:=x3; K[2]:=R3;
H[1]:=-1/2*(-x2^2+x3^2-R3^2+R2^2)/(x2-x3);
H[2]:=sqrt(R3^2-(H[1]-x3)^2) elif C[1]=D[1] then
x3:=C[1]; R3:=sqrt((x3-x1)^2-R1^2); H[1]:=x3; H[2]:=R3;
K[1]:=-1/2*(-x1^2+x3^2-R3^2+R1^2)/(x1-x3);
K[2]:=sqrt(R3^2-(K[1]-x3)^2) else if R1=R2 then
x3:=(x1+x2)/2 else x0:=solve((x1-x)/R1=(x2-x)/R2,x);
x_E:=x1-R1^2/(x1-x0); x_F:=x2-R2^2/(x2-x0);
x3:=(x_E+x_F)/2 fi; R3:=sqrt((x3-x1)^2-R1^2);
K[1]:=-1/2*(-x1^2+x3^2-R3^2+R1^2)/(x1-x3);
H[1]:=-1/2*(-x2^2+x3^2-R3^2+R2^2)/(x2-x3);
K[2]:=sqrt(R3^2-(K[1]-x3)^2);
H[2]:=sqrt(R3^2-(H[1]-x3)^2) fi;
H:=[H[1],H[2]]; K:=[K[1],K[2]]; d:=distance(H,K);
T:=plots[textplot]({[A[1],A[2],`a`], [B[1],B[2],`b`],
[C[1],C[2],`c`], [D[1],D[2],`d`], [H[1],H[2],`h`],
[K[1],K[2],`k`]}, align=ABOVE);
g.1:=segment(A,B); g.2:=segment(C,D); g.3:=segment(H,K);
RETURN(plots[display]([g.1, g.2, g.3, T],
title=convert(distance=d,string))) end:
```

```
> biortho([1,1], [5,3], [11,1], [14,3]);
```

The following generalizes the notion of common perpendicular.

Definition 15.2.1 The straight line L is called a *θ-transversal* for L_1 and L_2 if it meets the same angle θ at each of the straight lines.

The common perpendicular L to L_1, L_2 is the unique $\frac{\pi}{2}$-transversal. For every other value of θ, there are four θ-transversals (two alternate and two adjacent). The length t_θ of the segment of the transversal is equal to

$$\sinh(\tfrac{1}{2}d(L_1, L_2)) = \sinh(\tfrac{1}{2}\, t_\theta)\sin\theta, \quad \text{the alternate transversal,}$$

$$\cosh(\tfrac{1}{2}d(L_1, L_2)) = \cosh(\tfrac{1}{2}\, t_\theta)\sin\theta, \quad \text{the adjacent transversal.}$$

The procedure for θ-transversals can be the following.

```
> transversal:=proc(A,B,P,Q,theta)
  local x1,R1,x2,R2,x3,R3,H,K,g,s,d,i,j,text;
  if A[1]<>B[1] then
  x1:=solve((A[1]-x)^2+A[2]^2=(B[1]-x)^2+B[2]^2,x);
  R1:=sqrt((A[1]-x1)^2+A[2]^2) fi; if P[1]<>Q[1] then
  x2:=solve((P[1]-x)^2+P[2]^2=(Q[1]-x)^2+Q[2]^2,x);
  R2:=sqrt((P[1]-x2)^2+P[2]^2) fi;
```

```
for i from 0 to 1 do for j from 0 to 1 do
x3:=`x3`: R3:=`R3`: if A[1]=B[1] then
solve({(A[1]-x3)/R3=evalf((-1)^i*cos(theta)),
(R2^2+R3^2-(x2-x3)^2)/(2*R2*R3)=evalf((-1)^j*cos(theta)),
R3>0}, {x3,R3}); assign(%);
H[1]:=A[1]: H[2]:=sqrt(R3^2-(A[1]-x3)^2);
K[1]:=solve(R3^2-(x3-x)^2=R2^2-(x2-x)^2,x);
K[2]:=sqrt(R2^2-(x2-K[1])^2) elif P[1]=Q[1] then
solve({(P[1]-x3)/R3=evalf((-1)^i*cos(theta)),
(R1^2+R3^2-(x1-x3)^2)/(2*R1*R3)=evalf((-1)^j*cos(theta)),
R3>0}, {x3,R3}); assign(%);
K[1]:=P[1]: K[2]:=sqrt(R3^2-(P[1]-x3)^2);
H[1]:=solve(R3^2-(x3-x)^2=R1^2-(x1-x)^2,x);
H[2]:=sqrt(R1^2-(x1-H[1])^2) else solve(
{(R1^2+R3^2-(x1-x3)^2)/(2*R1*R3)=evalf((-1)^i*cos(theta)),
(R2^2+R3^2-(x2-x3)^2)/(2*R2*R3)=evalf((-1)^j*cos(theta)),
R3>0}, {x3,R3}); assign(%);
H[1]:=solve(R3^2-(x3-x)^2=R1^2-(x1-x)^2,x);
H[2]:=sqrt(R1^2-(x1-H[1])^2);
K[1]:=solve(R3^2-(x3-x)^2=R2^2-(x2-x)^2,x);
K[2]:=sqrt(R2^2-(x2-K[1])^2) fi; K:=[K[1],K[2]]; H:=[H[1],H[2]];
s[i+2*j]:=segment(H,K); d[i+2*j]:=distance(H,K) od od;
g.1:=segment(A,B); g.2:=segment(P,Q);
text:=plots[textplot]({[A[1],A[2],`a`], [B[1],B[2],`b`],
[P[1],P[2],`p`], [Q[1],Q[2],`q`]}, align=ABOVE);
plots[display]([g.1, g.2, seq(s[i], i=0..3), text],
title=convert(dist=[d[1],d[2]], string)) end:
```

```
> transversal([-4,0],[-1,0],[1,0],[3,1], Pi/3);
```

```
> transversal([2,2],[5,1],[7,5],[7,1], Pi/3);
```

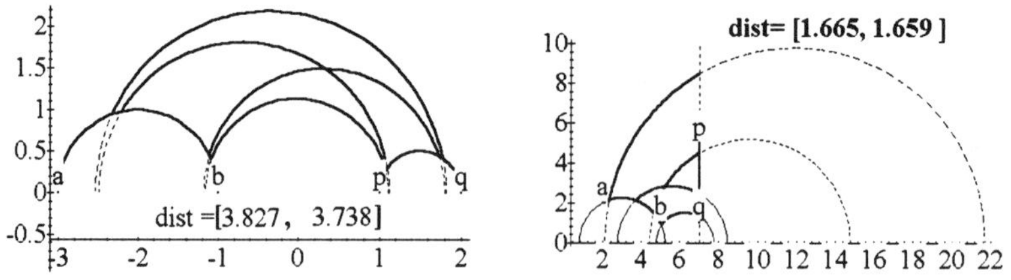

Figs. 15.9–15.10. Four θ-transversals to "super-parallel" lines

Example 15.2.9 *Pencils of lines and their orthogonal trajectories.*

The pencils of

(1) lines through a given point,
(2) parallel lines (in some direction), and
(3) lines orthogonal to a straight line,

are plotted, respectively, as *an elliptic, a parabolic* or *a hyperbolic* pencil of circles.

Let us plot these pencils of Euclidean circles.

The curves that are orthogonal to these sheaves are called in hyperbolic geometry *a circle, a horocycle,* and *equidistant*; they are analytic curves.

(1) The *circle* is drawn as a Euclidean circle that does not intersect the axis OY. The pencil of straight lines is represented by a family of Euclidean circles through the given point and its symmetric point with respect to the axis OY. The orthogonal trajectories of this elliptic pencil are seen as the family of Euclidean circles from the problem (with constant ratio of distances from given points). The figure is similar to the net of *bipolar coordinates*, see Fig. 6.14 d.

One can prove this using calculations. The area of the circle of radius R and the circumference of the circle are given by the formulas $S(R) = 4\pi \sinh^2(\frac{R}{2})$ and $L(R) = S(R)' = 2\pi \sinh^2(R)$. One can check that the angle subtended by the diameter of the circle is less than 90°.

Idea of the program. Let the circle of radius R with center at A intersect the vertical ray $\{x = A[1]\}$ at (diametric) points $y_1 < y_2$. Then by the distance formula we have $R = \ln \frac{A[2]}{y_1} = \ln \frac{y_2}{A[2]}$. From this it follows that $y_1 = A[2]\exp(R)$ and $y_1 = A[2]\exp(-R)$. The Euclidean center of the circle lies on the altitude $y_E = A[2](y_1+y_2)/2 = A[2]\cosh(R)$, and its Euclidean radius is equal to $R_1 = A[2](y_2 - y_1)/2 = A[2]\sinh(R)$.

```
> circle_L:=proc(A,R)    local P1,P2,yE,R1,AA,text;
  yE:=A[2]*cosh(R); R1:=A[2]*sinh(R);
  AA:=plottools[point](A, symbol=circle);
  P1:=plottools[circle]([A[1],yE], R1, thickness=2);
  P2:=segment([A[1]-A[2],0],[A[1]+A[2],0]);
  text:=plots[textplot]({[A[1],A[2],`a`]}, align=ABOVE);
  plots[display]([P1,AA,P2,text], scaling=constrained) end:
```

```
> plots[display]([seq(circle_L([3,1],i/5), i=1..5),
  seq(segment([3,1], [3+i/4,0]), i=-3..3)]); # Fig. 15.11
```

```
> n:=10: plots[display]([seq(circle_L([.3,1],i/n), i=1..n),
  seq(circle_L([-.3,1], i/n), i=1..n)]); # Fig. 15.12
```

Exercise 15.2.4 Plot the *circumcircle* for a given triangle. Note that a circumcircle does not always exist for a triangle, because through three points in the non-Euclidean half-plane that do not belong to the same line, we can plot one of three curves: circle, equidistant, or horocycle.

Plot the *inscribed circle* for a given triangle. This triangle always exists; its center lies at the intersection of the three angle bisectors.

(2) The *equidistant* is represented in the half-plane as the arc of the Euclidean circle that intersects the axis OY at two improper points for the straight line. It is a simple non-closed curve. The pencil of straight lines orthogonal to the given line, the axis OY, is represented as the family of Euclidean concentric half-circles whose center is at O. The orthogonal trajectories of this pencil are the Euclidean rays through the point O. The figure is similar to the net of *polar coordinates*; see Fig 6.14 a.

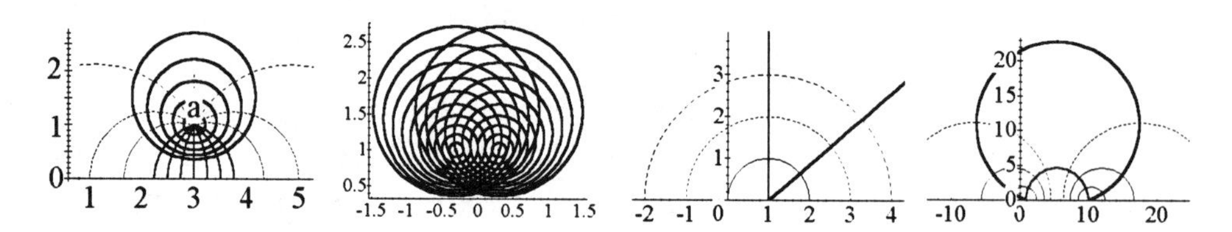

Figs. 15.11–15.14. Circles on a non-Euclidean river. Equidistant (2 cases)

Using inversion with respect to the Euclidean circle of radius R that is tangent to the axis OY, one maps the family of g rays (equidistants of OY) to the arcs of the hyperbolic pencil of Euclidean circles through the points (0, 0) and $(2R, 0)$. One can prove this using calculations.

```
> equid:=proc(A,B,d)  local x1,R,i,p,n,xi,Ri,ti,y1,y2,R2,T;
  n:=4;  if A[1]=B[1] then x1:=A[1];
  y1:=max(A[2], B[2])+1; y2:=2*arctan(exp(-d));
  p[0]:=plot([x1,t, t=0..y1], thickness=2);
  p[n]:=plot(tan(y2)*(x-x1), x=x1-y1..x1+y1, y=0..y1,
  thickness=3); for i from 1 to n-1 do Ri:=y1*i/n;
  p[i]:=plot([x1+Ri*cos(t),Ri*sin(t),t=0..Pi],linestyle=2)
  od else x1:=solve((B[1]-x)^2+B[2]^2=(A[1]-x)^2+A[2]^2,x);
  R:=sqrt((B[1]-x1)^2+B[2]^2); for i from 1 to n-1 do
  ti:=Pi*i/(2*n); xi:=x1+R*cos(ti)+R*sin(ti)*tan(ti);
  Ri:=sqrt((x1+R*cos(ti)-xi)^2+(R*sin(ti))^2);
  p[i]:=plot([[xi+Ri*cos(t), Ri*sin(t), t=0..Pi],
  [2*x1-xi+Ri*cos(t),Ri*sin(t), t=0..Pi]], linestyle=2) od;
  p[0]:=plot([x1+R*cos(t),R*sin(t), t=0..Pi], thickness=2);
  y1:=R*exp(d); y2:=(y1^2-R^2)/(2*y1); R2:=y1-y2;
  p[n]:=plot([x1+R2*cos(t), y2+R2*sin(t),
  t=-arcsin(y2/R2)..Pi+arcsin(y2/R2)], thickness=3) fi;
```

```
T:=plots[textplot] ({[A[1],A[2],`a`],[B[1],B[2],`b`]},
align=ABOVE); plots[display]([seq(p[i], i=0..n), T],
scaling=constrained) end:
```

```
> equid([1,1],[1,3], 1); equid([1,1],[2,3], 1.6);
```

```
>plots[display]({equid([1,1],[2,3],2),equid([1,1],[2,3],-2)});
```

(3) A *horocycle* is represented in the half-plane by a Euclidean circle that is tangent to the axis OY. It is a simple non-closed curve. A family of straight lines parallel to a given line is either represented by the parabolic pencil of Euclidean circles through the point on the axis OY, or by the family of rays parallel to the axis OY. In the first case their orthogonal trajectories form the set of Euclidean circles tangent to the axis OY at the point. The figure is similar to the net of *tangential coordinates*

```
> plots[coordplot](tangent, title=`Tangential`);
```

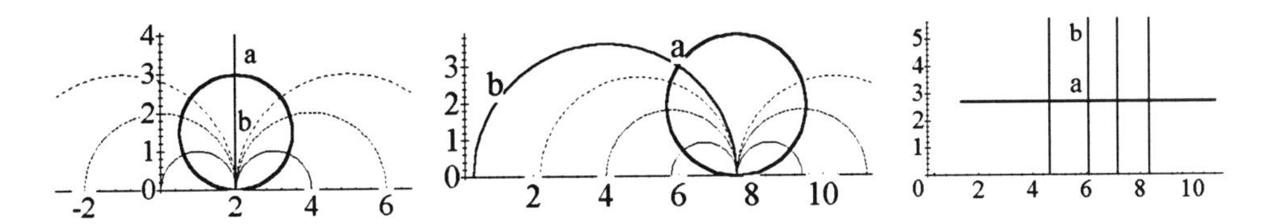

Figs. 15.15–15.17. Horocycle through A in the direction AB (3 cases)

In the second case, obviously their orthogonal trajectories form the family of Euclidean straight lines parallel to the axis OX. Try to prove this using calculations. The *broken horocycle* (obtained as the union of two parts of a mutually tangent pair of horocycles) is only a C^1-regular curve.

```
> horoc:=proc(A,B) local x0,x1,R,i,p,n,xi,Ri,ti,y2,R2,text;
  n:=4; if A[1]=B[1] then
  x1:=A[1]; R:=max(A[2],B[2])+1; R2:=A[2]/2;
  p[0]:=plot([x1,t,t=0..R],thickness=2) else
  x0:=solve((B[1]-x)^2+B[2]^2=(A[1]-x)^2+A[2]^2,x);
  R:=sqrt((B[1]-x0)^2+B[2]^2);
  x1:=x0+R; R2:=solve(y^2=(A[1]-x1)^2+(A[2]-y)^2,y);
  p[0]:=plot([x0+R*cos(t),R*sin(t),t=0..Pi],thickness=2) fi;
  p[n]:=plottools[circle]([x1,R2], R2, thickness=3);
  for i from 1 to n-1 do Ri:=R*i/n;
  p[i]:=plot([[x1+Ri+Ri*cos(t),Ri*sin(t), t=0..Pi],
  [x1-Ri+Ri*cos(t),Ri*sin(t),t=0..Pi]], linestyle=2) od;
  text:=plots[textplot]({[A[1],A[2],`a`], [B[1],B[2],`b`]},
  align=ABOVE); plots[display]([seq(p[i], i=0..n), text],
```

```
scaling=constrained) end:
> horoc([2,3],[2,1]); horoc([6,3],[1,2]);
> plots[display]({horoc([4,3],[1,2]), horoc([1,2],[4,3])});
```

Example 15.2.10 There is an interesting generalization of the construction of the equidistant.

Definition 15.2.2 Let S be the family of all straight lines intersecting a straight line at an angle $\alpha \in (0, \frac{\pi}{2})$. The orthogonal trajectory to the family S is called a *fifth line.*

Let us deduce the equations of a fifth line (a simple non-closed analytic curve) in the non-Euclidean half-plane and plot it.

Idea of the program. We deduce the differential equations of the fifth line in the case where the basic line coincides with the axis OY. The general case is obtained using an inversion (a rigid motion of the hyperbolic plane). A Euclidean circle intersecting the axis OY at the point $(t, 0)$ with angle $a \in (0, \pi)$ has the equation $(x-t)^2 + y^2 = t^2/\cos(a)^2$.

The parameter $t > 0$ of such a circle through the point (x, y) is equal to $t_1 = \frac{(x\cos(a)+\sqrt{x^2+y^2\sin(a)^2})\cos(a)}{-\sin(a)^2}$.

The normal vector $\vec{\mathbf{n}}(x, y)$ to the above circle at the point (x, y) is $[x - t_1,\ y] = \left[\frac{x+\cos(a)\sqrt{x^2+y^2\sin(a)^2}}{\sin(a)^2},\ y\right]$.

The differential equations of a fifth line are the following: $[x'(s),\ y'(s)] = \vec{\mathbf{n}}(x, y)$.

```
> fifth:=proc(A,B,a)  local n,i,p,x,s,P,q,x1,R1,x2,R2,T,F;
  q[0]:=segment(A,B); n:=8: if A[1]=B[1] then
  T:=solve((x(s)-t-A[1])^2+y(s)^2-t^2/cos(a)^2,t)[1];
  for i from 1 to n do
  p[i]:=dsolve({diff(x(s),s)=x(s)-T-A[1],diff(y(s),s)=y(s),
  x(0)=A[1],y(0)=i},{x(s),y(s)},type=numeric, method=classical);
  P[i]:=plots[odeplot](p[i],[x(s),y(s)],-3..2/i);
  q[i]:=plot([A[1]+i*cot(a)+i/sin(a)*cos(t), i/sin(a)*
  sin(t), t=0..Pi], linestyle=2) od;
  plots[display](seq(P[i],i=1..n), seq(q[i],i=0..n)) else
  x1:=solve((B[1]-x)^2+B[2]^2=(A[1]-x)^2+A[2]^2,x);
  R1:=sqrt((B[1]-x1)^2+B[2]^2);
  for i from 1 to n-1 do R2:=R1*sin(a*i/n)/sin(a-a*i/n);
  x2:=x1+R1*cos(a*i/n)+R1*sin(a*i/n)*cot(a-a*i/n);
  q[i]:=plot([x2+R2*cos(t),R2*sin(t),t=0..Pi],linestyle=2);
```

```
R2:=R1*sin(a+(Pi-a)*i/n)/sin((Pi-a)*i/n);
x2:=x1+R1*cos(a+(Pi-a)*i/n)+R1*sin(a+(Pi-a)*i/n)*
cot(-(Pi-a)*i/n);
q[-i]:=plot([x2+R2*cos(t),R2*sin(t),t=0..Pi]) od;
P[n]:=plots[polygonplot]([[x1+R1*cos(a),0],[x1+R1*cos(a),
4*R1*sin(a)]]);
T:=solve((x(s)-t-2*R1)^2+y(s)^2-t^2/cos(a)^2,t)[1];
F:=plottools[transform]((x,y)->[x/(x^2+y^2),y/(x^2+y^2)]);
for i from 2 to n-1 do
p[i]:=dsolve({diff(x(s),s)=x(s)-T-2*R1,diff(y(s),s)=y(s),
x(0)=-2*R1,y(0)=i},{x(s),y(s)}, type=numeric, method=classical);
P[i]:=plottools[translate](plottools[scale](F(plots
[odeplot](p[i],[x(s),y(s)],-3..1,scaling=constrained)),
4*R1^2,4*R1^2),x1-R1,0) od;
plots[display](seq(P[i],i=2..n), seq(q[i],i=1-n..n-1),
scaling=constrained) fi end:
> fifth([-5,0],[5,0],Pi/2-.4); fifth([5,0],[5,5],Pi/2-.4);
```

Definition 15.2.3 A partition of a metric space by a family of congruent and mutually equidistant sets is called a *metric fibration.*

Examples are parallel straight lines in the Euclidean plane and generators or parallels on a circular cylinder. What are all the metric fibrations (by congruent and mutually equidistant curves) of the hyperbolic plane?

Theorem 15.2.4 (see [EVGL]) *There are exactly three nontrivial metric fibrations of the hyperbolic plane:*

(1) the fibration by horocycles,
(2) the fibration by broken horocycles,
(3) the fibration by fifth lines.

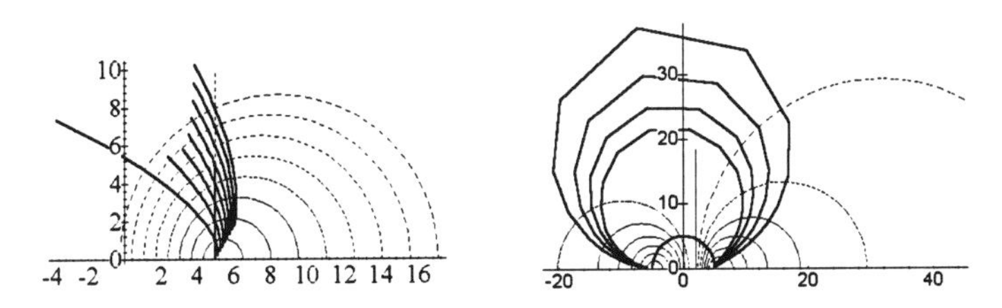

Figs. 15.18–15.19. Families of *fifth lines* (two cases)

16
Convex Hulls

A convex polygon is a particular case of a plane convex curve. In Sections 4.2, 14.3, 14.4.1 the notion of the *convex hull* (CH) of a finite planar set $\mathbf{P} = \{P_1, P_2, \ldots, P_n\}$ of points was used.

The derivation and plotting of CH is the key to a number of problems of *computational geometry*, in which algorithms of geometrical problems and their computational complexity are studied. Some algorithms for finding CH are well known. They differ in computational efficiency and the geometrical idea.

Here we consider one of them (see [Las]), based on the *gift wrapping* method. The idea is the following:

1. *Find the point $P[i_1]$ with minimal y-coordinate ($P[i_1]$ lies on CH).*

2. *Rotate a ray around $P[i_1]$ until the last point, say $P[i_2]$, of the set lies on it. (Obviously, $P[i_2]$ belongs to CH.)*

3. *Repeat this rotation for $P[i_2]$ and find $P[i_3]$, etc., until we return to the initial point $P[i_1]$.*

In the program, (a) the $(n + 1)$th element copies $P[i_1]$ and plays the role of the guard; (b) the final array L contains all given points $\mathbf{P}$, but the initial m elements of L are the vertices of CH; (c) global variables for convex layers are necessary in the program below.

```
> convex:=proc(P) local i,Y,r,j,C,B,a,T; global L,m,n;
  n:=nops(P); L:=array(1..n+1);
  for i from 1 to n do L[i]:=P[i] od;
  Y:=op(2,L[1]); a:=1; for i from 2 to n do
```

```
if op(2,L[i])<=Y then a:=i; Y:=op(2,L[i]) fi od;
L[n+1]:=L[a]; B:=[1,0]; for j from 1 to n+1 do
C:=-1; T:=L[a]; L[a]:=L[j]; L[j]:=T;
if a=n+1 or j=n+1 then m:=j; j:=n+1 fi;
for i from j+1 to n+1 do r:=L[i]-L[j];
if r[1]^2+r[2]^2>0 then Y:=evalf((r[1]*B[1]+r[2]*B[2])/
(sqrt(r[1]^2+r[2]^2)*sqrt(B[1]^2+B[2]^2)));
if Y>=C then C:=Y; a:=i fi fi od; B:=L[a]-L[j] od;
plot([seq(L[i],i=1..m)], thickness=2) end:
```

```
> N:=37: P:=[seq([rand(1..40)(),rand(1..40)()],j=1..N)]:
```

```
> PP:=plot(P, style=point,symbol=circle,color=blue):
plots[display]([PP, convex(P)]); # Fig. 16.1
```

Several problems are related to this, for example, finding the intersection of two convex polygons. Below we consider one problem with applications in statistics.

The *depth of a point A* in a finite planar set **P** is by definition the number of CH (or *convex layers*) that bound this point and that must be deleted before the point A can be deleted. The *depth of a finite planar set* is the number of CH (i.e., convex layers) obtained using the above procedure. For example, the depth of a triangle and its center (i.e., four points) is equal to two.

The following program uses the procedure `convex(P)` and returns the set of layers of the finite planar set **P**.

Program 16.0.1

```
> G.0:=plot(P, style=point,symbol=circle,color=blue):
G.1:=convex(P): for t from 2 while m<=n do
G.t:=convex([seq(L[i], i=m+1..n+1)]); k:=t od:
plots[display]([seq(G.i, i=0..k)]); # Fig. 16.2
```

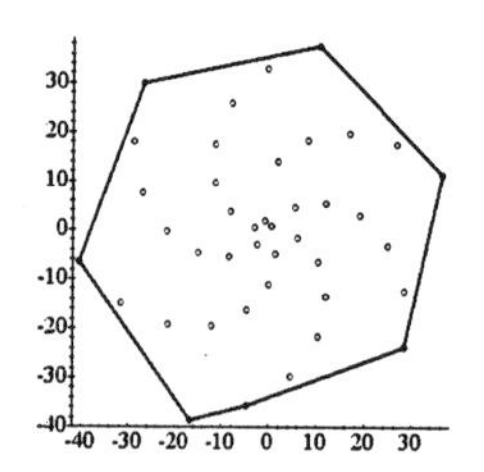

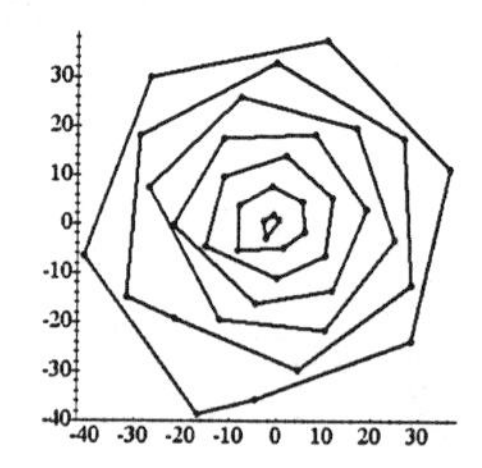

Figs. 16.1–16.2. The convex hull. The depth (layers) of the planar point set

Part III

Polyhedra with MAPLE

17
Regular Polyhedra

The studies of polyhedra with MAPLE are naturally organized in a sequence of themes: the notion of polyhedra, Platonic solids, regular star polyhedra, Archimedean solids, selected questions and problems.

MAPLE programs for plotting polyhedra have a similar structure: the list of vertices with their coordinates, the lists of vertices in faces, and the visualizing command (including rotation about coordinate axes).

In Section 17.1 we study basic definitions of polyhedra and describe simple programs for plotting the polygonal curves, prisms and pyramids.

In Section 17.2 we calculate and visualize Platonic solids (comparing them with MAPLE commands for Platonic solids) in rotation about coordinate axes. The reader can then experiment with more complicated polyhedra; putting objects through various transformations: rotation, animation, coloring, various operations (commands `cutin`, `cutout`, `stellate`), change of parameters in coordinates of vertices and faces, sections, intersections and other combinations of solids, obtaining real scenes with shadows and brightness, and discovering new polyhedra, and so on.

In Section 17.3 we recall star polygons and procedures. As a fun example of their use, we plot a (two-dimensional) Christmas tree. We then study star-shaped polyhedra: Poinsot's polyhedra and Kepler's octahedron.

In Chapter 17 the reader will become acquainted with the commands

`tetrahedron`, `octahedron`, `icosahedron`, `dodecahedron`, `hexahedron`, `polygonplot3d`, `stellate`, `POLYGONS`, `RegularPolygon`, `RegularStarPolygon`, `RegularPolyhedron`.

17.1 What Is a Polyhedron?

Becoming acquainted with polyhedra leaves a bright impression in the human mind, sometimes leading to creative work and discoveries. The importance of this theme in school geometry and for developing space imagination is well known.

The theory of polyhedra goes back to antiquity: Volume XII of Euclid's fundamental work *Elements* ... is devoted to the five regular polyhedra. Archimedes, in his work *On polyhedra*, described all the so-called semi-regular polyhedra.

A closed polygonal line without self-intersections that bounds a plane domain is called a *simple polygon*. A simple polygon is said to be *convex*, if for each side of the polygon, all the vertices that do not lie on the side, belong to one of the two half-planes defined by this side.

Definition 17.1.1 A *solid angle* is a conic surface whose directrix is a simple polygon.

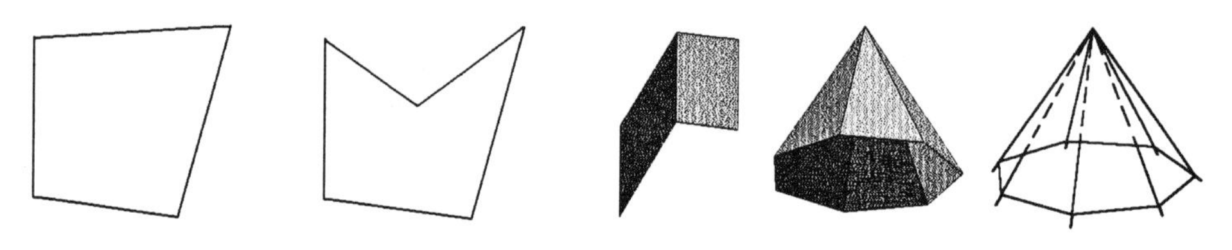

Figs. 17.1–17.5.
Convex quadrangle and non-convex pentagon. Adjacent faces. Solid angle.

Definition 17.1.2 A *polyhedron* is a figure in space that consists of a finite number of planar polygons (*faces* of the polyhedron) such that

(1) Each side of any of these faces meets exactly one other face (said to be *adjacent* to the first one) along a side.

(2) For faces α, β one can find a sequence of faces $\alpha_1, \alpha_2, \dots, \alpha_n$ such that face α is adjacent to α_1; face α_1 is adjacent to α_2; ...; and face α_n is adjacent to β.

(3) Let V be any vertex and let $F_1, F_2, \dots, F_n$ be the n faces that meet at V. It is possible to travel over the polygons F_i from one to another without passing through V.

Simple examples of polyhedra are a *prism* (in particular, the cube) and a *pyramid* (in particular, the tetrahedron).

If all faces are simple polygons, then by Definition 17.1.2, a polyhedron is a *surface* (see Section 19.1) in space. Moreover, condition (1) means that this surface is without a *boundary* and has singular edges, as in Figs. 19.4, 20.2, and 20.18. Condition (2) means that it is *connected* (i.e., has exactly one component). Condition (3) excludes singular vertices; see Figs. 19.15, 19.40, and 20.13.

To exclude self-intersecting polyhedra we give the following definitions:

Definition 17.1.3 A polyhedron is called *simple* if

- All its faces are simple polygons.
- Any two non-adjacent faces have no (interior or boundary) points in common except perhaps for a single vertex.
- Any two adjacent faces have only one common edge and do not have any other common points.

A simple polyhedron divides space into two regions, a finite one, called *the interior*, and another, which contains infinite straight lines, called *the exterior*.

Even Definition 17.1.3 does not exclude polyhedra (solids) with *tunnels*.

Definition 17.1.4 A closed curve in a surface is called *non-separating* if the surface remains in one piece when the curve is cut.

Some polyhedra and surfaces possess non-separating curves (see Figs. 19.3, 19.7, and 21.7); in others (*simply connected*), every curve separates the surface or polyhedron into two pieces (see Figs. 20.7, 20.8, and 20.11).

Non-separating curves (consisting of edges) in a polyhedron correspond to tunnels through a solid. One can define the number of tunnels as the maximum number of cuts (curves) which leave the polyhedron connected.

Example 17.1.1 Let us plot with MAPLE a plane triangle (in a similar way we may plot any polygon with arbitrary vertices) and a regular simple n-gon (inscribed in a circle).

```
> restart: # triangle in the plane, Fig. 17.6
> plots[polygonplot]([[0,0],[2,4],[5,3]], color=blue);
> n:=6: a:=i-> [cos(2*Pi*i/n),sin(2*Pi*i/n)]; # hexagon
> plots[polygonplot]([seq(a(i), i=1..n)]); # first method
> p1:=plot([seq(a(i), i=1..n+1)], thickness=2):
  p2:=plot([cos(t),sin(t), t=0..2*Pi], linestyle=2):
  plots[display]({p1,p2}); # circle and n-gon, Fig. 17.7
```

Then we plot a regular pyramid and a prism of height 2. Similarly, using the vector $[x_0, y_0, z_0]$, we can plot an oblique pyramid and a prism; see the program. Note that the tetrahedron and the cube are particular cases of a regular pyramid and a prism whose heights have special values. In Section 17.2 and Chapter 18 we plot these and other regular convex polyhedra by different methods and programs.

```
> n:=6: # pyramid with vertex S(0, 0, 2), Fig. 17.8
> a:=i -> [cos(2*Pi*i/n),sin(2*Pi*i/n),0]:
  for i from 1 to n do g[i]:=[a(i), a(i+1), [0,0,2]] od:
  g[n+1]:=[seq(a(i), i=1..n)];
```

$$g_7 := [\,[\tfrac{1}{2},\, \tfrac{1}{2}\sqrt{3},\, 0],\, [\tfrac{-1}{2},\, \tfrac{1}{2}\sqrt{3},\, 0],$$
$$[-1,\, 0,\, 0],\, [\tfrac{-1}{2},\, -\tfrac{1}{2}\sqrt{3},\, 0],\, [\tfrac{1}{2},\, -\tfrac{1}{2}\sqrt{3},\, 0],\, [1,\, 0,\, 0]\,]$$

```
> P:=j -> plots[polygonplot3d]([seq(g[i], i=1..n+1)],
  orientation=[20*j,50], style=PATCH):
> plots[display]([seq(P(j), j=1..18)], insequence=true);
```

```
> n:=6: # prism with vertical edge [0, 0, 2], Fig. 17.9
> a:=i -> [cos(2*Pi*i/n), sin(2*Pi*i/n), 0]:
  for i from 1 to n do
  g[i]:=[a(i),a(i+1),a(i+1)+[0,0,2],a(i)+[0,0,2]] od:
  g[n+1]:=[seq(a(i), i=1..n)]:
  g[n+2]:=[seq(a(i)+[0,0,2], i=1..n)]:
> P:=j -> plots[polygonplot3d]([seq(g[i], i=1..n+2)],
  orientation=[20*j,40], style=PATCH):
> plots[display]([seq(P(j), j=1..18)], insequence=true);
```

For plotting the pyramid and the prism without rotation, instead of the last two commands, one can use the command

```
> plots[polygonplot3d]([seq(g[i], i=1..n+1)], style=PATCH);
```

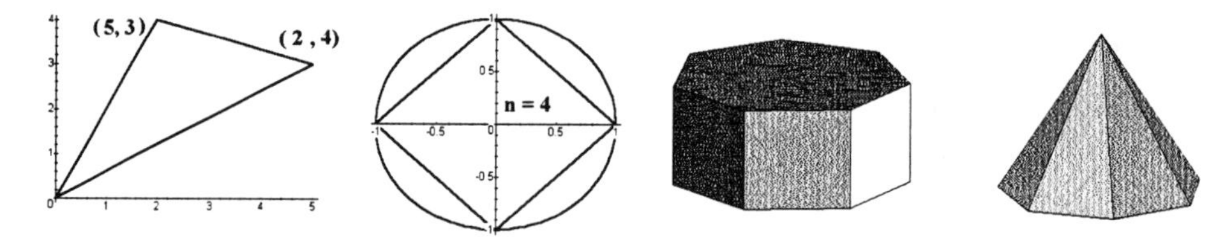

Figs. 17.6–17.9. Planar polygons. Prism and pyramid

Definition 17.1.5 The *development of a polyhedron* is the union of plane polygons according to a rule stating how one should glue them together along their sides and vertices to obtain the desired polyhedron. Moreover, the following conditions must be satisfied:

• Each side of the polygon is attached to at most one side of another polygon.

• Any two glued sides must have the same length.

• One can reach any polygon by starting from any other polygon and moving along a sequence of glued polygons.

One can apply the notion of development to curved surfaces (for example in descriptive geometry and drawing).

In particular, the development of the side of a cone is a circular sector.

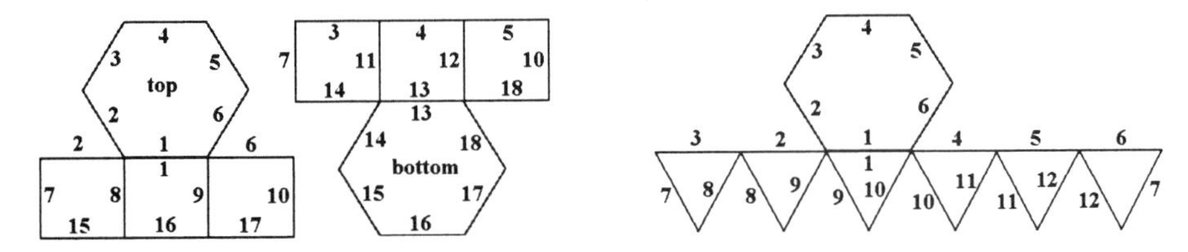

Figs. 17.10–17.11. Development of 6-faced prism and pyramid

Definition 17.1.6 A polyhedron is said to be *convex* if all vertices that do not belong to an arbitrary face of this polyhedron lie in one of the half-spaces defined by this face.

The faces of such a polyhedron are *convex polygons*. The interior of a convex polyhedron (CP) is a *convex body*.

The relation between the numbers of faces F, edges E and vertices V of a CP is given by *Euler's theorem* $F - E + V = 2$. In the case of a simple polyhedron M, the term $F - E + V$ is called the *Euler characteristic* and is denoted by $\chi(M)$. The integer $\chi(M) - 2$ is called the *genus of the polyhedron*.

The following theorems are important in the theory of polyhedra.

Theorem 17.1.1 (rigidity, A. Cauchy) *If two CPs are isometric to each other, then the second can be obtained from the first by a motion in* $\mathbb{R}^3$ *(of first or second order).*

Theorem 17.1.2 (existence, A. Alexandrov) *Necessary and sufficient conditions for gluing CP starting from the development of a polyhedron (containing only convex polygons) are the following:*

- *Euler's condition* $F - E + V = 2$;
- *the sum of the plane angles glued at any vertex is not more than* $360°$.

Theorem 17.1.3 (G. Minkowski)

(a) There exist CPs with arbitrary area of faces and arbitrary directions of the extrinsic normals to them, but the sum of the vectors in the directions of these normals with lengths equal to the area of corresponding faces must be zero, and not all the vectors are coplanar.

(b) A CP is completely defined by the area of its faces and directions of the extrinsic normals to the faces.

A *lattice of edges* of a CP is a graph consisting of its edges; such a lattice can be obtained in the plane, for instance, by projection from any extrinsic point near the center of an arbitrary face. Two polyhedra belong to the same type if their lattices of edges are topologically equal.

Theorem 17.1.4 (E. Steinitz) *There exists a CP with any lattice of its edges.*

17.2 Platonic Solids

Definition 17.2.1 A *regular polyhedron* is a polyhedron whose faces are regular polygons each equal to the other, with all solid angles equal.

As we can see from calculating the sum of the plane angles at a vertex, there are at most five regular convex CPs. These polyhedra, the so-called *Platonic solids — tetrahedron, cube, octahedron, dodecahedron, icosahedron—* were known in antiquity, and their existence was proved by Euclid.

Example 17.2.1 `MAPLE` has special commands for plotting *Platonic solids*; see the program below. (In `MAPLE`, Release 5 one can define a regular polyhedron by using the command `RegularPolyhedron(gon,[m,n],o,r)` where gon is the name of the polyhedron to be defined, `[m,n]` the Schlafli symbol, `o` the center of the polyhedron.) Later we give special programs for them, which are used for plotting more complicated polyhedra.

```
> restart: with(plottools): # Figs. 17.12–17.16
> f:=hexahedron([0,0,0], .8):plots[display](f,style=patch);
> f:=tetrahedron([0,0,0], .8): plots[display](f);
> f:=octahedron([0,0,0], .8): plots[display](f);
> f:=icosahedron([0,0,0], .8): plots[display](f);
> f := dodecahedron([0,0,0],0.8): plots[display](f);
```

Each of these polyhedra can be rotated using the commands

```
> P:=i ->display(f, style=patch, orientation=[10*i, 50],
  light=[75,75, 1,.9,.2], ambientlight=[.7, .7, .8]):
  plots[display]([seq(P(i), i=1..36)], insequence=true);
```

The other four *Platonic solids* can be constructed using the cube.

• *The cube and octahedron are dual, i.e., one can be obtained from the other if the centers of the faces of the first are considered the vertices of the second. Analogously, the dodecahedron and icosahedron are dual.*

• *Any four vertices of a cube taken in pairs, not adjacent along an edge, can be assumed to be the vertices of a tetrahedron.*

• *The icosahedron can be inscribed by a special method in a cube if its opposite edges are fixed in pairs on parallel faces of the cube by a special method.*

• *The dodecahedron is obtained from the cube by building roofs over its faces (method of Euclid).*

The above polyhedra admit a number of *symmetries* (self-transformation by a rigid motion of space): cyclic transposition of vertices on any face and cyclic transposition of faces that meet at any vertex.

A polyhedron is denoted by the *symbol* $\{p, q\}$, or p^q if every vertex is surrounded by q p-polygons.

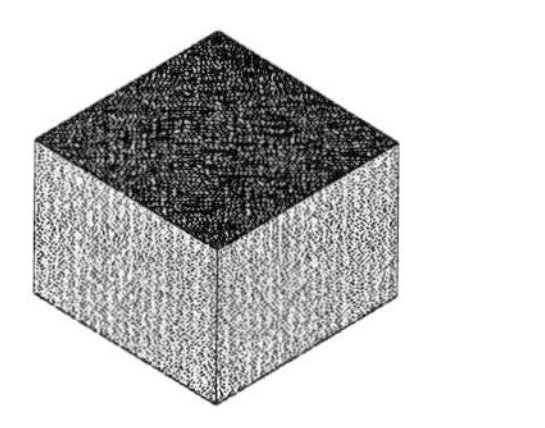
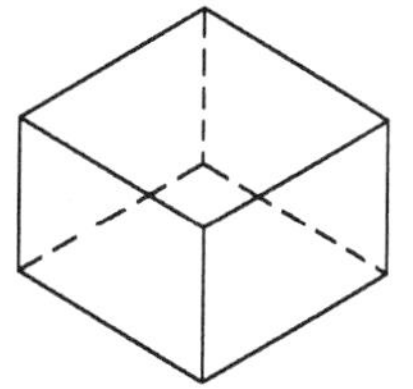

Fig. 17.12. Cube

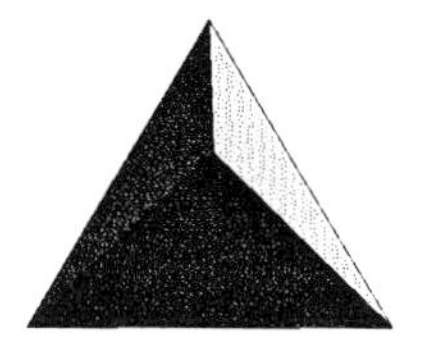
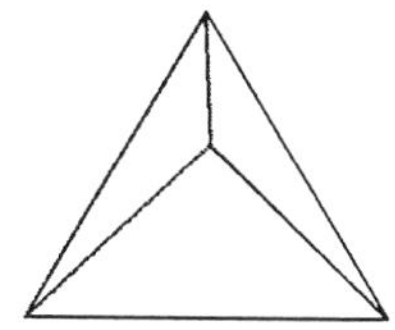
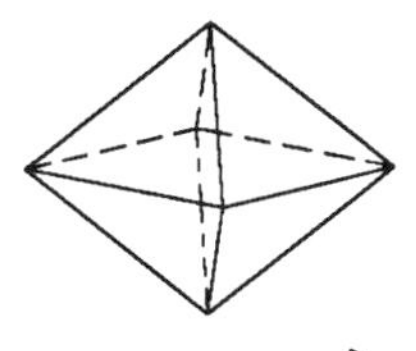
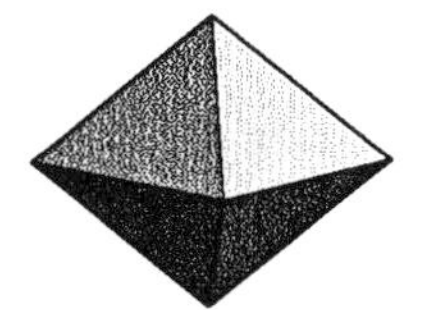

Figs. 17.13–17.14. Tetrahedron and octahedron

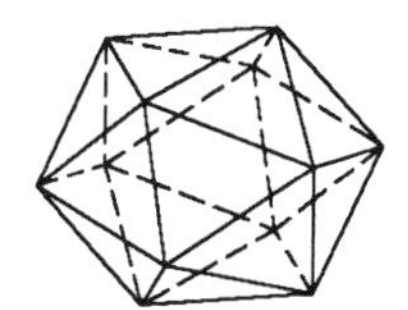
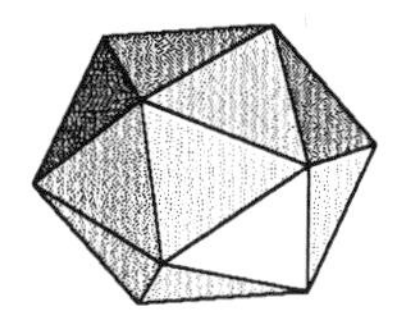
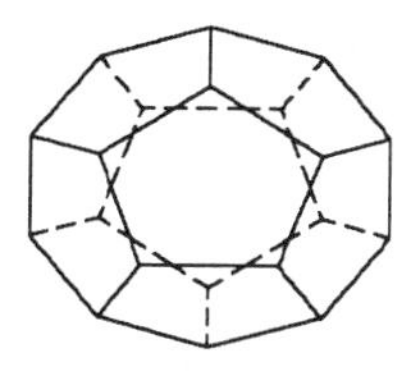
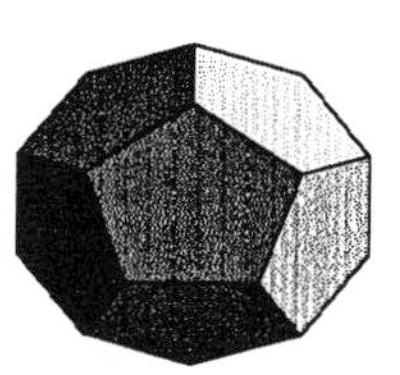

Figs. 17.15–17.16. Icosahedron and dodecahedron

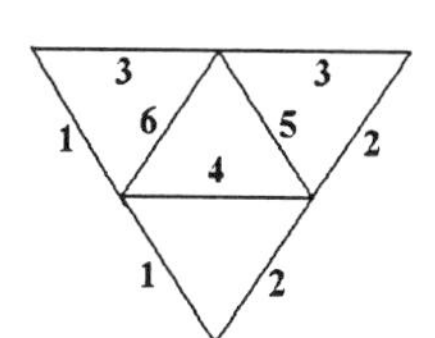

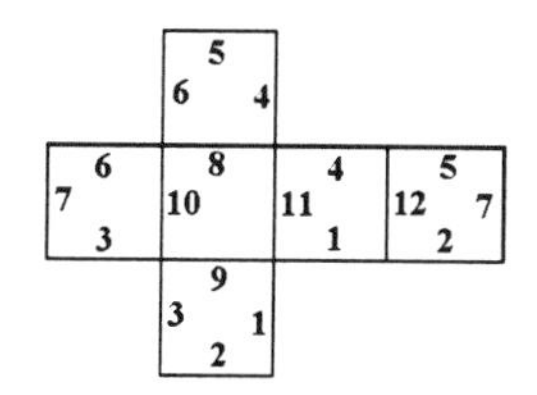

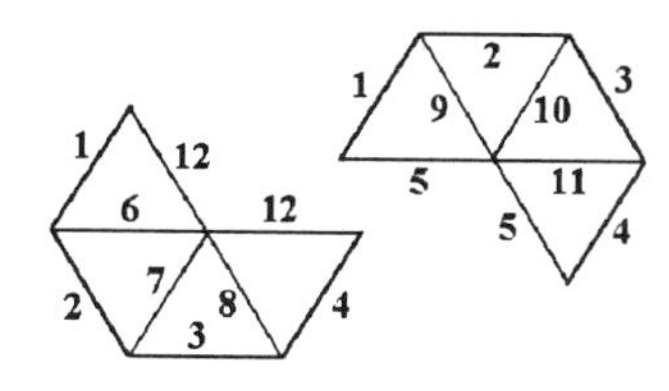

Figs. 17.17–17.18. Development of a tetrahedron, cube, and octahedron

The radii R, r of the circumscribed and inscribed spheres, and the volume of each regular CP with edge of length a are as follows:

tetrahedron $\{3, 3\}$: $R = a\frac{\sqrt{6}}{4}, r = a\frac{\sqrt{6}}{12}, V = a^3\frac{\sqrt{2}}{12},$

cube $\{4,3\}$: $R = a\frac{\sqrt{3}}{2}, r = \frac{1}{2}a, V = a^3,$

octahedron $\{4,4\}$: $R = a\frac{\sqrt{2}}{2}, r = a\frac{\sqrt{6}}{6}, V = a^3\frac{\sqrt{2}}{3},$

dodecahedron $\{5,3\}$:

$$R = \tfrac{a}{4}\sqrt{18+6\sqrt{5}},\ r = \tfrac{a}{2}\sqrt{(25+11\sqrt{5})/10},\ V = \tfrac{1}{4}a^3(15+7\sqrt{5}),$$

icosahedron $\{3,5\}$: $R = \frac{a}{4}\sqrt{10+2\sqrt{5}},\ r = \frac{a\sqrt{3}}{12}(3+\sqrt{5}),\ V = \frac{5}{12}a^3(3+\sqrt{5}).$

Consider the **cube** with edge $2c$ and the following vertices:

$A_1(c,c,c),\ A_2(-c,c,c),\ A_3(-c,-c,c),\ A_4(c,-c,c),$
$A_5(c,c,-c),\ A_6(-c,c,-c),\ A_7(-c,-c,-c),\ A_8(c,-c,-c).$

Let us find vertices of other regular CP using their relation to the cube.

Tetrahedron. We take diagonals of four faces of a cube in the role of edges of a tetrahedron. Then the vertices of the tetrahedron are the following:

$A_1(c,c,c),\ A_3(-c,-c,c),\ A_4(c,-c,c),\ A_8(c,-c,-c).$

The last four vertices of a cube also define a tetrahedron *dual* to the first one.

Octahedron. We connect the centers of the faces of a cube by line segments and obtain an octahedron with the following vertices:

$O_1(c,0,0),\ O_2(0,c,0),\ O_3(-c,0,c),\ O_4(0,-c,0),\ O_5(0,0,c),\ O_6(0,0,-c).$

Recall that the *golden section* (or *golden ratio*) is the division of a line segment AB by a point C, such that the ratio of the smaller part to the larger one is equal to the ratio of the larger part to the whole segment: $CB : AC = AC : AB$. Assuming $AB = a,\ AC = a\cdot\tau$, we have the equation $\frac{1-\tau}{\tau} = \frac{\tau}{1}$, from which it follows that $\tau^2+\tau-1=0$. The positive root of this equation $\tau = \frac{\sqrt{5}-1}{2} \approx 0.61803$ is the *golden section*.

Euclid made use of the golden section for plotting regular *pentagons* and *hexagons* and also for plotting the *icosahedron* and *dodecahedron*.

Icosahedron. We arrange six equal line segments with ratio of lengths to the edge of the cube equal to τ (see the first exercise below) on the faces of a cube in such a way that each segment is symmetric to itself with respect to the center of a certain face and parallel to its two opposite sides, and such that the line segments in neighboring faces are mutually orthogonal. Then we connect the endpoints of these line segments in a certain order and obtain an icosahedron inscribed in the cube. Hence our icosahedron has the following vertices (where $p = c\cdot\tau$):

$B_1(p,0,c),\ B_2(-p,0,c),\ B_3(p,0,-c),\ B_4(-p,0,-c),$
$B_5(c,-p,0),\ B_6(c,p,0),\ B_7(-c,-p,0),\ B_8(-c,p,0),$

$B_9(0, c, p)$, $B_{10}(0, c, -p)$, $B_{11}(0, -c, p)$, $B_{12}(0, -c, -p)$,

Dodecahedron. As in the case of the icosahedron, we choose six line segments of length $p = c\,\tau$ on the faces of a cube. Then we make a parallel translation of the distance p (see the second exercise below) of each of these segments in the direction of the extrinsic normal to the face. After connecting the twelve endpoints of the line segments so defined in a certain order with the eight vertices of the cube, we obtain a regular dodecahedron. Its vertices, in addition to the eight vertices of the cube, are the following:

$D_1(p, 0, c+p)$, $D_2(-p, 0, c+p)$, $D_3(p, 0, -c-p)$, $D_4(-p, 0, -c-p)$,
$D_5(c+p, -p, 0)$, $D_6(c+p, p, 0)$, $D_7(-c-p, -p, 0)$, $D_8(-c-p, p, 0)$,
$D_9(0, c+p, p)$, $D_{10}(0, c+p, -p)$, $D_{11}(0, -c-p, p)$, $D_{12}(0, -c-p, -p)$.

Example 17.2.2 Let us plot the image of a cube, tetrahedron, and octahedron using the coordinates of their vertices by the following program.

```
> restart:# hexahedron, Fig. 17.12
> h.1:=[[-1,-1,-1],[1,-1,-1],[1,-1,1],[-1,-1,1]]:
  h.2:=[[1,1,-1],[-1,1,-1],[-1,1,1],[1,1,1]]:
  h.3:=[[-1,-1,-1],[1,-1,-1],[1,1,-1],[-1,1,-1]]:
  h.4:=[[-1,-1,1],[1,-1,1],[1,1,1],[-1,1,1]]:
  h.5:=[[1,-1,-1],[1,1,-1],[1,1,1],[1,-1,1]]:
  h.6:=[[-1,-1,-1],[-1,1,-1],[-1,1,1],[-1,-1,1]]:
> P:=j -> plots[polygonplot3d]([seq(h.i, i=1..6)],
  orientation=[20*j, 50], style=PATCH):
> plots[display]([seq(P(j), j=1..18)], insequence=true);
```

```
> restart: # tetrahedron, dual tetrahedrons, Figs. 17.13, 17.33
> t1:=[[1,1,-1],[1,-1,1],[-1,1,1]]:
  t2:=[[1,1,-1],[1,-1,1],[-1,-1,-1]]:
  t3:=[[-1,-1,-1],[1,1,-1],[-1,1,1]]:
  t4:=[[-1,-1,-1],[1,-1,1],[-1,1,1]]:
> plots[polygonplot3d]({-t1,-t2,-t3,-t4}); # tetrahedron,
> P:=j ->plots[polygonplot3d]([t1,t2,t3,t4,-t1,-t2,-t3,-t4],
  orientation=[20*j,50], style=PATCH):
> plots[display]([seq(P(j), j=1..18)], insequence=true);
```

```
> restart: g[1]:=[[1,1,0],[-1,1,0],[0,0,1]]: # octahedron,
  g[2]:=[[1,1,0],[1,-1,0],[0,0,1]]:                # Fig. 17.14
  g[3]:=[[-1,-1,0],[1,-1,0],[0,0,1]]:
  g[4]:=[[-1,-1,0],[-1,1,0],[0,0,1]]:
```

```
  g[5]:=[[1,1,0],[-1,1,0],[0,0,-1]]:
  g[6]:=[[1,1,0],[1,-1,0],[0,0,-1]]:
  g[7]:=[[-1,-1,0],[1,-1,0],[0,0,-1]]:
  g[8]:=[[-1,-1,0],[-1,1,0],[0,0,1]]:
> P:=j -> plots[polygonplot3d]([seq(g[i], i=1..8)],
  orientation=[20*j, 50], style=PATCH):
> plots[display]([seq(P(j), j=1..18)], insequence=true);
```

Example 17.2.3 Plotting the icosahedron and dodecahedron using the coordinates of their vertices is based on some calculations (see Exercises 1, 2), and it can be done with the following program. Here for the icosahedron we directly describe its faces, but for the dodecahedron, we first describe all its vertices, and then give sequences of vertices for each face.

```
> restart: c:=1: p:=evalf(c*(sqrt(5)-1)/2): # icosahedron,
> g[1]:=[[p,0,c],[-p,0,c],[0,c,p]]:         # Fig. 17.15
  g[2]:=[[p,0,c],[-p,0,c],[0,-c,p]]:
  g[3]:=[[-p,0,c],[0,c,p],[-c,p,0]]:
  g[4]:=[[-p,0,c],[-c,p,0],[-c,-p,0]]:
  g[5]:=[[p,0,c],[c,p,0],[0,c,p]]:
  g[6]:=[[p,0,c],[c,-p,0],[c,p,0]]:
  g[7]:=[[c,p,0],[c,-p,0],[p,0,-c]]:
  g[8]:=[[c,p,0],[0,c,-p],[0,c,p]]:
  g[9]:=[[c,p,0],[0,c,-p],[p,0,-c]]:
  g[10]:=[[0,c,p],[0,c,-p],[-c,p,0]]:
  g[11]:=[[0,c,-p],[-c,p,0],[0,c,p]]:
  g[12]:=[[p,0,-c],[-p,0,-c],[0,c,-p]]:
  g[13]:=[[p,0,c],[c,-p,0],[0,-c,p]]:
  g[14]:=[[0,-c,p],[-p,0,c],[-c,-p,0]]:
  g[15]:=[[0,-c,p],[0,-c,-p],[-c,-p,0]]:
  g[16]:=[[p,0,-c],[-p,0,-c],[0,-c,-p]]:
  g[17]:=[[0,-c,-p],[-p,0,-c],[-c,-p,0]]:
  g[18]:=[[-p,0,-c],[-c,p,0],[-c,-p,0]]:
  g[19]:=[[c,-p,0],[0,-c,-p],[p,0,-c]]:
  g[20]:=([[0,-c,p],[0,-c,-p],[c,-p,0]]):
> P:=j -> plots[polygonplot3d]([seq(g[i], i=1..20)],
  orientation=[20*j,50], style=PATCH):
> plots[display]([seq(P(j), j=1..18)], insequence=true);
```

```
> restart: c:=1: p:=evalf(c*(sqrt(5)-1)/2): # dodecahedron,
> a[1]:=[p,0,c+p]: a[2]:=[-p,0,c+p]:         # Fig. 17.16
```

```
  a[3]:=[p,0,-c-p]: a[4]:=[-p,0,-c-p]:
  a[5]:=[c+p,-p,0]: a[6]:=[c+p,p,0]:
  a[7]:=[-c-p,-p,0]: a[8]:=[-c-p,p,0]:
  a[9]:=[0,c+p,p]: a[10]:=[0,c+p,-p]:
  a[11]:=[0,-c-p,p]: a[12]:=[0,-c-p,-p]:
  a[13]:=[c,c,c]: a[14]:=[c,c,-c]:
  a[15]:=[c,-c,c]: a[16]:=[c,-c,-c]:
  a[17]:=[-c,c,c]: a[18]:=[-c,c,-c]:
  a[19]:=[-c,-c,c]: a[20]:=[-c,-c,-c]:
> g[1]:=[a[2],a[17],a[8],a[7],a[19]]:
  g[2]:=[a[1],a[13],a[9],a[17],a[2]]:
  g[3]:=[a[17],a[9],a[10],a[18],a[8]]:
  g[4]:=[a[8],a[18],a[4],a[20],a[7]]:
  g[5]:=[a[7],a[20],a[12],a[11],a[19]]:
  g[6]:=[a[19],a[11],a[15],a[1],a[2]]:
  g[7]:=[a[6],a[5],a[15],a[1],a[13]]:
  g[8]:=[a[6],a[13],a[9],a[10],a[14]]:
  g[9]:=[a[14],a[10],a[18],a[4],a[3]]:
  g[10]:=[a[3],a[4],a[20],a[12],a[16]]:
  g[11]:=[a[16],a[12],a[11],a[15],a[5]]:
  g[12]:=[a[5],a[6],a[14],a[3],a[16]]:
> P:=j -> plots[polygonplot3d]([seq(g[i], i=1..12)],
  orientation=[20*j,50], style=PATCH):
> plots[display]([seq(P(j), j=1..18)], insequence=true);
```

There are other polyhedra besides the Platonic solids that are bounded by equal regular faces. In fact they are bounded by triangles, and together with the tetrahedron, the octahedron and the icosahedron they form the family of **8** (convex) *deltahedra* (see figures in [Cro, p. 76]).

Exercise 17.2.1 **1**. Find the length of the line segment on a face of the cube that we use for plotting the icosahedron.

2. Find the height and an edge of the roof over the face of the cube that we use for plotting the dodecahedron. The resulting pentagons must be planar and regular.

3. Prove the formulas for the radii of the inscribed and circumscribed spheres and for the volume of the *Platonic solids* as functions of their edges.

4. Write a program for plotting five non-regular deltahedra. (Two of them are triangular and pentagonal bipyramids).

17.3 Star-Shaped Polyhedra

Aside from the regular simple polygons there also exist *regular star* (or simply *star*) *polygons*: one for $n = 5$ (pentagram), none for $n = 6$ (non-simple cases decompose into two triangles), two for $n = 7$, one for $n = 8$ (other, non-simple, cases decompose into two squares), etc.

It is easy to show that n sides of a regular star (m, n)-gon are diagonals of a simple regular n-gon that define the same number m of sides.

If m is not relatively prime to n, then the star polygon has more than one component. For each natural number n, there exists as many distinct (not similar to each other) *disconnected* regular star n-gons as there are integers between 1 and $\frac{n}{2}$ relatively prime to n, i.e., $\frac{1}{2}\varphi(n)$, where $\varphi(n)$ is known from the theory of numbers as *Euler's φ-function.*

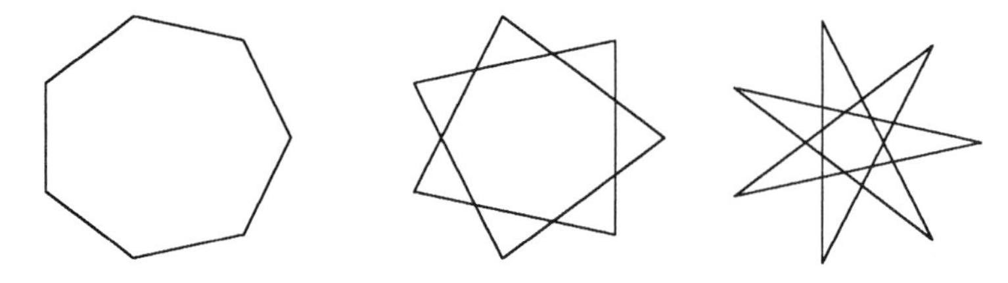

Figs. 17.19–17.21. Three regular heptagons

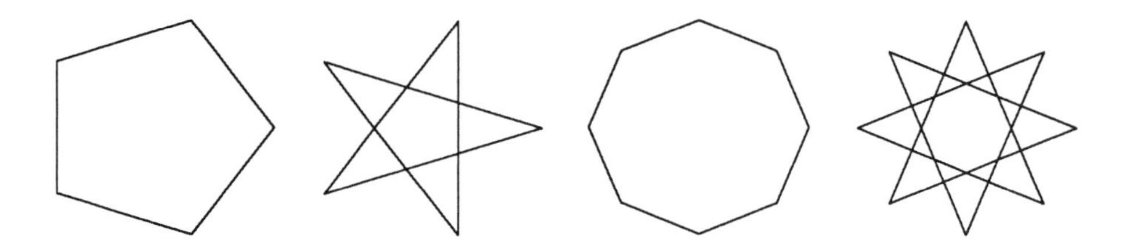

Figs. 17.22–17.25. Pairs of regular pentagons and octagons

Example 17.3.1 Let us plot a regular star polygon (inscribed in a circle) using the program for a simple regular polygon, Figs. 17.22-17.25. The commands `RegularPolygon(p,n,cen,rad)` and `RegularStarPolygon(p,n,cen,rad)` from the `geometry` package in `Maple V` Release 5 define regular (star) polygons.)

```
> n:=5: m:=2: # star pentagon, 2 methods
> a:=i -> [cos(2*Pi*i*m/n), sin(2*Pi*i*m/n)]:
> plots[polygonplot]([seq(a(i), i=1..n)]); # first method
> p1:=plot([cos(t),sin(t),t=0..2*Pi]): # second method
  p2:=plot([seq(a(i),i=1..n+1)]): plots[display]({p1,p2});
```

We plot five star-shaped polyhedra (SPs) over *Platonic solids* using `stellate` (stellar means starlike) with several values of the height parameter h less or greater than 1. For particular values of the parameter h (derive them) these polyhedra can be seen below as regular SPs.

```
> with(plottools): with(plots): # Figs. 17.26–17.30
  f:=display(stellate(tetrahedron(),5)): %;
  f:=display(stellate(hexahedron(),3)): %;
  f:=display(stellate(octahedron(),2)): %;
  f:=display(stellate(icosahedron(),2)): %;
  f:=display(stellate(dodecahedron(),2)): %;
```

Each of these polyhedra f can be rotated using the commands

```
> P:=i -> display(f, style=patch, orientation=[10*i, 50],
  light=[75,75, 1,.9,.2], ambientlight=[.7, .7, .8]):
  plots[display]([seq(P(i), i=1..36)], insequence=true);
```

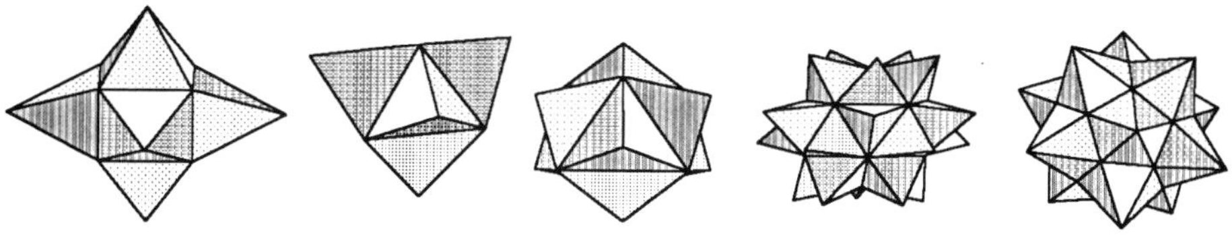

Figs. 17.26–17.30. Stars over Platonic solids

Example 17.3.2 Let us use polygons to plot a Christmas tree, Fig. 17.31. Generalize this program to the three-dimensional case.

```
> toy1:=proc(x,y,R) local P;  P.1:=plot([[x,y],[x,y-.3]]);
  P.2:=plottools[circle]([x,y-.3-R],R,color=blue,thickness
  =2); plots[display]([P.1,P.2],scaling=constrained) end:
> toy2:=proc(x,y,n,R) local P; P.1:=plot([[x,y],[x,y-.3]]);
  P.2:=plot([seq([x+R*sin(2*Pi*i/n), y+R*cos(2*Pi*i/n)
  -0.3-R], i=0..n)], thickness=2,axes=none):
  P.3:=plottools[circle]([x,y-.5],.2,color=green,thickness
  =2); plots[display]([P.1,P.2],scaling=constrained) end:
  ...
> star:=proc(x,y,n,m,R) local i,a; for i from 0 to n+1 do
  a[i]:=[x+R*cos(2*Pi*i*m/n),y+R*sin(2*Pi*i*m/n)] od:
  plot([seq(a[i],i=0..n)], thickness=2, color=red) end:
> tree:=proc() local S,Q,L,M,k,i,t,s;
  k:=4; s:=8; for i from 1 to s do t[i]:=4-i;
  S[i]:=plot([[0,t[i]],[k,t[i]-2]],color=green);
  Q[i]:=plot([[k,t[i]-2],[0,t[i]-1]],color=green);
  L[i]:=plot([[0,t[i]],[-k,t[i]-2]],color=green);
  M[i]:=plot([[-k,t[i]-2],[0,t[i]-1]],color=green) od;
```

```
plots[display]([seq(S[i],i=1..s),seq(Q[i],i=1..s),
seq(L[i],i=1..s),seq(M[i],i=1..s)],thickness=3) end:
> y:=-4.3: d:=.25: box:=plot([[-d,y],[d,y],[d,y-2],
[-d,y-2],[-d,y]], thickness=2,color=brown):
> k := 4: plots[display](star(0,3.4,7,2,.3),box,tree(),
seq([toy1(k+.1,2-i,.2),toy1(-k-.1,2-i,.2)],i=1..7),
seq([toy2(evalf(k/2+(-1)^i/2),2-i,3,.2),          # Christmas
toy2(evalf(-k/2+(-1)^i/2),2-i,4,.2)],i=1..6));   # tree
```

Impossible figures are those that seem to be projections of three-dimensional figures but actually do not correspond to any space objects. An example with stairs is shown in Fig. 17.32.

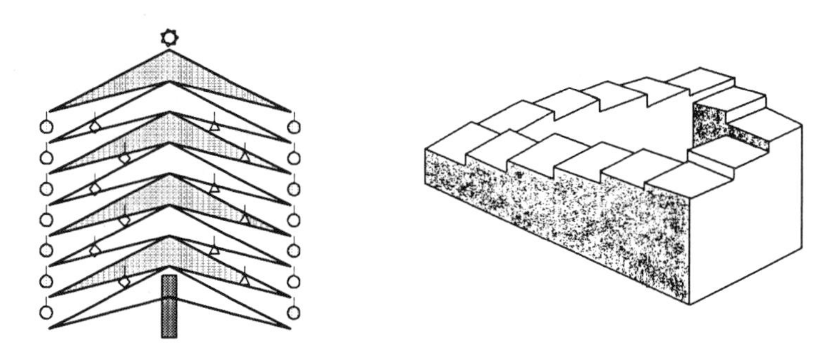

Figs. 17.31–17.32. Christmas tree. Impossible stairs

Aside from the *regular convex solid angles* there also exist *regular star-shaped solid angles*; in both cases their planar angles are equal to each other; the solid angles are equal as well.

The regular star 5-faced angle is seen in Fig. 17.36. Let O be the center of the star pentagon $A_1A_2A_3A_4A_5$. Plot the line OB perpendicular to the plane of pentagon. Then five plane angles A_1BA_2, A_2BA_3, A_3BA_4, A_4BA_5, and A_5BA_1 form a star pentagonal solid.

In the 17th century J. Kepler found two regular SPs. In 1810 L. Poinsot proved the existence of four such SPs, now called *Poinsot's polyhedra,* Figs. 17.35–17.38:

• The *small star dodecahedron* has 12 faces (regular star pentagons), 30 edges, 12 vertices.

• The *large star dodecahedron* has 12 faces (regular star pentagons), 30 edges, 20 vertices.

• The *large icosahedron* has 12 triangular faces, 30 edges, 12 vertices.

• The *large dodecahedron* has 12 faces (regular simple pentagons), 30 edges, 12 vertices.

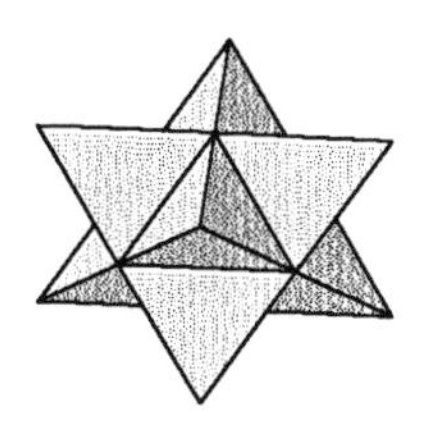 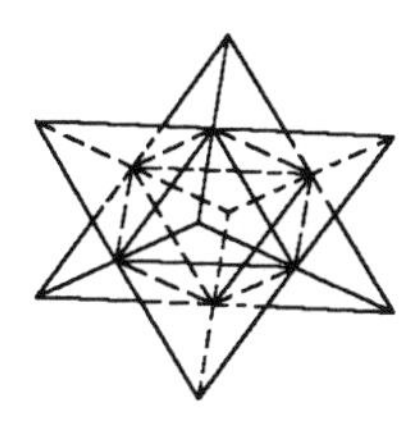 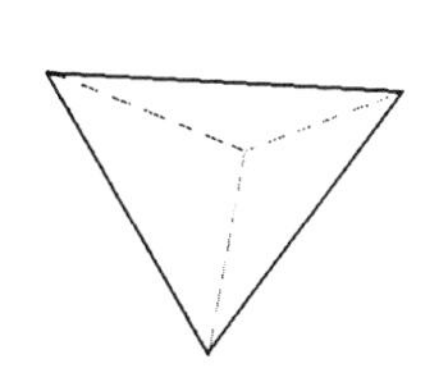 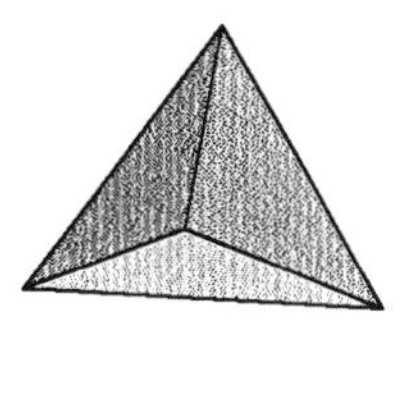

Figs. 17.33–17.34. The star octahedron consists of two tetrahedra

 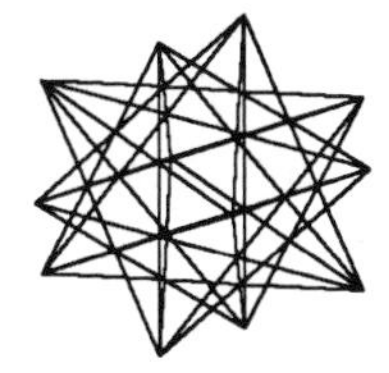 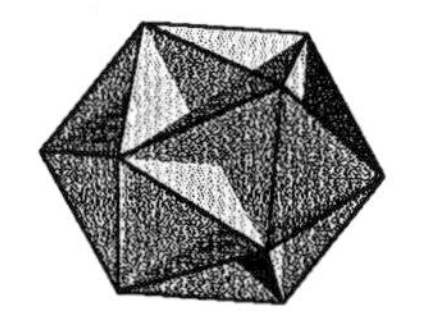 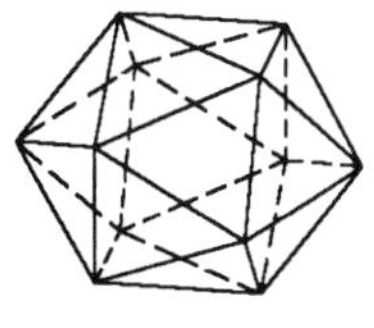

Figs. 17.35–17.36. Small and large star dodecahedra

In 1812 A. Cauchy proved that the set of all regular polyhedra contains only the five *Platonic solids*, the four *Poinsot bodies*, and the *star Kepler octahedron*; Fig. 17.33. The last polyhedron has 8 faces and decomposes into two tetrahedra.

The faces of an SP are either star polygons or mutually intersect in space. The supporting planes of the faces of a regular SP M divide space into some number of convex domains, one of which, the *kernel* of the polyhedron M, is a simple regular polyhedron M_0. The polyhedron M_0 has the same number of faces as M.

The *star octahedron* is obtained from the octahedron by continuing each of its faces until it intersects with three non-neighboring non-parallel faces. This polyhedron *decomposes into two tetrahedra*, "dual" to each other inscribed in a cube. The program for plotting a star octahedron is a short continuation of the program for the tetrahedron; see Example 17.2.2.

Continuing the edges of a dodecahedron, i.e., replacing each face α by a star pentagon with kernel α, leads to the *small star dodecahedron*. The vertices of the new face are the images of the vertices of the kernel face under homothety from its center with the coefficient $k = -(1 + \frac{2}{\sqrt{3}}\tau)$; see Exercise 17.3.1. In the program, which is a simple continuation of the one for the dodecahedron, we assemble each star-shaped face from five triangles but without including its kernel, which is a simple pentagon.

Continuing each face of the dodecahedron before its intersection with five non-neighbors and non-parallel faces leads to two possible cases.

• If we consider *simple* pentagons in the role of new faces, then we obtain the *large dodecahedron*. The `MAPLE` program given below is a short continuation of the program for the dodecahedron.

• If we consider *star* pentagons in the role of new faces, then we obtain the *large star dodecahedron*. Our MAPLE program is based on results of the previous program for the small star dodecahedron, using homothety with negative coefficient $k = -(1+\frac{2}{\sqrt{3}}\tau)$; see Exercise 17.3.1. Here we again assemble each star-shaped face from five triangles.

For calculating the vertices of the *large icosahedron*, we start from an icosahedron inscribed in a cube. The continuation of the faces of the icosahedron leads to only one case where the new polyhedron does not decompose. Each face must be continued until its intersection with the three faces neighboring its opposite face. By symmetry we see that each face of the large icosahedron is a triangle homothetic to the triangular face of the central icosahedron with some negative coefficient k.

For calculating k, consider the cross section of the polyhedron cut by the plane XZ, which contains the center O_1 of the face $B_1B_1B_9$, and medians of this face and the face $B_3B_4B_{10}$, which neighbors the parallel face $B_3B_4B_8$. Such medians $B_9B_{1,2}$, $B_{10}B_{3,4}$, where $B_{1,2} = \frac{1}{2}(B_1 + B_2) = (0, 0, c)$, $B_{3,4} = \frac{1}{2}(B_3 + B_4) = (0, 0, -c)$, intersect at the point $F(0, y, 0)$ on the axis OY, which is the center of an edge of the large icosahedron. Hence, F is the image of the point $B_{1,2}$ under homothety with the center O_1, i.e., $k = -\frac{O_1E}{O_1B_{1.2}}$. From similar triangles we find the value of y:

$$\frac{y}{c} = \frac{y-c}{p} \Rightarrow y = \frac{c^2}{c-p} = \frac{2}{3-\sqrt{5}}\,c.$$

Then we express the coefficient of homothety k through the ratio of projections of line segments: $k = -\frac{y-\frac{1}{3}c}{\frac{1}{3}c} = -\frac{3y-c}{c} = -\frac{3+\sqrt{5}}{3-\sqrt{5}} \approx -6.854$.

The program for plotting the *large icosahedron* with MAPLE given below is based on data of the program for the simple icosahedron.

Example 17.3.3 Let us plot regular star dodecahedra starting from the coordinates of their vertices and lists of sequences of vertices in all faces. Complete the program for the regular simple dodecahedron with some additional calculations; see Exercise 17.3.1. Here the vectors $c[i]$ contain the coordinates of the center of the ith face of the dodecahedron.

```
> restart: with(plottools):      # star forms of dodecahedron
> c:=1:p:=evalf(c*(sqrt(5)-1)/2):k:=evalf(1+2*cos(Pi/5)):
  a[1]:=[p,0,c+p]: ...  # see the program for the dodecahedron
  a[20]:=[-c,-c,-c]:
> g[1]:=[a[2],a[17],a[8],a[7],a[19]]:
  ... # see the program for the dodecahedron
   g[12]:=[a[5],a[6],a[14],a[3],a[16]]:
> for i from 1 to 12 do # continuation: large dodecahedron
```

```
  c[i]:=(sum('g[i][j]', 'j'=1..5)/5):
 p[i]:=[seq((c[i]-k*(g[i][j]-c[i])), j=1..5)] od:
> P:=j -> plots[polygonplot3d]([seq(p[i], i=1..12)],
  orientation=[20*j, 50], style=PATCH, lightmodel=light2):
> plots[display]([seq(P(j), j=1..18)], insequence=true);
> for i from 1 to 12 do
  b1[i]:=[op(1,g[i]), op(2,g[i]), op(4,p[i])]:
  b2[i]:=[op(2,g[i]), op(3,g[i]), op(5,p[i])]:
  b3[i]:=[op(3,g[i]), op(4,g[i]), op(1,p[i])]:
  b4[i]:=[op(4,g[i]), op(5,g[i]), op(2,p[i])]:
  b5[i]:=[op(5,g[i]), op(1,g[i]), op(3,p[i])] od:
> plots[polygonplot3d]({seq(b1[i],i=1..12), seq(b2[i],
  i=1..12),seq(b3[i],i=1..12),seq(b4[i],i=1..12), seq(b5[i],
  i=1..12)}, style=patch, lightmodel=light2);
  continuation: small star dodecahedron
> for i from 1 to 12 do
  d1[i]:=[seq((c[i]-k*(b1[i][j]-c[i])),j=1..3)]:
  d2[i]:=[seq((c[i]-k*(b2[i][j]-c[i])),j=1..3)]:
  d3[i]:=[seq((c[i]-k*(b3[i][j]-c[i])),j=1..3)]:
  d4[i]:=[seq((c[i]-k*(b4[i][j]-c[i])),j=1..3)]:
  d5[i]:=[seq((c[i]-k*(b5[i][j]-c[i])),j=1..3)] od:
> plots[polygonplot3d]({seq(d1[i],i=1..12), seq(d2[i],
  i=1..12),seq(d3[i],i=1..12),seq(d4[i],i=1..12), seq(d5[i],
  i=1..12)}, style=patch, lightmodel=light2); # first method
> plots[display](stellate(POLYGONS(seq(g[i],i=1..12)),2.3),
  style=patch, lightmodel=light2);                # second method
> plots[display](stellate(dodecahedron(),2.3),
  style=patch,lightmodel=light2);                 # third method
```

Example 17.3.4 Let us plot the regular star icosahedron by using the coordinates of its vertices and their order in all faces. The program is based on the one for the simple icosahedron with some additional calculations; see Exercise 17.3.1. Here the vectors $c[i]$ contain coordinates of the center of the ith face of the icosahedron. Note that this case is complicated for MAPLE V. We do not see on the display the complete picture of the intersections of all 20 of its triangular faces.

```
> c:=1: p:=evalf(c*(sqrt(5)-1)/2):
  g[1]:=[[p,0,c],[-p,0,c],[0,c,p]]:
  ... # see the program for the icosahedron
   g[20]:=([[0,-c,p],[0,-c,-p],[c,-p,0]]):
  k:=evalf((3+sqrt(5))/(3-sqrt(5))):
```

```
  for i from 1 to 20 do c[i]:=(sum('g[i][j]', 'j'=1..3)/3):
  b[i]:=[seq((c[i]-k*(g[i][j]-c[i])), j=1..3)] od:
> P:=j -> plots[polygonplot3d]([seq(b[i], i=1..20)],
  orientation=[20*j,50], style=PATCH):
> plots[display]([seq(P(j), j=1..18)], insequence=true);
```

Remark 17.3.1 One can define a regular star polyhedron in MAPLE by using the command `RegularPolyhedron(gon,[m,n],o,r);` or the command `PolyhedronName(gon,o,r);` (as for Platonic solids) where `PolyhedronName` is one of `GreatStellatedDodecahedron`, `SmallStellatedDodecahedron`, `GreatIcosahedron`, `GreatDodecahedron`.

Definition 17.3.1 A *compound polyhedron* is a set of distinct polyhedra, called the components of the compound, which are placed together so that their centers coincide. (The compound of two tetrahedra is shown in Fig. 17.33).

Compounds of Platonic solids in which all components are the same (see [Cro, Plates 11–16]) have a high degree of symmetry, which makes them very attractive. The idea of placing one polyhedron in another in different ways is used in plotting compounds. Another method to plot compounds uses matrix representation of finite symmetry groups in space.

Exercise 17.3.1

1. Prove that the vertices of any face of the small star dodecahedron are the images under homothety from the center of the "kernel" face with coefficient $k = -(1 + \frac{2}{\sqrt{3}}\tau)$, but in a different order.

Analogously, the vertices of the large star dodecahedron are the images under a homothety from the center of mass of the "kernel" face with the same coefficient k, this time in the same order.

2. Some of the most fascinating models to play with are compounds. Plot (rotating) compounds corresponding to

a) An octahedron and a cube coupled together so that their edges bisect each other at right angles. (The dodecahedron and the icosahedron can also be coupled together to form a compound polyhedron.)

b) Five different ways to inscribe a cube in a dodecahedron.

c) Three different ways to inscribe a cube in octahedron.

18
Semi-Regular Polyhedra

In Chapter 18 we recall how to construct Archimedean solids via transformations of Platonic solids (Section 18.1), and we present programs (Section 18.2) to plot them by coordinates of vertices (in a way that is similar to programs for the *Platonic solids* but with some additional calculations and also difficult preparatory work with the lists of vertices in the faces).

The way to define an Archimedean solid in MAPLE is to use the command `Archimedean(gon,sch,o,r);` or `PolyhedronName(gon,o,r);` from the library `geom3d` where `PolyhedronName` is one of

`TruncatedTetrahedron`, `TruncatedOctahedron`, `TruncatedHexahedron`, `TruncatedIcosahedron`, `TruncatedDodecahedron`, `SmallRhombicuboctahedron`, `SmallRhombiicosidodecahedron`, `GreatRhombicuboctahedron`, `TruncatedCuboctahedron`, `GreatRhombiicosidodecahedron`, `TruncatedIcosidodecahedron`, `SnubCube`, `cuboctahedron`, `icosidodecahedron`.

To access the information relating to an Archimedean solid gon, we use the corresponding function calls.

In MAPLE V Release 5, the dual of a given polyhedron `geom3d[duality](dgon,gon,s)` can be defined. For a given regular solid, its dual is also a regular solid. Archimedean solids are also included in this case.

18.1 What Are Semi-Regular Polyhedra?

Definition 18.1.1 An *isohedron* (*isogon*) is a CP whose rotation group (of the first and second orders) translates any face (vertex) to any other face (respectively, vertex).

Each isohedron corresponds to a dual isogon, and conversely. There are 13 different combinatorially special types and two infinite series of isohedra (isogons). Each of these isohedra can be realized in space in such a way that all faces are regular polygons, and we obtain semi-regular polyhedra from the following definition.

Definition 18.1.2 A polyhedron whose faces are the regular polygons (perhaps of different sizes and types) and all its solid angles are equal is called a *semi-regular polyhedron.*

Simple examples of such polyhedra are the *regular prisms* $4^2 \cdot n$ (i.e., each vertex belongs to two squares and one n-gon; in the following way we use this classic designation for semi-regular polyhedra), whose bottom faces are regular simple n-gons and whose lateral faces are squares, and *antiprisms* $3^3 \cdot n$, whose bottom faces are regular simple n-gons with $n \geq 3$ and lateral faces are two regular triangles. In particular, the 3-antiprism is the tetrahedron, and the 4-prism is the cube. Note that the n-antiprism can be obtained from the n-prism by rotating one of its bases about the center through an angle $\frac{\pi}{n}$ in the same plane, while at the same time decreasing the height so that distances between corresponding vertices is equal to their edges.

Long ago Archimedes proved that except for the two series of prisms and antiprisms, there exist 13 types of semi-regular CP, the so-called *Archimedean solids.* The geometer Pappus of Alexandria tells about the work of Archimedes and gives a short description of the Archimedean solids.

The complete theory of semi-regular polyhedra was discovered by Kepler in his book *Harmonices mundilibri quinque ...* (1619).

Some restrictions on polyhedra follow directly from the definition:

• *There are no semi-regular polyhedra bounded by more than three different types of faces.* (In fact, in view of the equality of their solid angles, each of them contains at least one planar angle of each type of face, but the minimum angles come from polygons with 3, 4, 5 and 6 sides, i.e., we have the inequality $60° + 90° + 108° + 120° > 360°$),

• *There are no semi-regular polyhedra whose vertices contain more than five faces.* (In fact, the assumption that a solid angle consists of at least six planar angles of regular polygons of two types leads to the inequality $5 \cdot 60° + 90° > 360°$).

The list of possible types of semi-regular CPs can be obtained by topological (combinatoric) reasoning using Euler's formula for a sphere.

The names of the semi-regular CPs are the following:

(1) *snub cube* $3^4 \cdot 4$: 32 triangles, 6 squares
(2) *cuboctahedron* $(3 \cdot 4)^2$: 8 triangles, 6 squares
(3) *rhombicuboctahedron* $3 \cdot 4^3$: 8 triangles, 18 squares
(4) *snub dodecahedron* $3^4 \cdot 5$: 80 triangles, 12 pentagons
(5) *icosidodecahedron* $(3 \cdot 5)^2$: 20 triangles, 12 pentagons
(6) *truncated tetrahedron* $3 \cdot 6^2$: 4 triangles, 4 hexagons
(7) *truncated cube* $3 \cdot 8^2$: 8 triangles, 6 octagons
(8) *truncated dodecahedron* $3 \cdot 10^2$: 20 triangles, 12 decagons
(9) *truncated octahedron* $4 \cdot 6^2$: 6 squares, 8 hexagons
(10) *truncated icosahedron* $5 \cdot 6^2$: 12 pentagons, 20 hexagons
(11) *rhombicosidodecahedron* $3 \cdot 4 \cdot 5 \cdot 4$: 20 triangles, 30 squares, 12 pentagons
(12) *truncated cuboctahedron* $4 \cdot 6 \cdot 8$: 12 squares, 8 hexagons, 6 octagons
(13) *truncated icosidodecahedron* $4 \cdot 6 \cdot 10$: 30 squares, 20 hexagons, 12 decagons
(14) *Ashkinuze solid* $3 \cdot 4^3$: 8 triangles and 18 squares.

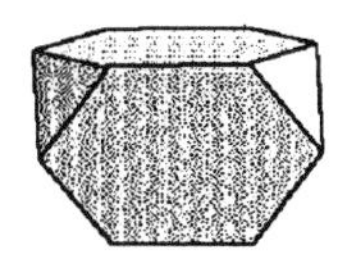
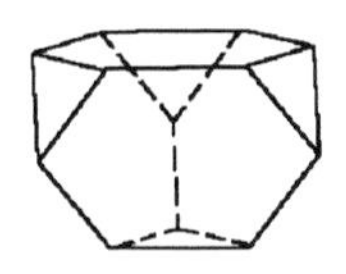
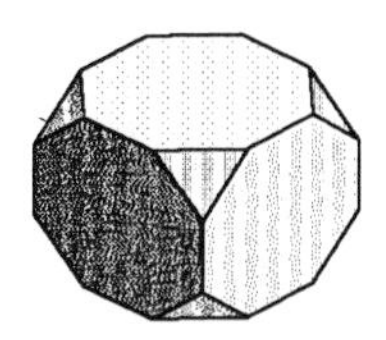
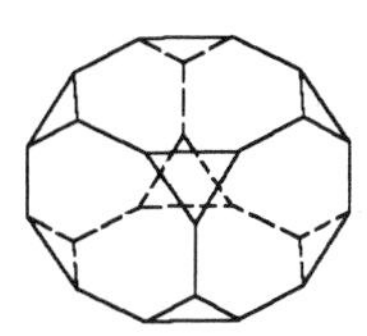

Figs. 18.1–18.2. (6) Truncated tetrahedron. (7) Truncated cube

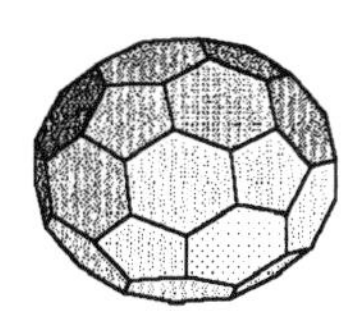
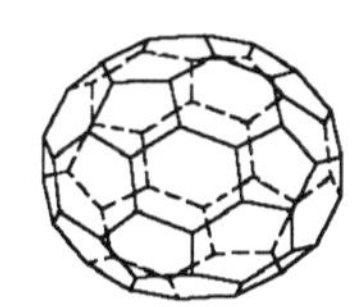

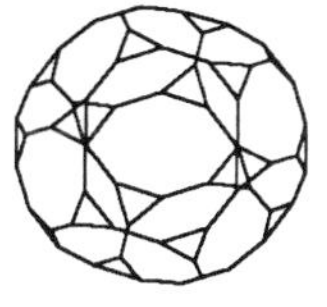

Figs. 18.3–18.4. (10) Truncated icosahedron. (8) Truncated dodecahedron

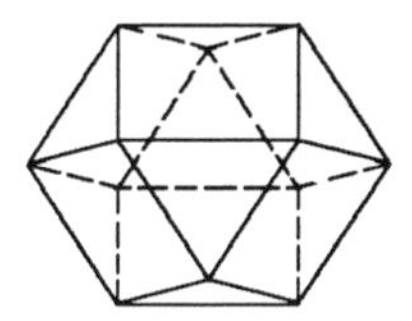
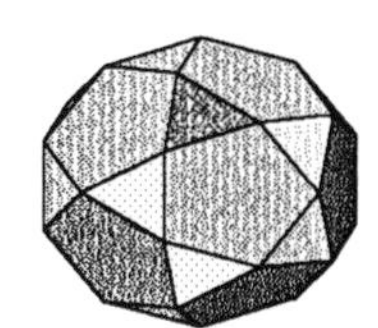
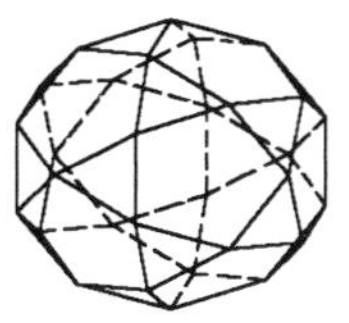

Figs. 18.5–18.6. (2) Cuboctahedron. (5) Icosidodecahedron

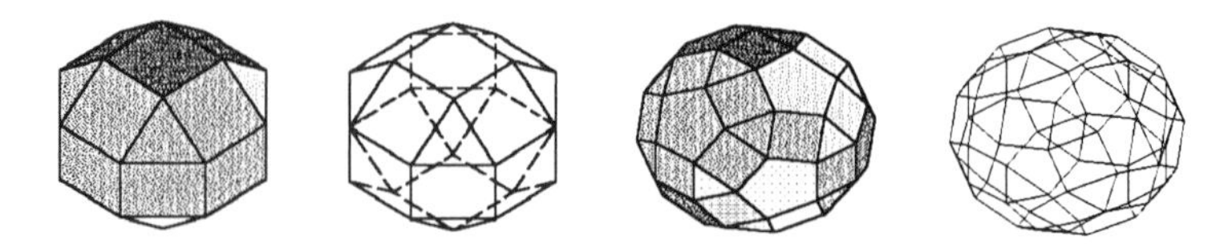

Figs. 18.7–18.8. (3) Rhombicuboctahedron. (11) Rhombicosidodecahedron

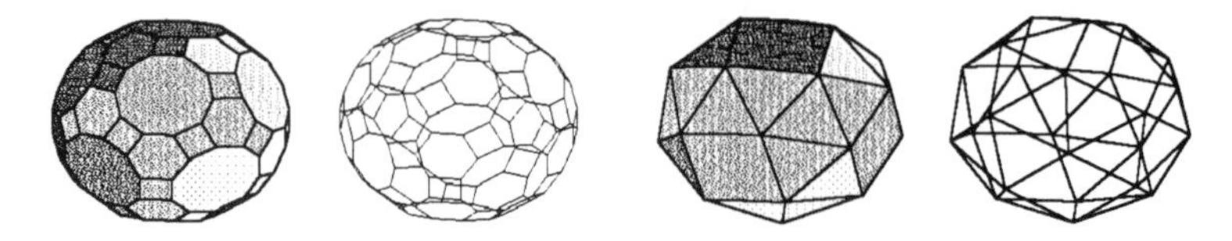

Figs. 18.9–18.10. (13) Truncated icosidodecahedron. (1) Snub cube

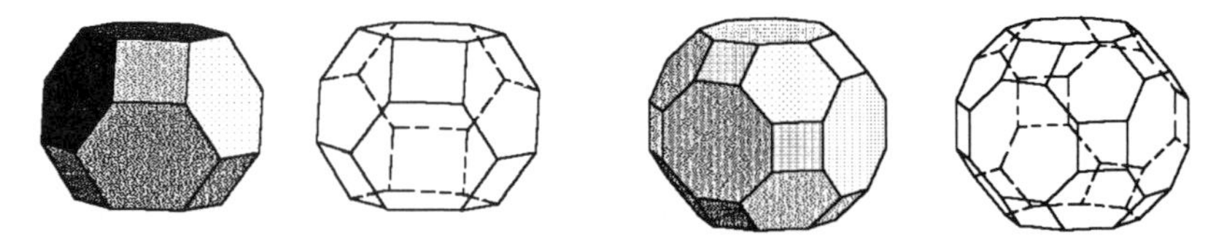

Figs. 18.11–18.12. (9) Truncated octahedron. (12) Truncated cuboctahedron

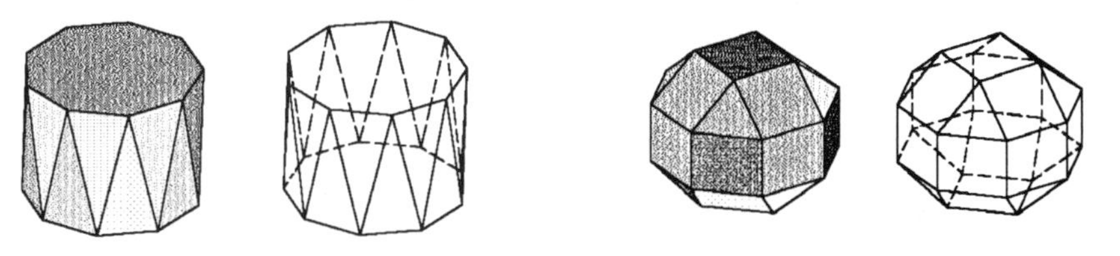

Figs. 18.13–18.14. n-antiprism. (14) Ashkinuze solid

Remark 18.1.1 In the 20th century V. Ashkinuze (and J. Miller) found the 14th semi-regular polyhedron (of type $3\cdot 4^3$), which differs from the rhombicuboctahedron, see case (3), only by rotating the whole upper part of the polyhedron, consisting of five squares and four triangles, through a 45° angle. For each of the 13 Archimedean solids, prisms, and antiprisms, any two vertices can be translated one to the other by a symmetry of the polyhedron, but for the *Ashkinuze solid* this does not hold.

One can plot the Archimedean solids using various methods. We consider one method using transformations of the five Platonic solids by cutting off the neighborhoods of vertices and edges by planes.

Let us start with the tetrahedron. We plot the planes at a distance $\frac{1}{3}$ from each edge meeting at a common vertex, cut off the resulting pyramid, and obtain the solid (6), bounded by regular triangles and hexagons.

An analogous transformation for the octahedron gives us the solid (9), and from the icosahedron we may obtain the solid (10).

Plotting the planes through the centers of certain edges of the cube or octahedron, we obtain the solid (2), i.e., the *cuboctahedron*. By the same method we obtain (from the icosahedron or dodecahedron) solid (5), i.e., the *icosidodecahedron*. If we employ such cutting planes in the cube so that its square faces are transformed into regular octagons, then we obtain the solid (7), i.e., the *truncated cube*. In an analogous way we obtain the solid (8) from the dodecahedron.

The transformations that proceed by cutting off neighborhoods of both vertices and edges are more complicated. If we cut off one after another, the edges having a common vertex of planes that define strips of equal height on the faces, we obtain four new trihedrals. Drawing this, we see that three of these vertices are located symmetrically around the third one. We cut off this last vertex by a plane that contains the first three vertices and thus obtain a triangle in the intersection.

Repeating the same procedure with the other edges and vertices, each time we obtain one triangle in place of a vertex of the cube and between them, instead of the edges of the cube, we obtain squares; the faces of the cube would also be transformed into new squares of smaller size.

To construct the regular solid (3), i.e., the *rhombicuboctahedron*, one must cut off by a plane the line segments $x = \frac{a}{2}(2 - \sqrt{2})$ and $y = \frac{a}{2}(\sqrt{2} - 1)$ on edges not parallel to it, where a is the length of the edge of the cube.

Employing the same transformation for the dodecahedron, we obtain the solid (11).

Now we will apply to the cube a transformation very similar to the one above. The difference is that we cut off not *one* of the four trihedrals, but *all* of them, and in the intersection we obtain not a triangle, but a (regular) hexagon. The result will be the solid (12).

Moreover, one can cut off by a plane the line segments $x = \frac{a}{14}(4 - \sqrt{2})$ and $y = \frac{a}{7}(2\sqrt{2} - 1)$ on non-parallel edges. Applying the same transformation to the dodecahedron gives us the solid (13).

Further, applying this new transformation to the cube and dodecahedron, we obtain two more analogous polyhedra. The resulting polyhedron for the cube is seen in Figure 18.10; it is (1), a *snub cube*. From the dodecahedron we obtain the *snub dodecahedron*, the solid (4).

There also exist *star semi-regular (uniform) polyhedra* (UPs). These are to the Archimedean solids what the regular SPs are to the Platonic solids. However, the faces of UPs are allowed to intersect each other. For a start we can form prisms and antiprisms with star polygons as bases. The decagon when the base for the prism is a pentagon is sometimes called a *pentacle*.

One can describe the list of 75 such regular UPs (see models in [Wen]); J. Skilling showed in 1975 using a computer that the list is complete.

Following Kepler, *half-regular polyhedra* are defined as in Definition 18.1.2 where we use half-regular (i.e., equilateral) polygons. Kepler restricted attention to half-regular 4-gons (rhombi) and found two such polyhedra, see figures in [Cro, p. 152]. The first (*rhombic dodecahedron*) is bounded by twelve rhombi whose diagonals are in the ratio of $1 : \sqrt{2}$. The second rhombic polyhedron (*rhombic triacontahedron*) is bounded by 30 rhombi whose diagonals are in the golden ratio.

A generalization of semi-regular polyhedra are the so-called *regular-faced polyhedra*, whose faces by definition are regular polygons. The constraints that restrict the number of such polyhedra are less topological and more metrical. Their classification requires length calculations.

Theorem 18.1.1 (see [Zal]) *Aside from the two series of prisms and antiprisms, there exist exactly* 92 *regular-faced polyhedra, but only* 28 *of them are indecomposable, i.e., cannot be broken by a plane into two regular-faced polyhedra.*

Examples of decomposable regular-faced polyhedra are the octahedron, icosahedron, Ashkinuze solid, and also among the Archimedean solids, the cuboctahedron, icosidodecahedron, rhombicuboctahedron, and rhombicosidodecahedron. Hence, only $8\ (= \frac{28-(5-2)-(13-4)}{2})$ regular-faced polyhedra are not regular or semi-regular polyhedra or parts of them.

Exercise 18.1.1
1. Plot the regular icosahedron, combining a pentagonal antiprism with two regular pentagonal pyramids having the same edge length.
2. What regular star polyhedra can be obtained from the *Platonic solids* by the MAPLE construction `stellate`?
Apply the construction `stellate` to *Archimedean solids*.
3. Rewrite the programs below in the form of procedures, and complete them with a program for plotting the *snub dodecahedron*.
4. Write a program for plotting two half-regular Kepler polyhedra.
5. Write a program for plotting UP: prisms and antiprisms.

18.2 Programs for Plotting Semi-Regular Polyhedra

```
> restart: c:=1: v:=0.7044022*c: u:=0.4524646*c: # 33334
> a[1]:=[-c+u,c,-c+v]:a[2]:=[-c+v,c,c-u]:
  a[3]:=[c-u,c,c-v]:a[4]:=[c-v,c,-c+u]:
  a[5]:=[c,c-u,-c+v]:a[6]:=[c,-c+v,-c+u]:
```

```
  a[7]:=[c,-c+u,c-v]:a[8]:=[c,c-v,c-u]:
  a[9]:=[c-v,c-u,c]:a[10]:=[c-u,-c+v,c]:
  a[11]:=[-c+v,-c+u,c]:a[12]:=[-c+u,c-v,c]:
  a[13]:=[-c+v,c-u,-c]:a[14]:=[-c+u,-c+v,-c]:
  a[15]:=[c-v,-c+u,-c]:a[16]:=[c-u,c-v,-c]:
  a[17]:=[-c,c-v,-c+u]:a[18]:=[-c,c-u,c-v]:
  a[19]:=[-c,-c+v,c-u]:a[20]:=[-c,-c+u,-c+v]:
  a[21]:=[c-u,-c,-c+v]:a[22]:=[-c+v,-c,-c+u]:
  a[23]:=[-c+u,-c,c-v]:a[24]:=[c-v,-c,c-u]:
> for i from 1 to 6 do g[i]:=[seq(a[4*(i-1)+j],j=1..4)] od:
> g[7]:=[a[4],a[5],a[3]]:g[8]:=[a[3],a[9],a[2]]:
  g[9]:=[a[2],a[18],a[1]]:g[10]:=[a[4],a[1],a[13]]:
  g[11]:=[a[8],a[3],a[5]]:g[12]:=[a[7],a[10],a[8]]:
  g[13]:=[a[6],a[21],a[7]]:g[14]:=[a[5],a[16],a[6]]:
  g[15]:=[a[9],a[8],a[10]]:g[16]:=[a[12],a[2],a[9]]:
  g[17]:=[a[11],a[19],a[12]]:g[18]:=[a[11],a[10],a[24]]:
  g[19]:=[a[13],a[17],a[14]]:g[20]:=[a[13],a[16],a[4]]:
  g[21]:=[a[16],a[15],a[6]]:g[22]:=[a[15],a[14],a[22]]:
  g[23]:=[a[24],a[23],a[11]]:g[24]:=[a[23],a[22],a[20]]:
  g[25]:=[a[21],a[15],a[22]]:g[26]:=[a[21],a[24],a[7]]:
  g[27]:=[a[19],a[18],a[12]]:g[28]:=[a[17],a[1],a[18]]:
  g[29]:=[a[17],a[20],a[14]]:g[30]:=[a[20],a[19],a[23]]:
  g[31]:=[a[9],a[3],a[8]]:g[32]:=[a[10],a[7],a[24]]:
  g[33]:=[a[11],a[23],a[19]]:g[34]:=[a[18],a[2],a[12]]:
  g[35]:=[a[4],a[16],a[5]]:g[36]:=[a[13],a[1],a[17]]:
  g[37]:=[a[22],a[14],a[20]]:g[38]:=[a[6],a[15],a[21]]:
> P:=j -> plots[polygonplot3d]([seq(g[i], i=1..38)],
  orientation=[20*j,50], style=PATCH):
  plots[display]([seq(P(j), j=1..18)], insequence=true);
> restart: c:=1: # 3434
> a[1]:=[c,c,0]: a[2]:=[c,0,-c]: a[3]:=[0,c,-c]:
  a[4]:=[-c,0,-c]: a[5]:=[-c,c,0]: a[6]:=[-c,0,c]:
  a[7]:=[-c,-c,0]: a[8]:=[0,c,c]: a[9]:=[0,-c,c]:
  a[10]:=[c,-c,0]: a[11]:=[c,0,c]: a[12]:=[0,-c,-c]:
> g[1]:=[a[1],a[2],a[3]]: g[2]:=[a[5],a[3],a[4]]:
  g[3]:=[a[7],a[4],a[12]]: g[4]:=[a[10],a[12],a[2]]:
  g[5]:=[a[8],a[11],a[1]]: g[6]:=[a[8],a[5],a[6]]:
  g[7]:=[a[9],a[6],a[7]]: g[8]:=[a[11],a[9],a[10]]:
  g[9]:=[a[1],a[8],a[5],a[3]]: g[10]:=[a[3],a[4],a[12],a[2]]:
  g[11]:=[a[12],a[7],a[9],a[10]]:
  g[12]:=[a[9],a[6],a[8],a[11]]:
  g[13]:=[a[2],a[10],a[11],a[1]]:
  g[14]:=[a[5],a[6],a[7],a[4]]:
> P:=j -> plots[polygonplot3d]([seq(g[i], i=1..14)],
  orientation=[20*j,50], style=PATCH):
  plots[display]([seq(P(j), j=1..18)], insequence=true);
> restart: # 3434-1
> c:=1: d:=evalf((c*sqrt(2))/(2+sqrt(2))):
> a[1]:=[d,d,c]: a[2]:=[-d,d,c]: a[3]:=[-d,-d,c]:
```

```
  a[4]:=[d,-d,c]: a[5]:=[c,d,d]: a[6]:=[c,d,-d]:
  a[7]:=[c,-d,-d]: a[8]:=[c,-d,d]: a[9]:=[-d,c,d]:
  a[10]:=[-d,c,-d]: a[11]:=[d,c,-d]: a[12]:=[d,c,d]:
  a[13]:=[d,-c,d]: a[14]:=[d,-c,-d]: a[15]:=[-d,-c,-d]:
  a[16]:=[-d,-c,d]: a[17]:=[-c,-d,d]: a[18]:=[-c,-d,-d]:
  a[19]:=[-c,d,-d]: a[20]:=[-c,d,d]: a[21]:=[d,d,-c]:
  a[22]:=[-d,d,-c]: a[23]:=[-d,-d,-c]: a[24]:=[d,-d,-c]:
> for i from 1 to 6 do g[i]:=[seq(a[4*(i-1)+j],j=1..4)] od:
> g[7]:=[a[1],a[4],a[8],a[5]]:
  g[8]:=[a[4],a[13],a[16],a[3]]:
  g[9]:=[a[3],a[17],a[20],a[2]]:
  g[10]:=[a[2],a[9],a[12],a[1]]:
  g[11]:=[a[12],a[5],a[6],a[11]]:
  g[12]:=[a[8],a[13],a[14],a[7]]:
  g[13]:=[a[16],a[17],a[18],a[15]]:
  g[14]:=[a[20],a[9],a[10],a[19]]:
  g[15]:=[a[6],a[7],a[24],a[21]]:
  g[16]:=[a[14],a[15],a[23],a[24]]:
  g[17]:=[a[18],a[23],a[22],a[19]]:
  g[18]:=[a[10],a[11],a[21],a[22]]:
  g[19]:=[a[1],a[12],a[5]]:
  g[20]:=[a[4],a[8],a[13]]:
  g[21]:=[a[3],a[16],a[17]]:
  g[22]:=[a[2],a[20],a[9]]:
  g[23]:=[a[6],a[11],a[21]]:
  g[24]:=[a[7],a[24],a[14]]:
  g[25]:=[a[15],a[23],a[18]]:
  g[26]:=[a[19],a[22],a[10]]:
> P:=j -> plots[polygonplot3d]([seq(g[i], i=1..26)],
  orientation=[20*j,50], style=PATCH):
  plots[display]([seq(P(j), j=1..18)], insequence=true);
> restart: # 3535
> c:=1:
  p:=evalf(c*((sqrt(5)-1)/2)): v:=p/2:
> a[1]:=[c/2,c+v,c/2+v]:a[2]:=[0,c+p,0]:
  a[3]:=[c/2,c+v,-c/2-v]:a[4]:=[c+v,c/2+v,-c/2]:
  a[5]:=[c+v,c/2+v,c/2]:a[6]:=[c/2+v,c/2,c+v]:
  a[7]:=[c/2+v,-c/2,c+v]:a[8]:=[c+v,-c/2-v,c/2]:
  a[9]:=[c+p,0,0]:a[10]:=[c/2+v,-c/2,-c-v]:
  a[11]:=[c/2,-c-v,-c/2-v]:a[12]:=[0,-c-p,0]:
  a[13]:=[c/2,-c-v,c/2+v]:a[14]:=[-c/2,-c-v,c/2+v]:
  a[15]:=[-c/2-v,-c/2,c+v]:a[16]:=[0,0,c+p]:
  a[17]:=[-c/2-v,c/2,c+v]:a[18]:=[-c/2,c+v,c/2+v]:
  a[19]:=[-c-v,c/2+v,c/2]:a[20]:=[-c-v,c/2+v,-c/2]:
  a[21]:=[-c/2,c+v,-c/2-v]:a[22]:=[-c/2-v,c/2,-c-v]:
  a[23]:=[0,0,-c-p]:a[24]:=[c/2+v,c/2,-c-v]:
  a[25]:=[-c/2-v,-c/2,-c-v]:a[26]:=[-c/2,-c-v,-c/2-v]:
  a[27]:=[-c-v,-c/2-v,-c/2]:a[28]:=[-c-p,0,0]:
  a[29]:=[-c-v,-c/2-v,c/2]:a[30]:=[c+v,-c/2-v,-c/2]:
> g[1]:=[a[1],a[2],a[18]]:g[2]:=[a[2],a[3],a[21]]:
```

```
  g[3]:=[a[1],a[6],a[5]]:g[4]:=[a[17],a[18],a[19]]:
  g[5]:=[a[7],a[6],a[16]]:g[6]:=[a[16],a[17],a[15]]:
  g[7]:=[a[7],a[13],a[8]]:g[8]:=[a[14],a[15],a[29]]:
  g[9]:=[a[4],a[24],a[3]]:g[10]:=[a[20],a[21],a[22]]:
  g[11]:=[a[14],a[12],a[13]]:g[12]:=[a[26],a[11],a[12]]:
  g[13]:=[a[5],a[9],a[4]]:g[14]:=[a[9],a[8],a[30]]:
  g[15]:=[a[19],a[20],a[28]]:g[16]:=[a[28],a[27],a[29]]:
  g[17]:=[a[23],a[24],a[10]]:g[18]:=[a[22],a[23],a[25]]:
  g[19]:=[a[25],a[26],a[27]]:g[20]:=[a[30],a[11],a[10]]:
  g[21]:=[a[6],a[1],a[18],a[17],a[16]]:
  g[22]:=[a[7],a[16],a[15],a[14],a[13]]:
  g[23]:=[a[3],a[24],a[23],a[22],a[21]]:
  g[24]:=[a[23],a[10],a[11],a[26],a[25]]:
  g[25]:=[a[3],a[2],a[1],a[5],a[4]]:
  g[26]:=[a[2],a[21],a[20],a[19],a[18]]:
  g[27]:=[a[27],a[26],a[12],a[14],a[29]]:
  g[28]:=[a[12],a[11],a[30],a[8],a[13]]:
  g[29]:=[a[8],a[9],a[5],a[6],a[7]]:
  g[30]:=[a[4],a[9],a[30],a[10],a[24]]:
  g[31]:=[a[20],a[22],a[25],a[27],a[28]]:
  g[32]:=[a[19],a[28],a[29],a[15],a[17]]:
> P:=j -> plots[polygonplot3d]([seq(g[i], i=1..32)],
  orientation=[20*j,50], style=PATCH):
  plots[display]([seq(P(j), j=1..18)], insequence=true);
> restart: c:=1: d:=c/3:   # 366
> a[1]:=[d,c,-d]: a[2]:=[c,d,-d]: a[3]:=[d,d,-c]:
  a[4]:=[-d,d,c]: a[5]:=[-d,c,d]: a[6]:=[-c,d,d]:
  a[7]:=[c,-d,d]: a[8]:=[d,-d,c]: a[9]:=[d,-c,d]:
  a[10]:=[-d,-d,-c]: a[11]:=[-c,-d,-d]: a[12]:=[-d,-c,-d]:
> g[1]:=[a[1],a[2],a[3]]:g[2]:=[a[4],a[5],a[6]]:
  g[3]:=[a[7],a[8],a[9]]:g[4]:=[a[10],a[11],a[12]]:
  g[5]:=[a[2],a[7],a[8],a[4],a[5],a[1]]:
  g[6]:=[a[2],a[7],a[9],a[12],a[10],a[3]]:
  g[7]:=[a[8],a[9],a[12],a[11],a[6],a[4]]:
  g[8]:=[a[10],a[11],a[6],a[5],a[1],a[3]]:
> P:=j -> plots[polygonplot3d]([seq(g[i], i=1..8)],
  orientation=[20*j,50], style=PATCH):
  plots[display]([seq(P(j), j=1..18)], insequence=true);
> restart: c:=1: d:=evalf((2*c)/(2+sqrt(2))): #388
> a[1]:=[c,c,c-d]: a[2]:=[c,c,d-c]: a[3]:=[-c,c,c-d]:
  a[4]:=[-c,c,d-c]: a[5]:=[c,-c,c-d]: a[6]:=[c,-c,d-c]:
  a[7]:=[-c,-c,c-d]: a[8]:=[-c,-c,d-c]: a[9]:=[c,c-d,c]:
  a[10]:=[c,d-c,c]: a[11]:=[c,c-d,-c]: a[12]:=[c,d-c,-c]:
  a[13]:=[-c,d-c,-c]: a[14]:=[-c,c-d,-c]: a[15]:=[-c,d-c,c]:
  a[16]:=[-c,c-d,c]: a[17]:=[c-d,c,-c]: a[18]:=[d-c,c,-c]:
  a[19]:=[c-d,c,c]: a[20]:=[d-c,c,c]: a[21]:=[c-d,-c,c]:
  a[22]:=[d-c,-c,c]: a[23]:=[c-d,-c,-c]: a[24]:=[d-c,-c,-c]:
> g[1]:=[a[17],a[2],a[11]]: g[2]:=[a[9],a[19],a[1]]:
  g[3]:=[a[12],a[23],a[6]]: g[4]:=[a[5],a[10],a[21]]:
```

```
g[5]:=[a[8],a[13],a[24]]: g[6]:=[a[7],a[22],a[15]]:
g[7]:=[a[4],a[14],a[18]]: g[8]:=[a[3],a[20],a[16]]:
g[9]:=[a[13],a[8],a[7],a[15],a[16],a[3],a[4],a[14]]:
g[10]:=[a[23],a[6],a[5],a[21],a[22],a[7],a[8],a[24]]:
g[11]:=[a[12],a[6],a[5],a[10],a[9],a[1],a[2],a[11]]:
g[12]:=[a[17],a[2],a[1],a[19],a[20],a[3],a[4],a[18]]:
g[13]:=[a[11],a[12],a[23],a[24],a[13],a[14],a[18],a[17]]:
g[14]:=[a[10],a[9],a[19],a[20],a[16],a[15],a[22],a[21]]:
```

```
> P:=j -> plots[polygonplot3d]([seq(g[i], i=1..14)],
  orientation=[20*j,50], style=PATCH):
  plots[display]([seq(P(j), j=1..18)], insequence=true);
```

```
> restart: c:=1: p:=evalf((sqrt(5)-1)*c/2): # 3-10-10
```

```
> m:=evalf((c*(sqrt(5)-1))/(2+2*sin(54*Pi/180))):
  k:=evalf(c*(sqrt(5)-1)-2*m):
  d:=evalf(p*(c*(sqrt(5)-1)-m)/(c*(sqrt(5)-1))):
  t:=evalf(m*p/(c*(sqrt(5)-1))):
  r:=evalf(m*cos(72*Pi/180)):
```

```
> x[1]: =c-k/2: y[1]:=c-r: z[1]:=c+t:
  x[2]:=k/2: y[2]:=p+r: z[2]:=c+d:
  x[3]:=0: y[3]:=p-m: z[3]:=c+p:
  x[4]:=0: y[4]:=-p+m: z[4]:=c+p:
  x[5]:=k/2: y[5]:=-p-r: z[5]:=c+d:
  x[6]:=c-k/2: y[6]:=r-c: z[6]:=c+t:
  x[7]:=c+t: y[7]:=-c+k/2: z[7]:=c-r:
  x[8]:=c+d: y[8]:=-k/2: z[8]:=p+r:
  x[9]:=c+d: y[9]:=k/2: z[9]:=p+r:
  x[10]:=c+t: y[10]:=c-k/2: z[10]:=c-r:
  x[11]:=c-r: y[11]:=c+t: z[11]:=c-k/2:
  x[12]:=p+r: y[12]:=c+d: z[12]:=k/2:
  x[13]:=p+r: y[13]:=c+d: z[13]:=-k/2:
  x[14]:=c-r: y[14]:=c+t: z[14]:=-c+k/2:
  x[15]:=c+t: y[15]:=c-k/2: z[15]:=r-c:
  x[16]:=c+d: y[16]:=k/2: z[16]:=-p-r:
  x[17]:=c+p: y[17]:=0: z[17]:=-p+m:
  x[18]:=c+p: y[18]:=0: z[18]:=p-m:
  x[19]:=c+d: y[19]:=-k/2: z[19]:=-p-r:
  x[20]:=c+t: y[20]:=-c+k/2: z[20]:=r-c:
  x[21]:=x[14]: y[21]:=-y[14]: z[21]:=z[14]:
  x[22]:=x[13]: y[22]:=-y[13]: z[22]:=z[13]:
  x[23]:=x[22]: y[23]:=y[22]: z[23]:=-z[22]:
  x[24]:=x[21]: y[24]:=y[21]: z[24]:=-z[21]:
  x[25]:=p-m: y[25]:=-c-p: z[25]:=0:
  x[26]:=-p+m: y[26]:=-c-p: z[26]:=0:
  x[27]:=-x[22]: y[27]:=y[22]: z[27]:=z[22]:
  x[28]:=-x[21]: y[28]:=y[21]: z[28]:=z[21]:
  x[29]:=-x[32]: y[29]:=y[32]: z[29]:=z[32]:
  x[30]:=-x[31]: y[30]:=y[31]: z[30]:=z[31]:
  x[31]:=x[5]: y[31]:=y[5]: z[31]:=-z[5]:
  x[32]:=x[6]: y[32]:=y[6]: z[32]:=-z[6]:
  x[33]:=x[1]: y[33]:=y[1]: z[33]:=-z[1]:
```

```
  x[34]:=x[2]: y[34]:=y[2]: z[34]:=-z[2]:
  x[35]:=-x[34]: y[35]:=y[34]: z[35]:=z[34]:
  x[36]:=-x[33]: y[36]:=y[33]: z[36]:=z[33]:
  x[37]:=-x[14]: y[37]:=y[14]: z[37]:=z[14]:
  x[38]:=-x[13]: y[38]:=y[13]: z[38]:=z[13]:
  x[39]:=x[26]: y[39]:=c+p: z[39]:=0:
  x[40]:=x[25]: y[40]:=c+p: z[40]:=0:
  x[41]:=x[3]: y[41]:=y[3]: z[41]:=-z[3]:
  x[42]:=x[4]: y[42]:=y[4]: z[42]:=-z[4]:
  x[43]:=-x[12]: y[43]:=y[12]: z[43]:=z[12]:
  x[44]:=-x[11]: y[44]:=y[11]: z[44]:=z[11]:
  x[45]:=-x[10]: y[45]:=y[10]: z[45]:=z[10]:
  x[46]:=-x[9]: y[46]:=y[9]: z[46]:=z[9]:
  x[47]:=-x[8]: y[47]:=y[8]: z[47]:=z[8]:
  x[48]:=-x[7]: y[48]:=y[7]: z[48]:=z[7]:
  x[49]:=-x[6]: y[49]:=y[6]: z[49]:=z[6]:
  x[50]:=-x[5]: y[50]:=y[5]: z[50]:=z[5]:
  x[51]:=-x[2]: y[51]:=y[2]: z[51]:=z[2]:
  x[52]:=-x[18]: y[52]:=0: z[52]:=z[18]:
  x[53]:=-x[17]: y[53]:=0: z[53]:=z[17]:
  x[54]:=-x[19]: y[54]:=y[19]: z[54]:=z[19]:
  x[55]:=-x[20]: y[55]:=y[20]: z[55]:=z[20]:
  x[56]:=x[27]: y[56]:=y[27]: z[56]:=-z[27]:
  x[57]:=x[28]: y[57]:=y[28]: z[57]:=-z[28]:
  x[58]:=-x[1]: y[58]:=y[1]: z[58]:=z[1]:
  x[59]:=-x[15]: y[59]:=y[15]: z[59]:=z[15]:
  x[60]:=-x[16]: y[60]:=y[16]: z[60]:=z[16]:
> for i from 1 to 60 do a[i]:=[x[i],y[i],z[i]] od:
> g[1]:=[a[1],a[2],a[3],a[4],a[5],a[6],a[7],a[8],
        a[9],a[10]]:
  g[2]:=[a[3],a[4],a[50],a[49],a[48],a[47],a[46],a[45],
        a[58],a[51]]:
  g[3]:=[a[15],a[16],a[19],a[20],a[32],a[31],a[42],a[41],
        a[34],a[33]]:
  g[4]:=[a[41],a[42],a[30],a[29],a[55],a[54],a[60],a[59],
        a[36],a[35]]:
  g[5]:=[a[7],a[8],a[18],a[17],a[19],a[20],a[21],a[22],
        a[23],a[24]]:
  g[6]:=[a[9],a[18],a[17],a[16],a[15],a[14],a[13],a[12],
        a[11],a[10]]:
  g[7]:=[a[59],a[60],a[53],a[52],a[46],a[45],a[44],a[43],
        a[38],a[37]]:
  g[8]:=[a[53],a[54],a[55],a[28],a[27],a[56],a[57],a[48],
        a[47],a[52]]:
  g[9]:=[a[14],a[33],a[34],a[35],a[36],a[37],a[38],a[39],
        a[40],a[13]]:
  g[10]:=[a[11],a[12],a[40],a[39],a[43],a[44],a[58],a[51],
        a[2],a[1]]:
  g[11]:=[a[29],a[30],a[31],a[32],a[21],a[22],a[25],a[26],
        a[27],a[28]]:
  g[12]:=[a[25],a[23],a[24],a[6],a[5],a[50],a[49],a[57],
```

```
        a[56],a[26]]:
  g[13]:=[a[1],a[10],a[11]]:  g[14]:=[a[2],a[51],a[3]]:
  g[15]:=[a[58],a[44],a[45]]: g[16]:=[a[46],a[52],a[47]]:
  g[17]:=[a[48],a[57],a[49]]: g[18]:=[a[4],a[50],a[5]]:
  g[19]:=[a[6],a[24],a[7]]: \  g[20]:=[a[8],a[18],a[9]]:
  g[21]:=[a[12],a[13],a[40]]: g[22]:=[a[43],a[39],a[38]]:
  g[23]:=[a[27],a[26],a[56]]: g[24]:=[a[22],a[23],a[25]]:
  g[25]:=[a[19],a[16],a[17]]: g[26]:=[a[15],a[33],a[14]]:
  g[27]:=[a[34],a[41],a[35]]: g[28]:=[a[36],a[59],a[37]]:
  g[29]:=[a[60],a[54],a[53]]: g[30]:=[a[55],a[29],a[28]]:
  g[31]:=[a[31],a[30],a[42]]: g[32]:=[a[21],a[32],a[20]]:
> P:=j -> plots[polygonplot3d]([seq(g[i], i=1..32)],
  orientation=[20*j,50], style=PATCH):
  plots[display]([seq(P(j), j=1..18)], insequence=true);
> restart: c:=1: t:=evalf(c/3): u:=evalf(c-t): #466
> a[1]:=[0,u,t]: a[2]:=[-t,u,0]: a[3]:=[0,u,-t]:
  a[4]:=[t,u,0]: a[5]:=[u,0,t]: a[6]:=[u,t,0]:
  a[7]:=[u,0,-t]: a[8]:=[u,-t,0]: a[9]:=[0,-u,t]:
  a[10]:=[t,-u,0]: a[11]:=[0,-u,-t]: a[12]:=[-t,-u,0]:
  a[13]:=[-u,0,t]: a[14]:=[-u,-t,0]: a[15]:=[-u,0,-t]:
  a[16]:=[-u,t,0]: a[17]:=[t,0,u]: a[18]:=[0,t,u]:
  a[19]:=[-t,0,u]: a[20]:=[0,-t,u]: a[21]:=[t,0,-u]:
  a[22]:=[0,t,-u]: a[23]:=[-t,0,-u]: a[24]:=[0,-t,-u]:
> for i from 1 to 6 do g[i]:=[seq(a[4*(i-1)+j],j=1..4)] od:
> g[7]:=([a[5],a[17],a[18],a[1],a[4],a[6]]):
   g[8]:=([a[10],a[9],a[20],a[17],a[5],a[8]]):
   g[9]:=([a[11],a[24],a[23],a[15],a[14],a[12]]):
   g[10]:=([a[23],a[15],a[16],a[2],a[3],a[22]]):
   g[11]:=([a[1],a[2],a[16],a[13],a[19],a[18]]):
   g[12]:=([a[9],a[20],a[19],a[13],a[14],a[12]]):
   g[13]:=([a[7],a[21],a[24],a[11],a[10],a[8]]):
   g[14]:=([a[6],a[7],a[21],a[22],a[3],a[4]]):
> P:=j -> plots[polygonplot3d]([seq(g[i], i=1..14)],
  orientation=[20*j,50], style=PATCH):
  plots[display]([seq(P(j), j=1..18)], insequence=true);
> restart:  # 566
  c:=1:t:=evalf(c/3): p:=evalf(c*(sqrt(5)-1)/2):
  d:=evalf(p/3): v:=evalf((c+2*p)/3): m:=evalf((2*c+p)/3):
> a[1]:=[d,0, c]:a[2]:=[2*d,t, m]:a[3]:=[v,d, c-t]:
  a[4]:=[v,-d,c-t]:a[5]:=[2*d,-t, m]:a[6]:=[d,0,-c]:
  a[7]:=[2*d,t,-m]:a[8]:=[v,d,t-c]:a[9]:=[v,-d,t-c]:
  a[10]:=[2*d,-t,-m]:a[11]:=[-d,0,c]:a[12]:=[-2*d,t,m]:
  a[13]:=[-v,d,c-t]:a[14]:=[-v,-d,c-t]:a[15]:=[-2*d,-t,m]:
  a[16]:=[-d,0,-c]:a[17]:=[-2*d,t,-m]:a[18]:=[-v,d,t-c]:
  a[19]:=[-v,-d,t-c]:a[20]:=[-2*d,-t,-m]:a[21]:=[-t,m,2*d]:
  a[22]:=[-d,c-t,v]:a[23]:=[d,c-t,v]:a[24]:=[t,m,2*d]:
  a[25]:=[0,c,d]:a[26]:=[-t,m,-2*d]:a[27]:=[-d,c-t,-v]:
  a[28]:=[d,c-t,-v]:a[29]:=[t,m,-2*d]:a[30]:=[0,c,-d]:
  a[31]:=[-t,-m,2*d]:a[32]:=[-d,t-c,v]:a[33]:=[d,t-c,v]:
```

```
a[34]:=[t,-m,2*d]:a[35]:=[0,-c,d]:a[36]:=[-t,-m,-2*d]:
a[37]:=[-d,t-c,-v]:a[38]:=[d,t-c,-v]:a[39]:=[t,-m,-2*d]:
a[40]:=[0,-c,-d]:a[41]:=[c-t,v,d]:a[42]:=[m,2*d,t]:
a[43]:=[c,d,0]:a[44]:=[m,2*d,-t]:a[45]:=[c-t,v,-d]:
a[46]:=[t-c,v,d]:a[47]:=[-m,2*d,t]:a[48]:=[-c,d,0]:
a[49]:=[-m,2*d,-t]:a[50]:=[t-c,v,-d]:a[51]:=[c-t,-v,d]:
a[52]:=[m,-2*d,t]:a[53:=[c,-d,0]:a[54]:=[m,-2*d,-t]:
a[55]:=[c-t,-v,-d]:a[56]:=[t-c,-v,d]:a[57]:=[-m,-2*d,t]:
a[58]:=[-c,-d,0]:a[59]:=[-m,-2*d,-t]:a[60]:=[t-c,-v,-d]:
> for i from 1 to 12 do g[i]:=[seq(a[5*(i-1)+j],j=1..5)] od:
> g[13]:=[a[11],a[12],a[22],a[23],a[2],a[1]]:
g[14]:=[a[5],a[33],a[32],a[15],a[11],a[1]]:
g[15]:=[a[23],a[24],a[41],a[42],a[3],a[2]]:
g[16]:=[a[42],a[43],a[53],a[52],a[4],a[3]]:
g[17]:=[a[52],a[51],a[34],a[33],a[5],a[4]]:
g[18]:=[a[51],a[55],a[39],a[40],a[35],a[34]]:
g[19]:=[a[60],a[56],a[31],a[35],a[40],a[36]]:
g[20]:=[a[56],a[57],a[14],a[15],a[32],a[31]]:
g[21]:=[a[14],a[57],a[58],a[48],a[47],a[13]]:
g[22]:=[a[12],a[13],a[47],a[46],a[21],a[22]]:
g[23]:=[a[21],a[46],a[50],a[26],a[30],a[25]]:
g[24]:=[a[24],a[25],a[30],a[29],a[45],a[41]]:
g[25]:=[a[28],a[7],a[8],a[44],a[45],a[29]]:
g[26]:=[a[44],a[8],a[9],a[54],a[53],a[43]]:
g[27]:=[a[9],a[10],a[38],a[39],a[55],a[54]]:
g[28]:=[a[10],a[6],a[16],a[20],a[37],a[38]]:
g[29]:=[a[36],a[37],a[20],a[19],a[59],a[60]]:
g[30]:=[a[58],a[59],a[19],a[18],a[49],a[48]]:
g[31]:=[a[50],a[49],a[18],a[17],a[27],a[26]]:
g[32]:=[a[28],a[27],a[17],a[16],a[6],a[7]]:
> P:=j -> plots[polygonplot3d]([seq(g[i], i=1..32)],
orientation=[20*j,50], style=PATCH):
plots[display]([seq(P(j), j=1..18)], insequence=true);
> restart: c:=1: p:=evalf(c*((sqrt(5)-1)/2)): # 3454
> a[1]:=[c/3,c/3,c+2*p/3]: a[3]:=[0,c+p/3,(2*c+p)/3]:
a[2]:=[(p+c)/3,(2*c+p)/3, 2*(c+p)/3]:
a[4]:=[-(p+c)/3,(c*2+p)/3,(2*p+2*c)/3]:
a[5]:=[-c/3,c/3,c+2*p/3]:a[6]:=[-(2*c+p)/3,0,c+p/3]:
a[7]:=[-(2*c+2*p)/3,(p+c)/3,(2*c+p)/3]:
a[8]:=[-c-2*p/3,c/3,c/3]:a[9]:=[-c-2*p/3,-c/3,c/3]:
a[10]:=[-(2*c+2*p)/3,-(c+p)/3,(2*c+p)/3]:
a[11]:=[-(2*c+p)/3,(2*c+2*p)/3,(c+p)/3]:
a[12]:=[-c/3,c+2*p/3,c/3]:a[13]:=[-c/3,c+2*p/3,-c/3]:
a[14]:=[-(2*c+p)/3,(2*c+2*p)/3,-(c+p)/3]:
a[15]:=[-c-p/3,(2*c+p)/3,0]: a[17]:=[c+p/3,(2*c+p)/3,0]:
a[16]:=[(2*c+p)/3,(2*c+2*p)/3,(c+p)/3]:
a[18]:=[(2*c+p)/3,(2*c+2*p)/3,-(c+p)/3]:
a[19]:=[c/3,c+2*p/3,-c/3]: a[20]:=[c/3,c+2*p/3,c/3]:
a[21]:=[-(c+p)/3,(2*c+p)/3,-(2*c+2*p)/3]:
a[22]:=[0,c+p/3,-(2*c+p)/3]: a[24]:=[c/3,c/3,-c-2*p/3]:
```

```
a[23]:=[(c+p)/3,(2*c+p)/3,-(2*c+2*p)/3]:
a[25]:=[-c/3,c/3,-c-2*p/3]: a[26]:=[(2*c+p)/3,0,-c-p/3]:
a[27]:=[-(2*c+2*p)/3,-(c+p)/3,-(2*c+p)/3]:
a[28]:=[-c-2*p/3,-c/3,-c/3]: a[29]:=[-c-2*p/3,c/3,-c/3]:
a[30]:=[-(2*c+2*p)/3,(c+p)/3,-(2*c+p)/3]:
a[31]:=[-c/3,-c/3,c+2*p/3]: a[33]:=[0,-c-p/3,(2*c+p)/3]:
a[32]:=[-(c+p)/3,-(2*c+p)/3,(2*c+2*p)/3]:
a[34]:=[(c+p)/3,-(2*c+p)/3,(2*c+2*p)/3]:
a[35]:=[c/3,-c/3,c+2*p/3]: a[36]:=[(2*c+p)/3,0,c+p/3]:
a[37]:=[(2*c+2*p)/3,-(c+p)/3,(2*c+p)/3]:
a[38]:=[c+2*p/3,-c/3,c/3]: a[39]:=[c+2*p/3,c/3,c/3]:
a[40]:=[(2*c+2*p)/3,(c+p)/3,(2*c+p)/3]:
a[41]:=[(2*c+p)/3,-(2*c+2*p)/3,(c+p)/3]:
a[42]:=[c/3,-c-2*p/3,c/3]: a[43]:=[c/3,-c-2*p/3,-c/3]:
a[44]:=[(2*c+p)/3,-(2*c+2*p)/3,-(c+p)/3]:
a[45]:=[c+p/3,-(2*c+p)/3,0]: a[46]:=[c+2*p/3,-c/3,-c/3]:
a[47]:=[(2*c+2*p)/3,-(c+p)/3,-(2*c+p)/3]:
a[48]:=[(2*c+p)/3,0,-c-p/3]: a[50]:=[c+2*p/3,c/3,-c/3]:
a[49]:=[(2*c+2*p)/3,(c+p)/3,-(2*c+p)/3]:
a[51]:=[-(c+p)/3,-(2*c+p)/3,-(2*c+2*p)/3]:
a[52]:=[-c/3,-c/3,-c-2*p/3]: a[53]:=[c/3,-c/3,-c-2*p/3]:
a[54]:=[(c+p)/3,-(2*c+p)/3,-(2*c+2*p)/3]:
a[55]:=[0,-c-p/3,-(2*c+p)/3]: a[56]:=[-c/3,-c-2*p/3,-c/3]:
a[58]:=[-(2*c+p)/3,-(2*c+2*p)/3,(c+p)/3]:
a[59]:=[-c-p/3,-(2*c+p)/3,0]: a[57]:=[-c/3,-c-2*p/3,c/3]:
a[60]:=[-(2*c+p)/3,-(2*c+2*p)/3,-(p+c)/3]:

> g[1]:=[a[4],a[11],a[7]]:g[2]:=[a[3],a[20],a[12]]:
g[3]:=[a[19],a[22],a[13]]:g[4]:=[a[14],a[21],a[30]]:
g[5]:=[a[8],a[15],a[29]]:g[6]:=[a[34],a[41],a[37]]:
g[7]:=[a[33],a[57],a[42]]:g[8]:=[a[38],a[45],a[46]]:
g[9]:=[a[44],a[54],a[47]]:g[10]:=[a[43],a[56],a[55]]:
g[11]:=[a[1],a[35],a[36]]:g[12]:=[a[16],a[2],a[40]]:
g[13]:=[a[50],a[17],a[39]]:g[14]:=[a[49],a[23],a[18]]:
g[15]:=[a[24],a[48],a[53]]:g[16]:=[a[52],a[26],a[25]]:
g[17]:=[a[51],a[60],a[27]]:g[18]:=[a[59],a[9],a[28]]:
g[19]:=[a[31],a[5],a[6]]:g[20]:=[a[58],a[32],a[10]]:
g[21]:=[a[30],a[11],a[10]]:
g[22]:=[a[1],a[2],a[3],a[4],a[5]]:
g[23]:=[a[6],a[7],a[8],a[9],a[10]]:
g[24]:=[a[11],a[12],a[13],a[14],a[15]]:
g[25]:=[a[16],a[17],a[18],a[19],a[20]]:
g[26]:=[a[21],a[22],a[23],a[24],a[25]]:
g[27]:=[a[26],a[27],a[28],a[29],a[30]]:
g[28]:=[a[31],a[32],a[33],a[34],a[35]]:
g[29]:=[a[36],a[37],a[38],a[39],a[40]]:
g[30]:=[a[41],a[42],a[43],a[44],a[45]]:
g[31]:=[a[46],a[47],a[48],a[49],a[50]]:
g[32]:=[a[51],a[52],a[53],a[54],a[55]]:
g[33]:=[a[56],a[57],a[58],a[59],a[60]]:
g[34]:=[a[2],a[16],a[20],a[3]]:
g[35]:=[a[3],a[12],a[11],a[4]]:
```

```
  g[36]:=[a[4],a[7],a[6],a[5]]:
  g[37]:=[a[7],a[11],a[15],a[8]]:
  g[38]:=[a[15],a[14],a[30],a[29]]:
  g[39]:=[a[14],a[13],a[22],a[21]]:
  g[40]:=[a[13],a[12],a[20],a[19]]:
  g[41]:=[a[19],a[18],a[23],a[22]]:
  g[42]:=[a[21],a[25],a[26],a[30]]:
  g[43]:=[a[29],a[28],a[9],a[8]]:
  g[44]:=[a[35],a[34],a[37],a[36]]:
  g[45]:=[a[34],a[33],a[42],a[41]]:
  g[46]:=[a[33],a[32],a[58],a[57]]:
  g[47]:=[a[57],a[56],a[43],a[42]]:
  g[48]:=[a[43],a[55],a[54],a[44]]:
  g[49]:=[a[44],a[47],a[46],a[45]]:
  g[50]:=[a[45],a[38],a[37],a[41]]:
  g[51]:=[a[38],a[46],a[50],a[39]]:
  g[52]:=[a[47],a[54],a[53],a[48]]:
  g[53]:=[a[55],a[56],a[60],a[51]]:
  g[54]:=[a[1],a[36],a[40],a[2]]:
  g[55]:=[a[16],a[40],a[39],a[17]]:
  g[56]:=[a[17],a[50],a[49],a[18]]:
  g[57]:=[a[23],a[49],a[48],a[24]]:
  g[58]:=[a[24],a[53],a[52],a[25]]:
  g[59]:=[a[26],a[52],a[51],a[27]]:
  g[60]:=[a[27],a[60],a[59],a[28]]:
  g[61]:=[a[9],a[59],a[58],a[10]]:
  g[62]:=[a[10],a[32],a[31],a[6]]:
  g[63]:=[a[5],a[31],a[35],a[1]]:
> P:=j -> plots[polygonplot3d]([seq(g[i], i=1..63)],
  orientation=[20*j,50], style=PATCH):
  plots[display]([seq(P(j), j=1..18)], insequence=true);
> restart: c:=1: d:=evalf(2*c*(3*sqrt(2)+2)/7): # 468
  m:=evalf(d*(3*sqrt(2)-4)/4):
  l:=evalf(m+d*(2-sqrt(2))/4): t:=evalf(d*(2-sqrt(2))/4):
> a[1]:=[c-l,-t,c]: a[2]:=[t,-c+l,c]: a[3]:=[-t,-c+l,c]:
  a[4]:=[-c+l,-t,c]: a[5]:=[-c+l,t,c]: a[6]:=[-t,c-l,c]:
  a[7]:=[t,c-l,c]: a[8]:=[c-l,t,c]: a[9]:=[c,t,c-l]:
  a[10]:=[c,-t,c-l]: a[11]:=[c,-c+l,t]: a[12]:=[c,-c+l,-t]:
  a[13]:=[c,-t,-c+l]: a[14]:=[c,t,-c+l]: a[15]:=[c,c-l,-t]:
  a[16]:=[c,c-l,t]: a[17]:=[c-l,c,t]: a[18]:=[c-l,c,-t]:
  a[19]:=[t,c,-c+l]: a[20]:=[-t,c,-c+l]: a[21]:=[-t,c-l,-c]:
  a[22]:=[t,c-l,-c]: a[23]:=[c-l,t,-c]: a[24]:=[-c+l,t,-c]:
  a[25]:=[-c+l,c,-t]: a[26]:=[-c,c-l,-t]: a[27]:=[-c,t,-c+l]:
  a[28]:=[-c,-t,-c+l]: a[29]:=[-c+l,-t,-c]: a[30]:=[-t,c,c-l]:
  a[31]:=[t,c,c-l]: a[32]:=[-c+l,c,t]: a[33]:=[-c,c-l,t]:
  a[34]:=[-c,t,c-l]: a[35]:=[-c,-t,c-l]: a[36]:=[-c,-c+l,t]:
  a[37]:=[-c+l,-c,t]: a[38]:=[-c+l,-c,-t]:
  a[39]:=[-c,-c+l,-t]: a[40]:=[-t,-c,-c+l]:
  a[41]:=[-t,-c+l,-c]: a[42]:=[t,-c+l,-c]:
  a[43]:=[t,-c,-c+l]: a[44]:=[c-l,-c,-t]: a[45]:=[c-l,-t,-c]:
  a[46]:=[t,-c,c-l]: a[47]:=[-t,-c,c-l]: a[48]:=[c-l,-c,t]:
```

```
> g[1]:=([a[1],a[2],a[3],a[4],a[5],a[6],a[7],a[8]]):
  g[2]:=([a[17],a[18],a[19],a[20],a[25],a[32],a[30],a[31]]):
  g[3]:=([a[22],a[23],a[45],a[42],a[41],a[29],a[24],a[21]]):
  g[4]:=([a[37],a[38],a[40],a[43],a[44],a[48],a[46],a[47]]):
  g[5]:=([a[9],a[10],a[11],a[12],a[13],a[14],a[15],a[16]]):
  g[6]:=([a[26],a[27],a[28],a[39],a[36],a[35],a[34],a[33]]):
  g[7]:=([a[5],a[6],a[30],a[32],a[33],a[34]]):
  g[8]:=([a[3],a[4],a[35],a[36],a[37],a[47]]):
  g[9]:=([a[1],a[2],a[46],a[48],a[11],a[10]]):
  g[10]:=([a[7],a[8],a[9],a[16],a[17],a[31]]):
  g[11]:=([a[20],a[25],a[26],a[27],a[24],a[21]]):
  g[12]:=([a[28],a[39],a[38],a[40],a[41],a[29]]):
  g[13]:=([a[14],a[23],a[22],a[19],a[18],a[15]]):
  g[14]:=([a[12],a[13],a[45],a[42],a[43],a[44]]):
  g[15]:=([a[6],a[7],a[31],a[30]]):
  g[16]:=([a[1],a[10],a[9],a[8]]):
  g[17]:=([a[2],a[3],a[47],a[46]]):
  g[18]:=([a[4],a[5],a[34],a[35]]):
  g[19]:=([a[32],a[25],a[26],a[33]]):
  g[20]:=([a[36],a[37],a[38],a[39]]):
  g[21]:=([a[11],a[12],a[44],a[48]]):
  g[22]:=([a[15],a[16],a[17],a[18]]):
  g[23]:=([a[19],a[20],a[21],a[22]]):
  g[24]:=([a[27],a[24],a[29],a[28]]):
  g[25]:=([a[40],a[41],a[42],a[43]]):
  g[26]:=([a[13],a[14],a[23],a[45]]):
> P:=j -> plots[polygonplot3d]([seq(g[i], i=1..26)],
  orientation=[20*j,50], style=PATCH):
  plots[display]([seq(P(j), j=1..18)], insequence=true);
> restart: # 4610
> c:=1: p:=evalf((sqrt(5)-1)*c/2): co:=evalf(cos(32)):
  m:=evalf((c*(sqrt(5)-1))/(2+2*sin(54*Pi/180))):
  k:=evalf(c*(sqrt(5)-1)-2*m): d:=evalf((2*p-m)/2): t:=m/2:
  r:=evalf(m*cos(72*Pi/180)): tau:=evalf((sqrt(5)-1)/2):
  l:=evalf((tau/(sqrt(3+tau)*(1+tau)+2*tau))):
> P:=j -> plots[polygonplot3d]([seq(g[i], i=1..62)],
  orientation=[20*j,50], style=PATCH):
  plots[display]([seq(P(j), j=1..18)], insequence=true);
> restart: # 3444-2
  c:=1: d:=evalf((c*sqrt(2))/(2+sqrt(2))):
> x[1]:=d: y[1]:=d: z[1]:=c:x[2]:=-d: y[2]:=d: z[2]:=c:
  x[3]:=-d: y[3]:=-d: z[3]:=c: x[4]:=d: y[4]:=-d: z[4]:=c:
  x[5]:=c: y[5]:=d: z[5]:=d: x[6]:=c: y[6]:=d: z[6]:=-d:
  x[7]:=c: y[7]:=-d: z[7]:=-d: x[8]:=c: y[8]:=-d: z[8]:=d:
  x[9]:=-d: y[9]:=c: z[9]:=d: x[10]:=-d: y[10]:=c: z[10]:=-d:
  x[11]:=d: y[11]:=c: z[11]:=-d: x[12]:=d: y[12]:=c: z[12]:=d:
  x[13]:=d: y[13]:=-c: z[13]:=d: x[14]:=d: y[14]:=-c: z[14]:=-d:
  x[15]:=-d: y[15]:=-c: z[15]:=-d: x[16]:=-d: y[16]:=-c: z[16]:=d:
  x[17]:=-c: y[17]:=-d: z[17]:=d: x[18]:=-c: y[18]:=-d: z[18]:=-d:
  x[19]:=-c: y[19]:=d: z[19]:=-d: x[20]:=-c: y[20]:=d: z[20]:=d:
```

```
  x[21]:=evalf(d*sqrt(2)): y[21]:=0: z[21]:=-c:
  x[22]:=0: y[22]:=evalf(d*sqrt(2)): z[22]:=-c:
  x[23]:=evalf(-d*sqrt(2)): y[23]:=0: z[23]:=-c:
  x[24]:=0: y[24]:=evalf(-d*sqrt(2)): z[24]:=-c:
> for i from 1 to 24 do a[i]:=[x[i],y[i],z[i]] od:
> for i from 1 to 6 do  g[i]:=[seq(a[4*(i-1)+j],j=1..4)] od:
> g[7]:=[a[1],a[4],a[8],a[5]]:
  g[8]:=[a[4],a[13],a[16],a[3]]:
  g[9]:=[a[3],a[17],a[20],a[2]]:
  g[10]:=[a[2],a[9],a[12],a[1]]:
  g[11]:=[a[12],a[5],a[6],a[11]]:
  g[12]:=[a[8],a[13],a[14],a[7]]:
  g[13]:=[a[16],a[17],a[18],a[15]]:
  g[14]:=[a[20],a[9],a[10],a[19]]:
  g[15]:=[a[11],a[22],a[21],a[6]]:
  g[16]:=[a[7],a[21],a[24],a[14]]:
  g[17]:=[a[15],a[24],a[23],a[18]]:
  g[18]:=[a[10],a[19],a[23],a[22]]:
  g[19]:=[a[1],a[12],a[5]]:g[20]:=[a[4],a[8],a[13]]:
  g[21]:=[a[3],a[16],a[17]]:g[22]:=[a[2],a[20],a[9]]:
  g[23]:=[a[6],a[21],a[7]]:g[24]:=[a[14],a[24],a[15]]:
  g[25]:=[a[18],a[23],a[19]]:g[26]:=[a[10],a[22],a[11]]:
> P:=j -> plots[polygonplot3d]([seq(g[i], i=1..26)],
  orientation=[20*j,50], style=PATCH):
  plots[display]([seq(P(j), j=1..18)], insequence=true);
> restart: # 333n
> n:=9: r:=1: u:=evalf(Pi/n):
  c:=evalf((r*sqrt(3*sqrt(sin(Pi/n))-sqrt(1-cos(Pi/n))))/2):
> for i from 1 to n+1 do
  x[2*i-1]:= evalf(r*cos(i*2*Pi/n)):
  y[2*i-1]:=evalf(r*sin(i*2*Pi/n)): z[2*i-1]:=c:
  x[2*i]:=evalf(r*cos(u+i*2*Pi/n)):
  y[2*i]:=evalf(r*sin(u+i*2*Pi/n)): z[2*i]:=-z[2*i-1] od:
> for i from 1 to 2*n+2 do a[i]:=[x[i],y[i],z[i]] od:
> for i from 1 to 2*n do g[i]:=[a[i], a[i+1], a[i+2]] od:
  g[2*n+2]:=[seq(a[2*i], i=1..n)]:
  g[2*n+1]:=[seq(a[2*i-1], i=1..n)]:
> P:=j -> plots[polygonplot3d]([seq(g[i], i=1..2*n+2)],
  orientation=[20*j, 50], style=PATCH):
  plots[display]([seq(P(j), j=1..18)], insequence=true);
> restart: # 44n
> n:=7: r:=1: c:=evalf(r*sin(Pi/n)): # n-prism
> a:=i->[evalf(r*cos(i*2*Pi/n)),evalf(r*sin(i*2*Pi/n)),0]:
> for i from 1 to n do
     g[i]:=[a(i),a(i+1),a(i+1)+[0,0,2*c],a(i)+[0,0,2*c]] od:
  g[n+1]:=[seq(a(i), i=1..n)]:
  g[n+2]:=[seq(a(i)+[0,0,2*c], i=1..n)]:
> P:=j -> plots[polygonplot3d]([seq(g[i], i=1..n+2)],
```

```
orientation=[20*j,50], style=PATCH):
plots[display]([seq(P(j), j=1..18)], insequence=true);
```

Part IV

Surfaces with MAPLE

19
Surfaces in Space

In Sections 19.1 and 19.2 we consider some basic notions of a parametrized surface and a regular surface (analogous definitions for curves were studied in Section 5.1). In Section 19.3 we use a number of MAPLE commands to produce surfaces by various methods. In Section 19.4 we calculate and plot tangent planes and normal vectors of a surface. As an application we solve the conditional extremum problems in space (see the two-dimensional case in Section 5.6). In Section 19.5 we use changes in coordinates and linear transformations in space to calculate and plot an osculating paraboloid at the point of a surface. This elementary approach is given only for methodical reasons. In Section 19.6 we consider parametrized and implicitly defined surfaces with singularities.

In Chapter 19 the reader will become acquainted with the commands

```
implicitplot3d, coordplot3d, plot3d, display3d, surfdata,
complexplot3d, listplot3d, matrixplot, PDEplot,
hilbert, toeplitz, multiply, transpose, coeftayl, mtaylor.
```

19.1 What Is a Surface?

Euclid defines a *surface* intuitively, as a two-dimensional figure swept by the path of a moving curve, bounding a solid body.

Planes, some polyhedra, and also some curved surfaces, each defined in some practical way, are studied in school. The mathematically correct definition of a surface is based on notions from *topology*, but it starts from the key notion of an *elementary surface*, which can be imagined as an *elementary do-*

main of the plane after continuous deformation that is stretched in and out in space.

Definition 19.1.1 The set $G \subset \mathbb{R}^n$ $(n = 2, 3)$ is called an *open set* if for each point $P \in G$ there exists $\varepsilon > 0$ such that the ball $B(P, \varepsilon)$ of radius ε with center P (the disk, when $n = 2$) lies inside of G.

The complement $F = \mathbb{R}^3 \setminus G$ to the open set G is called a *closed set*.

A *connected* open set (i.e., one that is not the union of two non-empty and non-intersecting open sets) is called a *domain*.

Domains are convenient for working with continuous and differentiable (vector) functions defined on such sets. An arbitrary open set is the union of a finite or countable number of disjoint domains.

Definition 19.1.2 A domain in the plane (in space) that is homeomorphic to a disk (respectively, a ball) is called an *elementary domain*.

The following two elementary domains in the plane with coordinates u, v are often used: the disc $u^2 + v^2 < R^2$ of radius R and the rectangle with sides a, b parallel to the coordinate axes.

As in the definition of an elementary curve, we use the notion of a *topological map* (*homeomorphism*) from one geometric figure to another, i.e., a one-to-one map that is continuous and has a continuous inverse.

Definition 19.1.3 A set M in space is called an *elementary surface* if it is the image of a planar elementary domain G under a topological map $\vec{\mathbf{r}} : G \to \mathbb{R}^3$.

If we fix rectangular coordinates (with the orthonormal base $\{\mathbf{i}, \mathbf{j}, \mathbf{k}\}$) with origin O in space $\mathbb{R}^3$, then the coordinates x, y, z of a point on an elementary surface given by the map $\vec{\mathbf{r}}$ are functions of the coordinates u, v of a point in the domain G:

$$x = x(u, v), \quad y = y(u, v), \quad z = z(u, v). \tag{19.1}$$

In other words, the position vector $\vec{\mathbf{r}} = \vec{OP}$ of the point $P(x, y, z)$ on the surface M is the following vector-valued function:

$$\vec{\mathbf{r}} = \vec{\mathbf{r}}(u, v) = x(u, v)\,\mathbf{i} + y(u, v)\,\mathbf{j} + z(u, v)\,\mathbf{k}, \qquad (u, v) \in G. \tag{19.1 a}$$

We define functions $x(u, v)$, $y(u, v)$, $z(u, v)$ and then use the following command in MAPLE:

```
> r:=[x(u,v), y(u,v), z(u,v)];
```

If the initial point of the continuous vector $\vec{\mathbf{r}}\,(u, v)$ coincides with the origin O, then its endpoint, as a rule, moves along a *surface* in $\mathbb{R}^3$ called the *hodograph* of the vector-valued function. For example, a plane in space is the hodograph of a linear vector-valued function in the variables u, v:

$$\vec{\mathbf{r}}(u, v) = [a_1 u + b_1 v + c_1,\ a_2 u + b_2 v + c_2,\ a_3 u + b_3 v + c_3].$$

Definition 19.1.4 The system of equalities (19.1) or the vector equation (19.1 a), representing a map $\vec{\mathbf{r}} : G \to \mathbb{R}^3$ from the planar domain G into $\mathbb{R}^3$ is called the *parametric equations of the surface* $M = \vec{\mathbf{r}}(G)$: the pair of real numbers (u, v) is called the *curvilinear coordinates* of the point $P(x, y, z)$ on the surface.

Let us first take $v = v_0$ as a constant and $u = t$ as varying. Then we obtain the coordinate *u-curve* on the surface M, which also is the space curve $\vec{\mathbf{r}}(t, v_0)$. Similarly, for a constant $u = u_0$ and $v = t$ varying, we get the *v-curve* on the surface M, that is, the space curve $\vec{\mathbf{r}}(u_0, t)$. These two families of curves, (v-curves and u-curves), form the *coordinate net* of the surface, Fig. 19.1.

An arbitrary *curve* γ *on the surface* M is defined locally by equations for the curvilinear coordinates $u = u(t)$, $v = v(t)$ $(t \in I)$. The equations for t γ have the following form in space:

$$\tilde{x}(t) = x(u(t), v(t)), \quad \tilde{y}(t) = y(u(t), v(t)), \quad \tilde{z}(t) = z(u(t), v(t)).$$

```
> subs({u=u(t), v=v(t)}, r);
```

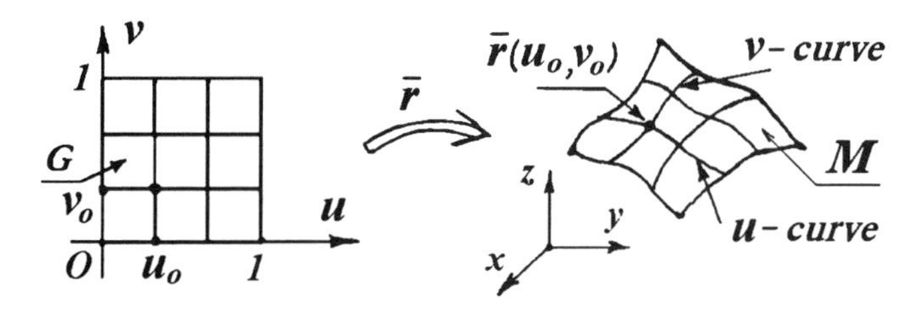

Fig. 19.1. Parametrized (elementary) surface

Since we will also be interested in *self-intersecting* surfaces, we give the following definitions:

Definition 19.1.5 A set M in $\mathbb{R}^3$ is called a *surface* if it can be covered by a finite or countable number of elementary surfaces.

Definition 19.1.6 A surface M is called a *simple surface* if it has no points of self-intersection, i.e., for any point $P \in M$ there exists $\varepsilon > 0$ such that the set $U(P, \varepsilon) = B(P, \varepsilon) \cap M$ (the intersection of M with the ball of the radius ε) is an elementary surface.

An elementary surface is a simple one. Examples of surfaces that are simple but not elementary are the following: the *sphere*, Fig. 20.7 (which can be covered by two disks); the *circular cylinder*, Fig. 19.5 (which can be covered by two cylindrical strips, each homeomorphic to a rectangle); the *torus*, Fig. 19.3 (which can be covered by four rectangles). The following surfaces are not simple: the *Klein bottle*, Fig. 19.7 (whose points of self-intersection form a circle); the *cone*, Fig. 19.15 (whose vertex is a *singular point*); a *pair of intersecting planes* (the self-intersection points form a straight line); the *pinched torus*, Fig. 19.4 (whose self-intersection points form two line segments). A more complicated example of a surface is the *union of a countable number of cylinders with the common generatrix* OZ: $M = \bigcup_{n\in\mathbb{N}} \{x^2 + (y-n)^2 = n^2\}$. It cannot be covered by a finite number of elementary surfaces. Note that for *local studies* of surfaces (i.e., in a neighborhood of some point) we need only the notion of an elementary surface.

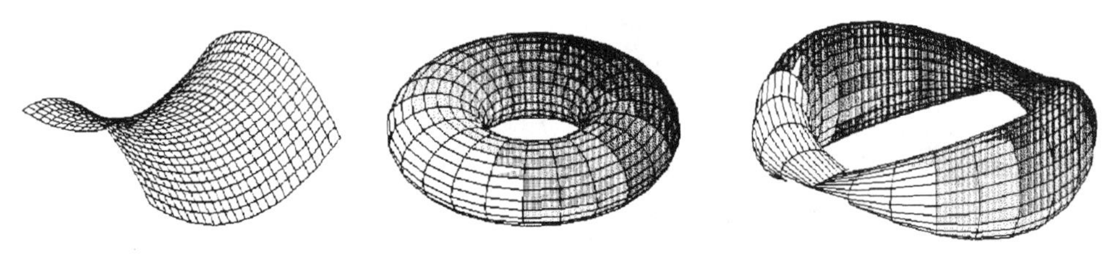

Figs. 19.2–19.4. An elementary surface, the torus, and a pinched torus

Remark 19.1.1 Sometimes we consider surfaces with a *boundary*: in the case of elementary surfaces, a boundary is the image of the boundary of an elementary domain and is homeomorphic to the circle. The boundary of the *half-plane* is a straight line, the boundary of part of a *cylinder* between its parallels is a pair of circles (Fig. 19.5), the boundary of the Möbius band (Fig. 19.6). Recall that a *Möbius band* is obtained when we twist a thin strip of paper through a $180°$ angle and then glue its lateral sides.

A remarkable property of the Möbius band is that it has only *one side*. Moreover, a surface that "contains" a Möbius band is called *one-sided*. Another example of a one-sided surface is the *Klein bottle*.

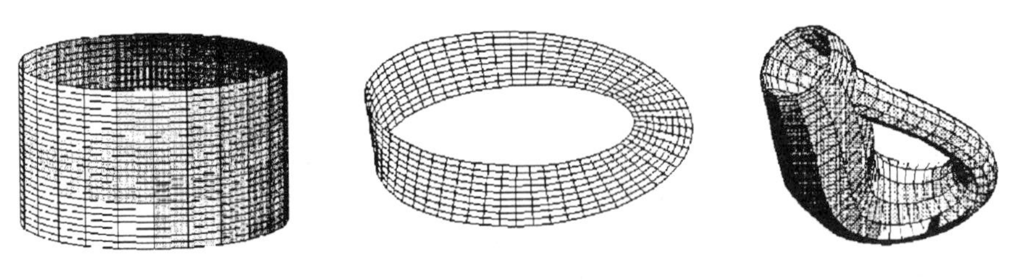

Figs. 19.5–19.7. Circular cylinder, the Möbius band, and the Klein bottle

Definition 19.1.7 A simple surface M is called *complete* if the limit point of any sequence of its points which converges in $\mathbb{R}^3$ belongs to M. If a complete surface does not have a *boundary*, it is called *closed*. A closed and bounded surface is called *compact* — recall that a body in $\mathbb{R}^3$ is called *bounded* if it is contained inside some ball of finite radius. This last notion has deep topological generalizations.

The paraboloid, the hyperboloid, and cylinders of the second order are examples of closed surfaces. The sphere, the torus, and the Klein bottle are examples of compact surfaces. If we omit a closed set from a complete surface, then we will obtain an incomplete surface.

Remark 19.1.2 We can derive in MAPLE some calculus-types of operations on vector-valued functions $\vec{\mathbf{r}}(u, v) = [x(u, v),\ y(u, v),\ z(u, v)]$ of the class C^k. The key is in using coordinates: the *limit* $\lim\limits_{(u,v)\to(u_0,v_0)} \vec{\mathbf{r}}(u, v)$,

```
> s:={u=u0, v=v0}:
  r0:=[subs(s,x(u,v)), subs(s,y(u,v)), subs(s,z(u,v))];
```

and the *partial derivatives of the first order*

$$\vec{\mathbf{r}}_u = \lim_{\Delta u\to 0} \frac{\vec{\mathbf{r}}(u+\Delta u, v) - \vec{\mathbf{r}}(u, v)}{\Delta u}, \qquad \vec{\mathbf{r}}_v = \lim_{\Delta v\to 0} \frac{\vec{\mathbf{r}}(u, v+\Delta v) - \vec{\mathbf{r}}(u, v)}{\Delta v}.$$

```
> ru:=[Diff(x(u,v),u), Diff(y(u,v),u), Diff(z(u,v),u)];
```

Since $\vec{\mathbf{r}}_u = [x'_u, y'_u, z'_u]$, $\vec{\mathbf{r}}_v = [x'_v, y'_v, z'_v]$ hold, where $x'_u = \frac{\partial x(u,v)}{\partial u}$, then the differential of the vector-valued function $\vec{\mathbf{r}}(u, v)$ has the form $d\vec{\mathbf{r}} = [dx, dy, dz]$, and hence

$$d\vec{\mathbf{r}} = [x'_u\, du + x'_v\, dv,\ y'_u\, du + y'_v\, dv,\ z'_u\, du + z'_v\, dv] = \vec{\mathbf{r}}_u\, du + \vec{\mathbf{r}}_v\, dv.$$

Analogously, we can calculate the second partial derivatives $\vec{\mathbf{r}}_{uu}$, $\vec{\mathbf{r}}_{vv}$, $\vec{\mathbf{r}}_{uv}$, the *Taylor decomposition* (see example in Section 19.5.2), and so on.

19.2 Regular Parametrized Surface

By definition, in a neighborhood of any its points, a simple surface can be defined by a parametric equation:

$$\vec{\mathbf{r}} = \vec{\mathbf{r}}(u, v) \iff x = x(u, v),\ y = y(u, v),\ z = z(u, v).$$

Suppose that the coordinate functions $x(u, v)$, $y(u, v)$, $z(u, v)$ are defined on some planar domain G and are of class C^k in the variables u and v. In this case the u- and v-curves (space curves) are said to be of class C^k, and we can find their velocity vectors:

$$\vec{\mathbf{r}}_u := x'_u(u, v)\,\mathbf{i} + y'_u(u, v)\,\mathbf{j} + z'_u(u, v)\,\mathbf{k} \ \text{(of the } u\text{-curve)},$$
$$\vec{\mathbf{r}}_v := x'_v(u, v)\,\mathbf{i} + y'_v(u, v)\,\mathbf{j} + z'_v(u, v)\,\mathbf{k} \ \text{(of the } v\text{-curve)}.$$

Definition 19.2.1 A surface M is called *regular of class C^k* (resp. C^∞ or C^ω) if each of its points has a neighborhood with a parametrization $\vec{\mathbf{r}}(u, v) = [x(u, v), y(u, v), z(u, v)]$, where the functions $x(u, v), y(u, v), z(u, v)$, defined on the elementary domain $G \in \mathbb{R}^2$, are of class C^k (resp. C^∞ or C^ω); moreover, we require that the vectors $\vec{\mathbf{r}}_u(u, v)$ and $\vec{\mathbf{r}}_v(u, v)$ be linearly independent at every point $(u, v) \in G$. For $k = 1$, such a surface is called *smooth.*

Each of the following conditions is equivalent to the property of linear independence of vectors $\vec{\mathbf{r}}_u(u, v)$ and $\vec{\mathbf{r}}_v(u, v)$ and is often used for checking this geometrical property:

(a) *The matrix* $\begin{pmatrix} x_u & y_u & z_u \\ x_v & y_v & z_v \end{pmatrix}$ *at the point* (u, v) *has rank* 2.

(b) *The cross product* $\vec{\mathbf{n}} := \vec{\mathbf{r}}_u \times \vec{\mathbf{r}}_v$ *is nonzero at the point* (u, v).

Definition 19.2.2 A point P on a simple surface M is called a *regular (smooth) point* if some neighborhood P admits a C^k-parametrization $\vec{\mathbf{r}}(u, v)$ such that the vectors $\vec{\mathbf{r}}_u$ and $\vec{\mathbf{r}}_v$ are linearly independent at P; otherwise, P is called a *singular point* (for example, the vertex on the top of a circular cone or the bottom fold of such a cone). A curve on a surface, all of whose points are singular, is called a *singular curve* (for example, two line segments on the pinched torus, Fig. 19.4; a singular generatrix on a cylinder over a plane curve with one singular point, Fig. 19.38).

A surface without singular points is smooth; its small open neighborhoods look like plane disks slightly deformed in space.

Although a regular surface admits a number of parametrizations locally, there does not exist a "natural" way to parameterize it, as in the case of curves. We consider the transition from one parametrization to another.

Lemma 19.2.1 *Let* $\mathbf{f} : G \to M$ *be a* C^k*-regular parametrization of the surface* M *and let* $\mathbf{h} : G_1 \to G$ *be an injective map of class* C^k *on the elementary domain* G_1 *into some domain* G *with nonzero Jacobian* J_h. *Then the composition* $\mathbf{f}_1 = \mathbf{f} \circ \mathbf{h} : G_1 \to M$ *is also a* C^k*-regular parametrization of a domain on* M*;* Fig. 19.8.

Proof. Let the map $\mathbf{f}$ be defined by the vector-valued function $\vec{\mathbf{r}} = [x(u, v), y(u, v), z(u, v)]$, and let $\mathbf{h}$ be given by the functions $u = \varphi(\alpha, \beta)$, $v = \psi(\alpha, \beta)$. Then the composition $\mathbf{f}_1$ is given by the function

$$\tilde{\mathbf{r}}(\alpha, \beta) = [x(u(\alpha, \beta), v(\alpha, \beta)),\ y(u(\alpha, \beta), v(\alpha, \beta)),\ z(u(\alpha, \beta), v(\alpha, \beta))],$$

and by standard theorems from calculus, the map $\mathbf{f}_1 : \tilde{\mathbf{r}} = \vec{\mathbf{r}}(\varphi(\alpha, \beta), \psi(\alpha, \beta))$ is of class C^k. To prove the regularity of the map $\mathbf{f}_1$ we start with $\tilde{\mathbf{r}}_\alpha = \vec{\mathbf{r}}_u \varphi_\alpha +$

$\vec{\mathbf{r}}_v\psi_\alpha, \quad \tilde{\mathbf{r}}_\beta = \vec{\mathbf{r}}_u\varphi_\beta + \vec{\mathbf{r}}_v\psi_\beta$. Moreover, $|\vec{\mathbf{r}}_u \times \vec{\mathbf{r}}_v| \neq 0$ holds in view of the regularity of the parametrization **f**. Hence

$$|\tilde{\mathbf{r}}_\alpha \times \tilde{\mathbf{r}}_\beta| = |\vec{\mathbf{r}}_u \times \vec{\mathbf{r}}_v| \cdot |\varphi_\alpha\psi_\beta - \varphi_\beta\psi_\alpha| = |\vec{\mathbf{r}}_u \times \vec{\mathbf{r}}_v| \cdot J_h \neq 0. \square$$

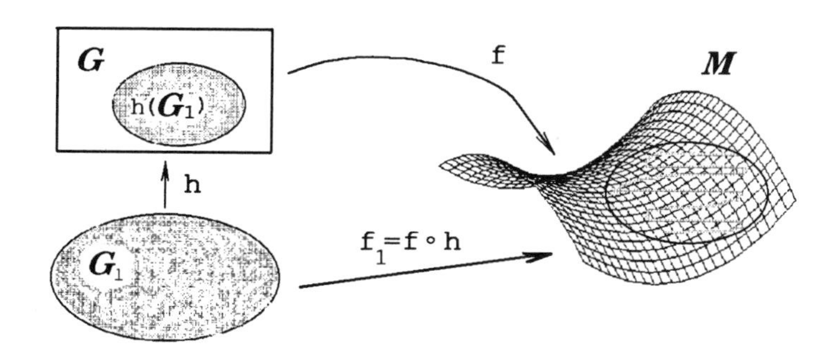

Figs. 19.8. A change of variables on a surface

Exercise 19.2.1

1. Prove that the vector-valued function

(i) $\vec{\mathbf{r}}(u, v) = [u + v, u - v, u^2 + v^2]$ is a C^∞-regular parametrization of the paraboloid $z = \frac{1}{2}(x^2 + y^2)$,

(ii) $\vec{\mathbf{r}}(u, v) = [u^2, uv, v^2]$, where $u > 0, \ v > 0$, defines a C^∞-regular coordinate system on part of the hyperbolic paraboloid $y^2 = xz$;

(iii) $\vec{\mathbf{r}}(u, v) = \left[\frac{a(uv+1)}{u+v}, \frac{b(u-v)}{u+v}, \frac{uv-1}{u+v}\right]$ is a C^∞-regular parametrization of part of a hyperboloid of one sheet.

What are the coordinate curves of these parametrizations?

Solution. (i) First check the regularity condition **(ii)**

```
> r:=[u+v, u-v, u^2+v^2]:
> ru:=diff(r,u): rv:=diff(r,v): linalg[crossprod](ru, rv);
```

$[2u + 2v, \ 2u - 2v, \ -2]$ # we obtain a nonzero vector.

Then substitute the vector-valued function in the explicit equation:

```
> r1:=simplify(subs({x=u+v,y=u-v,z=u^2+v^2} , z-(x^2+y^2)/2));
```

0

2. Find which of the following surfaces are compact, and plot them using the command `implicitplot3d` (see Section 19.3.2).

$$(a)\ x^2 - y^4 + z^6 = 1, \qquad (b)\ x^2 - 2x + y^2 + z^4 = 1,$$

$$(c)\ \ x^2 + y^2z^2 = 1, \qquad (d)\ x^2 + y^4 + z^6 = 1.$$

```
> plots[implicitplot3d](x^2+y^4+z^6=1,
  x=-2..2, y=-2..2, z=-2..2);   # etc.
```

3. A net of curves on a surface M is said to be *regular* if

• *Every point of the surface belongs to only one curve from each of two families of net curves.*

• *Any curve of one family of the net intersect every curve from another family at exactly one point.*

Check that

(a) The net of straight lines (two families) on the hyperboloid of one sheet is regular locally (i.e., on a neighborhood of an arbitrary point), but it is not regular on the whole surface.

(b) Parallels and meridians on the torus of revolution form a regular net.

Find domains in $\mathbb{R}^2$ where the (polar, parabolic, elliptic, hyperbolic, bipolar) coordinate nets, see Section 6.1, are regular nets.

19.3 Methods of Generating Surfaces

19.3.1 Graphs and Level Sets of Functions

Definition 19.3.1 The *graph* of the function $f : G \subset \mathbb{R}^2 \to \mathbb{R}$ in two real variables (x, y) is the set in space

$$\Gamma_f = \{(x, y, z) \in \mathbb{R}^3 : z = f(x, y), \quad (x, y) \in G\}.$$

Among surfaces of the second order, the elliptic and hyperbolic paraboloids $z = \frac{x^2}{p} \pm \frac{y^2}{q}$ can be expressed as graphs.

Every function in several variables is related to a number of *level sets* (forming the *chart of the function*), which are useful for studying the qualitative properties of the functions.

Definition 19.3.2 Let the function $f(x, y)$ be given on the domain $G \subset \mathbb{R}^2$. The *level curve* with height c of the function f is the set $f^{-1}(c) = \{(x, y) \in G : f(x, y) = c\}$ (the inverse image of c), for example, isobars and isoterms on geographical maps. For every function $F(x, y, z)$ defined on the domain $G \subset \mathbb{R}^3$, the analogous set $F^{-1}(c) = \{(x, y, z) \in G : F(x, y, z) = c\}$ (the inverse image of c) is called the *level surface* of height c of F. In this notation the real number c is called the *height* of the level set.

Note that the level set of height c for the function $f(x)$ consists of roots of the equation $f(x) = c$. In this way, space (or the plane) is fibered by level surfaces (or level curves) of the function F. Obviously, the graph of $f(x, y)$ is the level surface of height 0 for the function $z - f(x, y)$ in three real variables. If we intersect the graph Γ_f of the function $f(x, y)$ with the horizontal plane $z = c$ and project the intersection onto the plane XY, then we obtain the level curve $f^{-1}(c)$ of the given function; see Figs. 19.9–19.14. For most functions, it is complicated to draw their graphs in space. It is more convenient sometimes to visualize their behavior by drawing the charts $\{f^{-1}(c)\}$ $(c \in \mathbb{R})$. If we fix level curves through equal intervals of values $0, \pm d, \pm 2d, \ldots$, then by the density of level curves, we can judge the steepness of the graph. Level curves lie closer together where the slope of the graph to the horizontal plane is larger. Such *projections with real marks* are used in the construction of earthworks and are studied in drawing. Level sets are also useful in problems with extrema; see Section 19.4.2.

19.3.2 Methods of Generating Surfaces

Rectangular coordinates are familiar to the reader. Recall that the location of a point P in space can also be described by other curvilinear coordinate systems; see Section 8.1.3. The `plots[coordplot3d]` function plots a graphical representation of most of the three-dimensional coordinate systems currently supported by `MAPLE`. The coordinate systems that can be plotted are bipolarcylindrical, bispherical, cardiodal, cardiodcylindrical, casscylindrical, confocalellip, confocalparab, conical, cylindrical, ellcylindrical, ellipsoidal, hypercylindrical, invcasscylindrical, invellcylindrical, invoblspheroidal, invprospheroidal, logcoshcylindrical, logcylindrical, maxwellcylindrical, oblatespheroidal, paraboloidal, paraboloidal2, paracylindrical, prolatespheroidal, rosecylindrical, sixsphere, spherical, tangentcylindrical, tangentsphere, and toroidal.

```
> plots[coordplot3d](cylindrical, style=`PATCH`);
> plots[coordplot3d](spherical, style=`PATCH`); # etc.
```

A surface graph defined by the term `expr` is plotted by the command `plots[ <3d-system>](expr,var1=a..b,var2=c..d,<options>)`, where `<3d-system>` is `cylinderplot`, `sphereplot`, and so on.

In addition to *parametrized equations* of the surface $M \subset \mathbb{R}^3$, there are other analytic methods to represent it by formulas:

(1) as the *graph* of a function in two variables: $z = f(x, y)$,

(2) in *implicit* form, as the level set of a function in three variables: $F(x, y, z) = 0$.

Similar formulas are used for surfaces in the cases of *cylindrical, spherical,* and other *curvilinear coordinates in space.* Three analytical ways of generating simple surfaces are equivalent to one another in a *local* sense, i.e., considering neighborhoods of points on surfaces, but they are not equivalent for surfaces in the *large*. This can be seen from the examples of *surfaces of second order*. One component of the *hyperboloid of two sheets* $\frac{x^2}{a^2}+\frac{y^2}{b^2}-\frac{z^2}{c^2}=-1$ is defined as the graph $z=c\sqrt{1+\frac{x^2}{a^2}+\frac{y^2}{b^2}}$, and hence admits a parametrized form, but on the whole this surface intersects "almost every" straight line at two points and hence cannot be a graph; the sphere $x^2+y^2+z^2=R^2$ is not a graph for the same reasons; also, the following simple surfaces are not graphs: the circular cylinder, the hyperboloid of one sheet, the Möbius band.

Plot graphs of functions in two variables using the command `plot3d`.

```
> plot3d(sin(x)*cos(y), x=-Pi..Pi, y=-Pi..Pi);
> plot3d (x^2+y^2, x=-3..3, y=-sqrt(9-x^2)..sqrt(9-x^2));
> plot3d(x^2-y^2, x=-3..3, y=-3..3);        # Figs. 19.9–19.12
```

Plot level curves of these functions using the command `contourplot`.

```
> with(plots): contourplot(x^2+y^2, x=-3..3, y=-3..3);
> contourplot(sin(x)*cos(y), x=-Pi..Pi, y=-Pi..Pi);
> contourplot(x^2-y^2,x=-3..3,y=-3..3); # Figs. 19.12–19.14
```

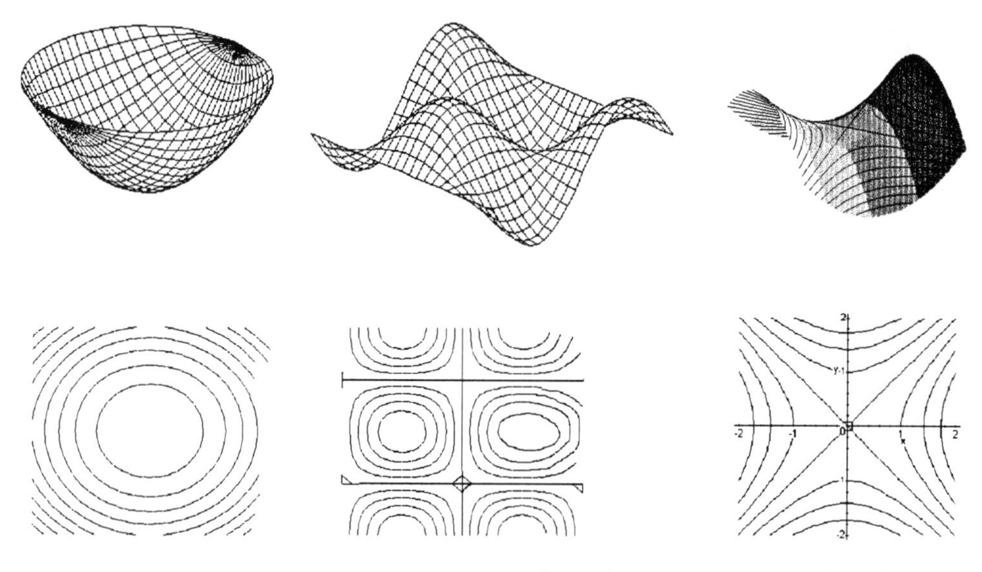

Figs. 19.9–19.14. Graphs of the functions x^2+y^2, $\sin(x)\cos(y)$, x^2-y^2 and their charts (level sets)

An implicitly given surface can be visualized using the command `implicitplot3d`, but this method is coarse, as can be seen from the example of the cone $x^2+y^2-z^2=0$.

```
> plots[implicitplot3d](x^2+y^2-z^2=0, x=-3..3, y=-3..3,
```

```
z=-5..5, style=patchcontour, orientation=[40,80]);
```

A graph is a particular case of a parametrized surface, because the equation $z = f(x, y)$ is equal to $\vec{\mathbf{r}}(u, v) = u\,\mathbf{i} + v\,\mathbf{j} + f(u, v)\,\mathbf{k}$:

```
> r:=[u, v, f(u,v)];
```

Starting with parametric equations, we plot both (up and down) sheets of a cone using the commands `plot3d` and `display3d`.

```
> r:=[u*cos(v), u*sin(v), u]:
> P1:=plot3d(r, u=0..3, v=0..2*Pi): %;
> P2:=plot3d(r, u=-3..0, v=0..2*Pi): %;
> plots[display3d]({P1, P2}, style=patchcontour,
  orientation=[45,75],axes=framed);    # Fig. 19.15
```

Plot the hyperboloid of one sheet using the above program with

```
> S:=signum(u)*sqrt(u^2+1): r:=[S*cos(v),S*sin(v),u]: #Fig.19.17
```

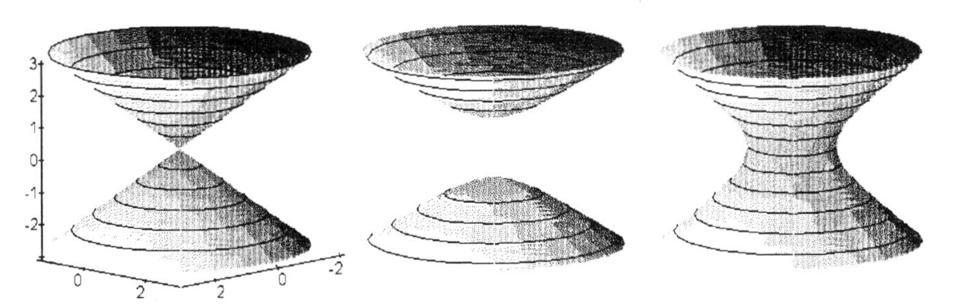

Figs. 19.15–19.17. Cone and hyperboloids

Plot graphs for complex expressions using the command `complexplot3d`

```
> plots[complexplot3d](GAMMA(z), z= -Pi-Pi*I..Pi+Pi*I,
  view=0..5, grid=[30,30], orientation=[-120,45],
  axes=frame, style=patchcontour, thickness=2);
```

Sufficient conditions for the smoothness of a surface graph are analogous to the case of the smooth plane curve.

Theorem 19.3.1 *Let $f(x, y)$ be a function of class C^k in the domain G. Then the equation $z = f(x, y)$ defines a C^k-regular surface graph.*

Proof. Take $x = u$, $y = v$ as the parameters of the vector-valued function $\vec{\mathbf{r}}(u, v) = [u,\ v,\ f(u, v)]$. We obtain linearly independent vectors $\vec{\mathbf{r}}_u = [1, 0,\ f'_u(u, v)]$ and $\vec{\mathbf{r}}_v = [0, 1,\ f'_v(u, v)]$. □

The notion of an implicitly defined surface also has its difficulty. An equation for continuous function in (x, y, z) without additional requirements can define an arbitrary closed set in space (for example, a ball). However, there are obvious sufficient conditions for the implicit equation $F(x, y, z) = c$ to define a surface, analogous to the case of an implicitly defined planar curve. Recall that the *gradient* of a function $F(x, y, z)$ is the vector field $\nabla F = [\frac{\partial F}{\partial x}, \frac{\partial F}{\partial y}, \frac{\partial F}{\partial z}]$.

Theorem 19.3.2 *Let $F(x, y, z)$ be a function of class C^k on the domain $G \subset \mathbb{R}^3$, and suppose that the gradient ∇F is nonzero on the level set $M_c = \{(x, y, z) \in G : F(x, y, z) = c\}$ of height c. Then M_c is a C^k-regular surface.*

Proof. Take the point $P(x_0, y_0, z_0) \in M_c$. If, for instance, $F_z'(P) \neq 0$ holds, than by the implicit function theorem, there exists a function $f(x, y)$ of class C^k such that $F(x, y, f(x, y)) \equiv 0$ holds on some neighborhood of the point (x_0, y_0). Hence the surface M_c in a neighborhood of the point P is defined by some explicit equation $z = f(x, y)$ and by Theorem 19.3.1, the surface is C^k-regular. □

Example 19.3.1 For the function $F = x^2 + y^2 - z^2$ the gradient $\nabla F = [2x, 2y, -2z]$ is zero only at the point $O(0, 0, 0)$,

```
> linalg[grad](x^2+y^2+z^2, [x,y,z]);
```

which is a singular point (the vertex) on the cone M_0. Other level sets M_c ($c \neq 0$) are smooth surfaces: hyperboloids (of one sheet when $c > 0$ and of two sheets when $c < 0$); Figs. 19.15–19.17. We see in a similar way that the ellipsoid is a smooth surface.

In addition to the above analytical methods there are other ways to represent a surface, for example, using a *table with coordinates of an array of points.* Also, *spline surfaces* that are continuously glued from patches parametrized by vector polynomials are used.

Let us consider an example with the command surfdata to plot a surface by a table of points. The function *wind chill factor* is defined as *how cold it feels (for example, $WC = 32°$ F) for a given wind speed W (5 miles per hour) and a given air temperature ($T = 35°$ F).*

```
> with(plots): L:=[[[5,35,32],[10,35,22],[15,35,16],
  [20,35,12],[25,35,8],[30,35,6],[35,35,4],
  ....................................
  [35,-10,-67],[40,-10,-69],[45,-10,-70]]]:
> P1:=surfdata(L): P2:=plot3d(-20, x=5..45, y=-10..35):
> display3d({P1,P2}, style=PATCHCONTOUR, labels=[W, T, WC]);
```

Here L is a 9×10 matrix where some of the data are represented in the program by dots. The intersection of the surface P1 with the horizontal plane

$WC = -20$, Fig. 19.18, is the level curve where one feels the coldness at $-20^\circ F$.

$w\backslash^t$	35	30	25	20	15	10	5	0	−5	−10
5	32	27	22	16	11	6	0	−5	−10	−15
10	22	16	10	3	−3	−9	−15	−22	−27	−34
15	16	9	2	−5	−11	−18	−25	−31	−38	−45
20	12	4	−3	−10	−17	−24	−31	−39	−46	−53
25	8	1	−7	−15	−22	−29	−36	−44	−51	−59
30	6	−2	−10	−18	−25	−33	−41	−49	−56	−64
35	4	−4	−12	−20	−27	−35	−43	−52	−58	−67
0	3	−5	−13	−21	−29	−37	−45	−53	−60	−69
45	2	−6	−14	−22	−30	−38	−46	−54	−62	−70

Table of the function *Wind Chill*

The commands `listplot3d` and `matrixplot` allow one to obtain surfaces defined by finite lists or matrices of coordinates of their points.

```
> plots[listplot3d]([seq([seq(sin((i-15)*(j-10)/Pi/20),
  i=1..30)], j=1..20)]);
> with(linalg): with(plots):   # Fig. 19.19
  A:=hilbert(6): B:=toeplitz([1,2,3,-3,-2,-1]):
> matrixplot(A+B,heights=histogram,gap=0.25,style=patch);
```

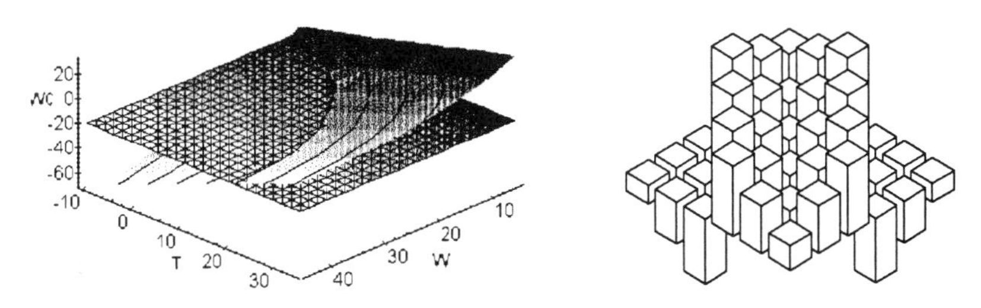

Figs. 19.18–19.19. Graph of the function *Wind Chill* and the level curve $WC = -20^0\, F$. Command `matrixplot`.

One can also plot surfaces (as the solutions for first-order linear or nonlinear partial differential equations for given initial conditions) using the `PDEplot` command from the library `PDETools`.

19.4 Tangent Planes and Normal Vectors

19.4.1 Main Equations and Properties

Definition 19.4.1 A *tangent line of a smooth surface M* is defined as a straight line that is tangent to some smooth curve on M.

Let $\vec{\mathbf{r}}(u, v)$ be a smooth parametrization of the surface M and let $\gamma : u = u(t),\ v = v(t),\ (t \in I)$ be some smooth curve on M through the point $P = \gamma(t_0)$. Write the equation of γ as the space curve $\tilde{\mathbf{r}}(t) = \vec{\mathbf{r}}(u(t), v(t))$ $(t \in I)$ and find its velocity vector as the linear combination

$$\tilde{\mathbf{r}}'(t) = \vec{\mathbf{r}}_u \cdot u'(t) + \vec{\mathbf{r}}_v \cdot v'(t)$$

of vectors $\vec{\mathbf{r}}_u,\ \vec{\mathbf{r}}_v$. By definition, the straight line through P that is parallel to the vector $\vec{\mathbf{r}}_u(u(t_0), v(t_0)) \cdot u'(t_0) + \vec{\mathbf{r}}_v(u(t_0), v(t_0)) \cdot v'(t_0)$ is tangent to the surface M at the point P.

Definition 19.4.2 The plane containing all tangent lines to the smooth surface M at the point P is called the *tangent plane* to M at P.

In fact, the linearly independent vectors $\vec{\mathbf{r}}_u,\ \vec{\mathbf{r}}_v$ define the plane that is tangent to the surface M at the point P. It is denoted by T_pM. The tangent plane T_pM contains the velocity vector $\gamma'(0)$ of every smooth curve $\gamma(t)$ on M with the initial condition $\gamma(0) = P$. Let $M^2 \subset \mathbb{R}^3$ be a surface, $P \in M^2$ a point, σ any plane through P, $Q \in M^2$ any point near P, $d = |PQ|$ the length of the segment between points P and Q, and $h = \rho(Q, \sigma)$ the distance between the point Q and the plane σ.

Theorem 19.4.1 *The tangent plane T_pM of a smooth surface M^2 is characterized among all planes through the point P by the condition* $\lim\limits_{Q \to P} \frac{h}{d} = 0$.

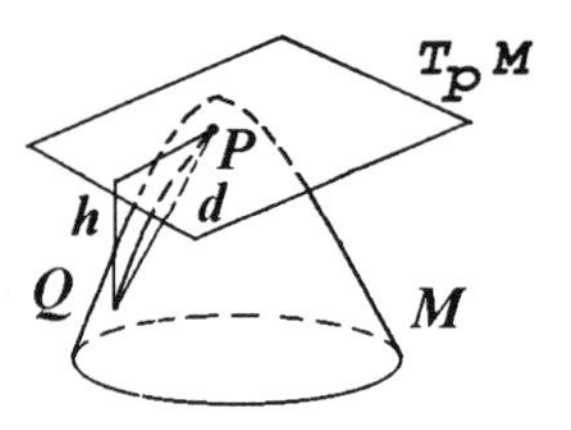

Fig. 19.20. Tangent plane at a point of a surface

Let $\vec{\mathbf{r}}(u, v) = [x(u, v),\ y(u, v),\ z(u, v)]$ be a smooth parametrization of the surface, where $(u, v) \in G \subset \mathbb{R}^2$. Consider points $P(u, v)$ and $A(X, Y, Z)$ on the tangent plane σ. The plane σ has the vector equation

$$\vec{\mathbf{r}} = \vec{\mathbf{r}}(u, v) + a\vec{\mathbf{r}}_u(u, v) + b\vec{\mathbf{r}}_v(u, v) \qquad (a, b \in \mathbb{R}). \tag{19.2}$$

Using scalar products, we write the equation of the tangent plane given by a point and two vectors:

$$(\vec{AP}, \vec{\mathbf{r}}_u, \vec{\mathbf{r}}_v) = \begin{vmatrix} X - x(u, v) & Y - y(u, v) & Z - z(u, v) \\ x_u(u, v) & y_u(u, v) & z_u(u, v) \\ x_v(u, v) & y_v(u, v) & z_v(u, v) \end{vmatrix} = 0. \tag{19.3}$$

In particular, when the surface is a graph $z = f(x, y)$, we obtain

$$\begin{vmatrix} x - x_0 & y - y_0 & z - f(x_0, y_0) \\ 1 & 0 & -f_x' \\ 0 & 1 & -f_y' \end{vmatrix}$$

$$= z - f(x_0, y_0) f_x'(x_0, y_0)(x - x_0) - f_y'(x_0, y_0)(y - y_0) = 0,$$

i.e., the tangent plane at $P(x_0, y_0, f(x_0, y_0))$ is given by the equation

$$z = f(x_0, y_0) + f_x'(x_0, y_0)(x - x_0) - f_y'(x_0, y_0)(y - y_0). \tag{19.4}$$

Plot the tangent plane to the surface graph using the following program:

```
> f:=x^2+y^2:                          # use any function f(x, y)
> fx:=diff(f, x): fy:=diff(f, y):
> TM:=f+fx*(u-x)+fy*(v-y);             # any tangent plane
```

$$TM := x^2 + y^2 + 2x(u - x) + 2y(v - y)$$

```
> TMP:=subs({x=1, y=2}, TM);           # tangent plane at P
```

$$TMP := -5 + 2u + 4v$$

```
> p1:=plot3d(f, x=-sqrt(16-y^2)..sqrt(16-y^2), y=-4..4):
> p2:=plot3d(TMP, u=-1..4, v=-1..3):
> plots[display3d]({p1, p2});
```

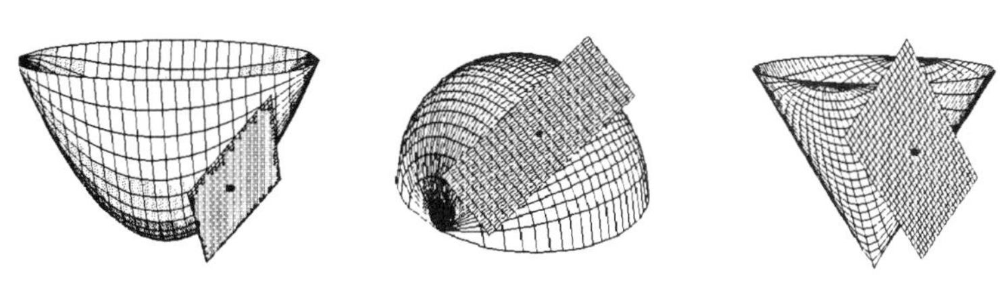

Figs. 19.21–19.23. Tangent planes to a paraboloid, sphere, and cone

For the implicitly defined smooth surface $M^2 : F(x, y, z) = 0$, the vector gradient $\nabla F = (F_x', F_y', F_z')$ is orthogonal to the surface. Thus, the equation of the tangent plane at $P(x_0, y_0, z_0) \in M$ is the following:

$$F_x'(x_0, y_0, z_0)(x - x_0) + F_y'(x_0, y_0, z_0)(y - y_0) + F_z(x_0, y_0, z_0)(z - z_0) = 0. \tag{19.5}$$

Definition 19.4.3 The vector $\vec{\mathbf{n}} = \frac{\vec{\mathbf{r}}_u \times \vec{\mathbf{r}}_v}{|\vec{\mathbf{r}}_u \times \vec{\mathbf{r}}_v|}$ is called the *unit normal* to the surface M^2 at the point $P(u, v)$. A straight line through a point of the surface orthogonal to the tangent plane is called a *normal line*.

Check that the vector-valued function $\vec{\mathbf{n}} := \frac{\vec{\mathbf{r}}_u \times \vec{\mathbf{r}}_v}{|\vec{\mathbf{r}}_u \times \vec{\mathbf{r}}_v|}$ (a unit normal vector to the surface) does not depend on the parametrization of the surface and that it changes its direction only when the sign of the Jacobian changes under coordinate displacement (see Lemma 19.2.1). If a surface bounds a body in space, then it has *exterior* and *interior* sides. If the surface is the graph of a function $z = f(x, y)$, one can define its *top* and *bottom* sides. Such surfaces have two sides. Moreover, the surface has two sides if and only if there exists a continuous field of unit normal vectors continuous on the whole surface. A surface that does not admit a continuous field of unit normal vectors has exactly one side (see Section 19.1).

Various equations for the normal to the surface:

$\vec{\mathbf{r}}(u, v) + \frac{\vec{\mathbf{r}}_u \times \vec{\mathbf{r}}_v}{|\vec{\mathbf{r}}_u \times \vec{\mathbf{r}}_v|} t$ at the point (u, v) of the parametrized surface;

$\frac{x-x_0}{F'_x(x_0,y_0,z_0)} = \frac{y-y_0}{F'_y(x_0,y_0,z_0)} = \frac{z-z_0}{F'_z(x_0,y_0,z_0)}$ at the point (x_0, y_0, z_0) of the implicitly defined surface $F(x, y, z) = 0$,

$\frac{x-x_0}{f'_x(x_0,y_0)} = \frac{y-y_0}{f'_y(x_0,y_0)} = \frac{z-f(x_0,y_0)}{-1}$ at the point (x_0, y_0) of the graph $z = f(x, y)$.

Exercise 19.4.1

1. Write the equations of the tangent plane and the normal vector of the ellipsoid $\frac{x^2}{a^2} + \frac{y^2}{b^2} + \frac{z^2}{c^2} = 1$ at the point $P(x_0, y_0, z_0)$, and plot them for some points P.

2. Prove that if the smooth surface M and the plane α have only one common point P, then α is the tangent plane to the surface at P.

3. Write equations of the tangent plane and the normal vector of the surfaces satisfying

- $\vec{\mathbf{r}}(u, v) = [u + v,\ u - v,\ uv]$ *at the point* $M(2, 1)$,
- $\vec{\mathbf{r}}(u, v) = [u \cos v,\ u \sin v,\ av]$ *at an arbitrary point,*
- $x^2 + 2y^2 - 3z^2 - 4 = 0$ *at the point* $M(3, 1, -1)$.

4. Prove that all tangent planes of the surface $z = x \cdot f\left(\frac{y}{x}\right)$ pass through the coordinate center. Plot an example.

5. Prove (using symbolic calculations with MAPLE) that the tangent planes of the surface $xyz = a^3$ form tetrahedra of constant volume with the three coordinate planes. Which property of tangent lines of the hyperbola does this problem generalize?

Solution. Denote by TM the tangent plane at the point (x, y, z) on the surface. Let $x_1,\ y_1,\ z_1$ be the segments of the intersection of TM with the coordinate axes:

```
> F:=x*y*z-a^3: gr:=linalg[grad](F,[x,y,z]):
> TM:=linalg[dotprod](gr, [xx-x,yy-y,zz-z],'orthogonal'):
> x1:=solve(subs({yy=0, zz=0}, TM), xx):
> y1:=solve(subs({xx=0, zz=0} ,TM), yy):
> z1:=solve(subs({yy=0, xx=0}, TM), zz):
> V:=subs(z=a^3/(x*y), x1*y1*z1);
```

The volume $V = 27\,a^3$ does not depend on the point on the surface. □

6. Prove that if all normal vectors on a surface pass through the same point, then this surface is a region on the sphere.

7. Prove that a surface containing a closed curve along which the unit normal vector $\vec{\mathbf{n}}$ continuously transforms to the opposite vector $-\vec{\mathbf{n}}$ is one-sided.

8. Prove that a one-sided surface cannot be represented as a graph $z = f(x, y)$ or in implicit form by $F(x, y, z) = 0$.

9. A curve on a surface $z = f(x, y)$ that forms the maximal angle with the plane XY at an arbitrary point is called a *maximal sloping curve*. Prove that a tangent line to such curve and a normal vector of the surface belong to the same plane that passes through the given point and is parallel to the axis OZ.

10. Let $f(x, y) \in C^1$ be a function, with $\nabla f \neq 0$. Prove that:

(a) Maximal sloping curves of the surface $z = f(x, y)$ (see Exercise 9) and the curve of the intersection of this surface with the planes $\{z = \text{const}\}$ form an orthogonal net.

(b) Projections of these curves onto the plane XY form an orthogonal net.

19.4.2 Extrema of Functions Defined on Surfaces

An *extremum* is a value of a continuous function that is a local maximum or minimum. Extrema of a function defined on a surface graph or on an implicitly given surface are related by notions of the tangent plane and the normal vector. It is known from calculus that a necessary condition for an extremum of a smooth function $z = f(x, y)$ at the point (x_0, y_0) in the domain G is the vanishing of all partial derivatives of the function at this point or, equivalently, that the gradient ∇f is zero at this point:

$$\{f'_x(x_0, y_0) = 0, \quad f'_y(x_0, y_0) = 0\} \iff \nabla f(x_0, y_0) = 0. \tag{19.6}$$

The geometrical sense of these conditions is that *the tangent plane (19.3) of the surface graph of the function f at $(x_0, y_0, f(x_0, y_0))$ is parallel to the horizontal coordinate plane XY*. For the function $f(x, y)$ of class C^2, we use the notations

$$a_{11} = f''_{xx}(x_0, y_0), \quad a_{12} = f''_{xy}(x_0, y_0), \quad a_{22} = f''_{yy}(x_0, y_0),$$
$$D = a_{11}a_{22} - a_{12}^2.$$

We have the following:

- *If $D > 0$, then at the point (x_0, y_0) the function f has a local extremum (maximum when $a_{11} < 0$ and minimum when $a_{11} > 0$).*

- *If $D < 0$, then (x_0, y_0) is a saddle point and the function f does not have a local extremum.*

- *If $D = 0$, then the point (x_0, y_0) might be an extremum of the function f.*

Problem 19.4.1 Find the local extrema of the function $z = 3x^2y - x^3 - y^4$.

Solution. First plot the graph and the chart of level curves of f:

```
> f:=(x,y) -> 3*x^2*y-x^3-y^4:
> plot3d(f(x,y), x=-6..10, y=-4..5);
> plots[contourplot](f(x,y), x=-6..10, y=-4..5);
```

Then calculate the partial derivatives f'_x, f'_y

```
> fx:=diff(f(x,y),x); fy:=diff(f(x,y),y);
```

$$f_x = -3x^2 + 6xy \qquad f_y = 3x^2 - 4y^3$$

and finally solve the system $\{\, f'_x = 0,\ f'_y = 0\,\}$

```
> sol:=solve({fx=0, fy=0}, {x, y});
```

$sol = \{x = 0,\ y = 0\},\ \{x = 0,\ y = 0\},\ \{x = 0,\ y = 0\},\ \{x = 6,\ y = 3\}$

Two points are possible extrema: $P_1(0, 0)$ and $P_2(6, 3)$. We calculate second partial derivatives of the given function and the term D

```
> fxx:=diff(fx,x); fxy:=diff(fx,y);
  fyy:=diff(fy,y); Delta:=fxx*fyy-fxy^2;
```

$$f_{xx} = -6x - 6y, \quad f_{xy} = 6x,\ f_{yy} = -12y^2 \quad \text{Delta} = 72xy^2 + 36y^3 - 36x^2$$

At the point P_1 we have

```
> subs({x=0,y=0}, [fxx, fxy, fyy, Delta]);
```

$$[0, 0, 0, 0]$$

Since $a_{11} = 0,\ a_{12} = 0,\ a_{22} = 0$ hold, then $D = 0$ and the point P_1 requires additional investigation.

The value of the function f at this point is zero: $f(0, 0) = 0$. Further, for $x < 0,\ y = 0$, we have $f(x, y) = -x^3 > 0$, and for $x = 0,\ y \neq 0$, we have

$f(x, y) = -y^4 < 0$. Consequently, in any neighborhood of the point P_1, the function $f(x, y)$ has values larger than $f(P_1)$ and smaller than $f(P_1)$. Hence, the function $f(x, y)$ does not have an extremum at the point P_1. Analogous calculations at the point P_2 yields $a_{11} = -18$, $a_{12} = 36$, $a_{22} = -108$, and hence $D = 648 > 0$. Since $a_{11} < 0$, the function has a local maximum at the point P_2. □

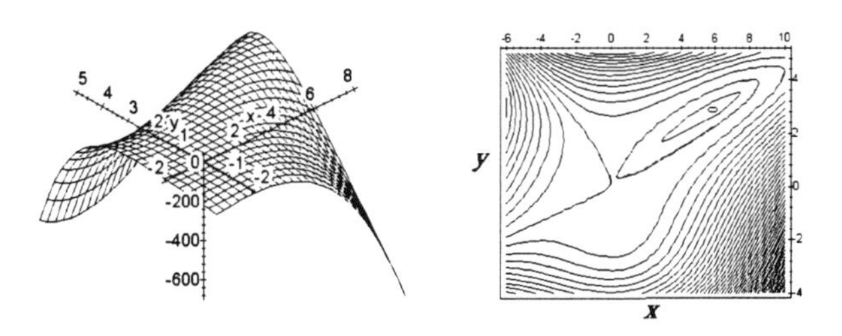

Figs. 19.24–19.25. Extrema of the function $z = 3x^2y - x^3 - y^4$: the graph and the chart (of level curves)

The general form of the conditional extremum problem is the following: *what is the maximal or minimal value of the continuous function $g : \mathbb{R}^3 \to \mathbb{R}$ on a given surface M?* As a rule, *the maximum (minimum) is reached at points where the surface M is tangent to some level surface of the function g.*

Theorem 19.4.2 *Let $F(x, y, z)$ be a smooth function on some domain $G \subset \mathbb{R}^3$, and let $M_c = F^{-1}(c)$ be the level surface (of height c). Assume M_c is smooth, i.e., $\nabla F \neq 0$ holds along M_c. Suppose that $g : G \to \mathbb{R}$ is a smooth function and $P \in M_c$ is an extremal point for g on M_c. Then the tangent plane of the surface M_c at P is orthogonal to the vector gradient $\nabla g(P)$, i.e., there exists a real number λ such that $\nabla g(P) = \lambda \nabla F(P)$ holds.*

The real number λ in Theorem 19.4.2 is called a *Lagrange multiplier*.

Remark 19.4.1 The condition in Theorem 19.4.2 is equivalent to the following system of four equations:

$$\begin{cases} F(x, y, z) = & c, \\ F'_x(x, y, z) = & \lambda g'_x(x, y, z), \\ F'_y(x, y, z) = & \lambda g'_y(x, y, z), \\ F'_z(x, y, z) = & \lambda g'_z(x, y, z). \end{cases}$$

If the surface M_c is compact (closed and bounded in space), then every smooth (simply continuous) function g takes its maximum and minimum on M_c. This means that Theorem 19.4.2 can be used for the selection of possible candidates (among all critical points of the function) for these extremal points.

For example, the *height function* $g(x, y, z) := z$ on the torus defined by revolution of the circle $\gamma : \{y = 0,\ x^2 + (z-2)^2 = 1\}$ around the axis OX has four critical points: in addition to two extrema (maximum at $P_1(0,0,3)$ and minimum at $P_2(0,0,-3)$) there are two saddle points $P_3(0,0,1)$ and $P_4(0,0,-1)$; see Fig. 19.26. If M_c is not compact, there may be no extrema (for instance, the height function $g(x, y, z) := z$ on the cylinder $M_1 : x^2 + y^2 = 1$). An analogous theorem holds for extrema of the function $g(x, y)$ along the plane curve γ; see Section 5.4.

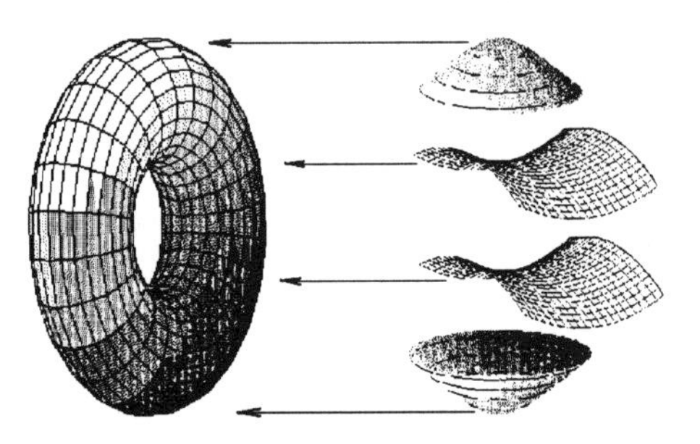

Fig. 19.26. Critical points of the height function on the torus

Proof. The condition $\nabla g(P) = \lambda \cdot \nabla F(P)$ means that any vector $\vec{\mathbf{b}}$ that is tangent to M_c at the point P is orthogonal to the vector $\nabla g(P)$. To see this, take any curve $\gamma : \vec{\mathbf{r}}(t),\ (t \in I)$ on the surface M_c passing through the point P in the direction $\vec{\mathbf{b}}$; i.e., $\vec{\mathbf{r}}(0) = P,\ \vec{\mathbf{r}}'(0) = \vec{\mathbf{b}}$ holds. Since the point P is an extremum of g on M_c, the point $t = 0$ is an extremum of the composed function $\tilde{g}(t) = g \circ \vec{\mathbf{r}}(t)$, defined on the real interval I. Thus we have $0 = \tilde{g}'(0) = \nabla g(P) \cdot \vec{\mathbf{r}}'(0) = \nabla g(P) \cdot \vec{\mathbf{b}}$; i.e., $\nabla g(P) \perp \vec{\mathbf{b}}$ holds for any vector $\vec{\mathbf{b}} \in T_p\{M_c\}$. □

Next consider a problem with a conditional extremum.

Problem 19.4.2 Find the dimensions of the cuboid of largest volume that can be fitted inside the ellipsoid $\frac{x^2}{a^2} + \frac{y^2}{b^2} + \frac{z^2}{c^2} = 1$, assuming that each edge is parallel to a coordinate axis.

Solution. Let $P(x, y, z)$ be the vertex of the cuboid in the first octant, so the edges of the cuboid are $2x,\ 2y,\ 2z$. We therefore wish to optimize the volume function $V = 8xyz$ under the constraint $F(x, y, z) = 1$, where $F(x, y, z) = \frac{x^2}{a^2} + \frac{y^2}{b^2} + \frac{z^2}{c^2} = 1$. It is sufficient to solve the system grad $(V) = \lambda \cdot$ grad (F) in variables x, y, z, and λ under the above constraint.

```
> with(linalg): V:=8*x*y*z: F:=x^2/a^2+y^2/b^2+z^2/c^2:
> GradV:=grad(V, [x,y,z]); GradF:=grad(F, [x,y,z]);
```

$$\text{GradV} := [8yz,\ 8xz,\ 8xy] \qquad \text{GradF} := [\tfrac{2x}{a^2},\ \tfrac{2y}{b^2},\ \tfrac{2z}{c^2}]$$

```
> sol:=solve({F=1, seq(GradV[i]=lambda*GradF[i], i=1..3)},
```

```
{x,y,z,lambda});
```

sol:= $\{z = c,\ y = 0,\ x = 0,\ \lambda = 0\},\ \{z = -c,\ y = 0,\ x = 0,\ \lambda = 0\},$
$\{z = RootOf(3\_Z^2 - 1)c,\ \lambda = -4RootOf(3\_Z^2 - 1)bac,$
$x = -RootOf(3\_Z^2 - 1)a,\ y = RootOf(3\_Z^2 - 1)b\},$
$\{z = RootOf(3\_Z^2 - 1)c,\ x = -RootOf(3\_Z^2 - 1)a,$
$\lambda = 4RootOf(3\_Z^2 - 1)bac,\ y = -RootOf(3\_Z^2 - 1)b\},$
...

```
> allvalues(RootOf(3*_Z^2- 1));
```

$\frac{1}{3}3^{1/2},\ -\frac{1}{3}3^{1/2}$

```
> x0:=allvalues(RootOf(3*_Z^2- 1))[1]*a;
  y0:=allvalues(RootOf(3*_Z^2- 1))[1]*b;
  z0:=allvalues(RootOf(3*_Z^2- 1))[1]*c;
```

$x_0 := \frac{1}{3}3^{1/2}a \qquad y_0 := \frac{1}{3}3^{1/2}b \qquad z_0 := \frac{1}{3}3^{1/2}c$

```
> V0:=subs({x=x0,y=y0,z=z0}, V); subs({x=x0,y=y0,z=z0}, F);
```

$V_0 := \frac{8}{9}ab3^{1/2}c \qquad 1$

Conclusion: Taking the positive roots, we find that the maximal cuboid has edges $x_0 = y_0 = z_0 = \frac{1}{\sqrt{3}}$ and the volume $V_0 = \frac{8\sqrt{3}}{9}abc$. □

Exercise 19.4.2

1. Find the local extrema of the function $z = \frac{1}{3}x^3 + 9y^3 - 4xy$.
Solution.

```
> f:=(x,y)->x^3/3+9*y^3-4*x*y;
> plot3d(f(x,y), x=-1..2, y=-0.5..1);
> plots[contourplot](f(x,y),x=-1..2,y=-0.5..1, contours=35);
> fx:=diff(f(x,y),x); fy:=diff(f(x,y),y);
```

$f_x := x^2 - 4y \quad f_y := 27y^2 - 4x$

```
> sol:=solve({fx=0, fy=0}, {x, y});
```

sol:= $\{y = 0, x = 0\}, \{x = \frac{4}{3}, y = \frac{4}{9}\}, \{x = \frac{4}{3}RootOf(\_Z^2 + \_Z + 1),$
$y = -\frac{4}{9} - \frac{4}{9}RootOf(\_Z^2 + \_Z + 1)\}$

```
> fxx:=diff(fx,x); fyy:=diff(fy,y);
  fxy:=diff(fx,y); Delta:=fxx*fyy-fxy^2;
```

$f_{xx} := 2x \qquad f_{yy} := 54y \qquad f_{xy} := -4 \qquad \text{Delta} := 108xy - 16$

```
> subs(sol[1], [fxx, Delta, f(x,y)]);
```

$[0, -16, 0]$

```
> subs(sol[2], [fxx,Delta,f(x,y)]);
```

$$[\frac{8}{3}, 48, -\frac{64}{81}]$$

Conclusion: The function has the minimum $f_{\min} = -\frac{64}{81}$ at $P_1(\frac{4}{3}, \frac{4}{9})$; the point $P_2(0, 0)$ is a saddle.

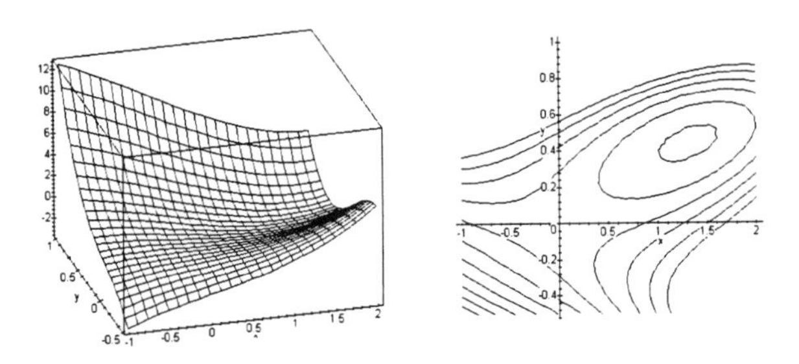

Figs. 19.27–19.28. Extrema of the function $z = \frac{1}{3}x^3 + 9y^3 - 4xy$: graph and chart of level curves

2. Find extrema of the functions: (a) $xy + 2x - \log(x^2y)$, (b) $xy + \frac{1}{2x+2y}$, (c) $\exp x + 2y(x^2 - y^2)$, (d) $xy\log(x^2 + y^2)$, (e) $\frac{x}{y} + \frac{1}{x} + y$.

3. Let $M \subset \mathbb{R}^3$ be a smooth surface and P_0 a point that does not belong to this surface. Prove that the shortest segment from P_0 to M (if it exists) is orthogonal to M.

Hint: Use Theorem 19.4.2 on Lagrange multipliers for the *distance function* $g_{P0} : Q \to \rho(P_0, Q)$.

19.5 The Osculating Paraboloid and a Type of Smooth Point

An osculating paraboloid that helps us visualize the shape of a surface near an arbitrary point is derived and plotted in Section 19.5.2 by the "elementary" method of Section 19.5.1. Using parallel transport and rotation (composition of two rotations about a coordinate axes) leads to the simple standard situation where the tangent plane at the given point, which coincides with the origin of coordinates, is horizontal; using the second-order Taylor formula, we then derive the osculating paraboloid, and finally return to the initial coordinates and plot its image.

19.5.1 Properties of the Osculating Paraboloid

The second-order equation

$$z = a_{11}x^2 + 2a_{12}xy + a_{22}y^2 \tag{19.7}$$

defines a paraboloid in $\mathbb{R}^3(x, y, z)$, excluding the degenerate cases of the parabolic cylinder and the plane. Rotation of the coordinate system in the plane XY by a suitable angle leads to the canonical form of equation (19.7):

$$z = k_1x^2 + k_2y^2. \tag{19.8}$$

We can f ind k_1 and k_2 as roots of the characteristic polynomial:

```
> sol:=solve((a11-t)*(a22-t)-a12*a12, t):
> k1:=sol[1]; k2:=sol[2];
```

Depending on the coefficients k_1, k_2, the *paraboloid* (19.8) has one of the following types: $\begin{cases} \textit{elliptic} & \text{for } k_1 k_2 > 0, \\ \textit{hyperbolic} & \text{for } k_1 k_2 < 0, \\ \textit{parabolic cylinder} & \text{for } k_1 = 0,\ k_2 \neq 0, \\ \textit{planar} & \text{for } k_1 = k_2 = 0. \end{cases}$

Let M be a regular surface and P a point of M. Let F be a paraboloid with vertex at P and axis parallel to the normal vector of the surface at P. Take the point Q on the surface close to P, and denote by d the distance $|PQ|$ between these points; let $h = \operatorname{dist}(Q, F)$ and denote the (vertical) distance from the point Q to the paraboloid F.

Definition 19.5.1 A paraboloid F through a point P on a surface M tangent to the plane T_pM and lying "closest" to the surface in a small neighborhood of the given point, that is, the condition $\lim_{Q\to P} \frac{h}{d^2} = 0$ holds, is called an *osculating paraboloid* at the point P.

In other words, an osculating paraboloid gives the best second-order approximation of the shape of the surface near the given point P, with respect to the distance from the point P. Thus we can use its shape for the classification of points on the C^2-regular surface.

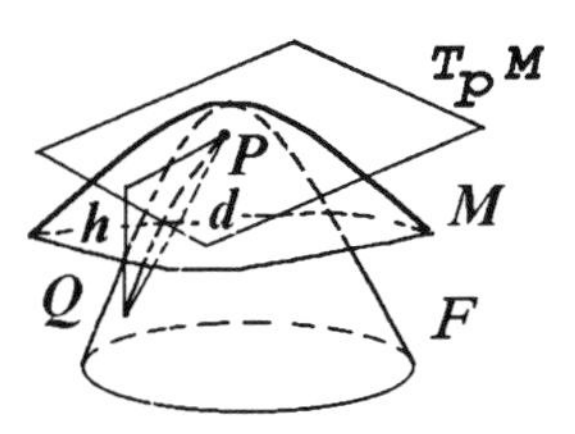

Fig. 19.29. Osculating paraboloid at a point on a surface

We fix rectangular coordinates in $\mathbb{R}^3$ such that the origin coincides with P, the axis OZ is parallel to the normal vector $\vec{\mathbf{n}}$ of the surface at the point P, and the coordinate plane XY coincides with the tangent plane of the surface at P. Now, the surface near $O = P$ is the graph of a function $z = f(x, y)$. Moreover, $f(0,0) = 0$ holds because O belongs to the surface, and $f_x(0,0) = f_y(0,0) = 0$ holds because XY is the tangent plane to the surface at O.

Using the Taylor decomposition of the function $f(x, y)$ at $(0, 0)$, we obtain the equation of the surface

$$z = \tfrac{1}{2}(rx^2 + 2sxy + ty^2) + o(x^2 + y^2)$$

where $r = f_{xx}(0,0),\ s = f_{xy}(0,0),\ t = f_{yy}(0,0)$. The paraboloid F at the point P is defined by the equation $z = \frac{1}{2}(rx^2 + 2sxy + ty^2)$, or in canonical form (with coordinate axes directed along the symmetry axes of the osculating paraboloid) by $z = k_1x^2 + k_2y^2$, where k_1, k_2 are eigenvalues of the quadratic form $rx^2+2sxy+ty^2$ at the point P. This differs from the surface near P by an infinitesimal magnitude of the second order. In fact, $d^2 = x^2 + y^2 + f^2(x, y)$ with $h \le |o(x^2 + y^2)|$. Hence we have $\lim\limits_{Q\to P} \frac{h}{d^2} = \lim\limits_{(x,y)\to 0} \frac{|o(x^2+y^2)|}{x^2+y^2+f^2(x,y)} = 0.$

Thus the following theorem has been proved.

Theorem 19.5.1 *For any point of a C^2-regular surface there exists a unique osculating paraboloid that can degenerate in special cases to a parabolic cylinder or plane.*

Definition 19.5.2 Depending on the shape of the osculating paraboloid, the point P of the surface has one of the following types:

$$\begin{cases} \textit{elliptic} & \text{for } \ k_1k_2 > 0, \\ \textit{hyperbolic} & \text{for } \ k_1k_2 < 0, \\ \textit{parabolic} & \text{for } \ k_1 = 0,\ k_2 \ne 0, \\ \textit{spherical} & \text{for } \ k_1 = k_2 \ne 0, \\ \textit{planar} & \text{for } \ k_1 = k_2 = 0. \end{cases}$$

The first three cases are general. In the last two cases F is a paraboloid of revolution or a plane. Plot examples of surfaces with $P = O$ of the above types:

```
> plot3d(x^2+9*y^2, x=-.1...1, y=-.1...1); # Fig. 19.30
> plot3d(x^2-y^2, x=-1..1, y=-1..1);        # Fig. 19.31
> plot3d(-x^2+y^7, x=-1..1, y=-1..1);       # Fig. 19.32
> plot3d(x^2+y^2, x=-1..1, y=-1..1);        # Fig. 19.33
> plot3d(-x^7-y^6, x=-1..1, y=-1..1);       # Fig. 19.34.
```

Example 19.5.1 All points in the plane are of planar type; all points on the sphere are of spherical type; all points on the cylinder or cone are of parabolic type; all points on the ellipsoid are of elliptical type. This explains the terminology. A torus of revolution (with the axis OZ) has points of the three main

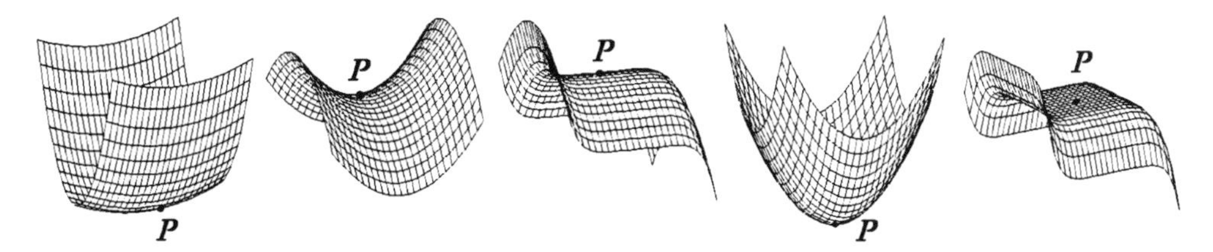

Figs. 19.30–19.34. Five types of smooth points on surfaces

types. The parabolic points fill two parallels (with maximal and minimal height z) and divide the torus into two domains: one with elliptical and another with hyperbolic points. We can imagine the elliptic and hyperbolic points for an arbitrary surface as also separated by curves consisting of parabolic points.

Since the osculating paraboloid at the point P does not depend on the choice of the parametrization of the surface in a neighborhood of the given point, one can define *asymptotic* and *principal directions* at a point of a surface coinciding respectively with the asymptotic directions of the osculating paraboloid and with its symmetry axes. Obviously, at each non-spherical and non-planar point there exist exactly two mutually orthogonal principal directions. We shall not discuss these notions any further. One can also define the *normal curvature at a point on a surface in a given direction* and the Gaussian and mean curvatures of a surface, but these topics are beyond the scope of this elementary volume. We note below only the following basic properties.

Lemma 19.5.1 *There exist two asymptotic directions at a hyperbolic point, one at a parabolic point, and none at an elliptical point. All directions at a planar point are asymptotic and also principal.*

19.5.2 Program for Plotting an Osculating Paraboloid

The arguments used in the proof of Theorem 19.5.1 allow us to write a program for deriving the coefficients k_1 and k_2 and the equations of an osculating paraboloid. Recall that parallel translations and rotations about axes, and also their compositions, are rigid motions of space, i.e., they preserve the shape and the relative position of figures. We use 3×3 matrices to represent rotations about the axes OX and OY:

$$R_x = \begin{vmatrix} 1 & 0 & 0 \\ 0 & \cos\psi & \sin\psi \\ 0 & -\sin\psi & \cos\psi \end{vmatrix}, \qquad R_y = \begin{vmatrix} \cos\varphi & 0 & -\sin\varphi \\ 0 & 1 & 0 \\ \sin\varphi & 0 & \cos\varphi \end{vmatrix}.$$

Let $z = f(x, y)$ be the equation of the surface M, and S the osculating paraboloid for M at $P(x_0, y_0, z_0)$. The program is divided into four steps and is tested for the following data: $z = ax^2 + by^2$ $(a, b > 0)$ with $a = 1$, $b = 2$ at the point $x_0 = 1$, $y_0 = 0$.

(1) Parallel translation along the vector $\overline{PO}$ congruently moves the surface M and the paraboloid S into a surface M_1, with the osculating paraboloid S_1 at the origin.

```
> with(linalg): f:=a*x^2+b*y^2: a:=1: b:=2:   # assume (a>0,b>0):
> x0:=1: y0:=0: z0:=subs({x=x0, y=y0}, f);
  f1:=simplify(subs({x=x+x0, y=y+y0}, f)-z0);
```

$z_0 := 1 \qquad f_1 := x^2 + 2x + 2y^2$.

(2) Assume that the common normal vector $\vec{\mathbf{n}} = [a_1, b_1, c_1]$ to M_1 and S_1 at O is not parallel to the plane XY. Denote by $\vec{\mathbf{n}}_1 = [0, b_1, c_1]$ the projection of this vector onto the coordinate plane YZ, and by ψ the angle between $\vec{\mathbf{n}}_1$ and the axis OZ. Obviously, $\cos\psi = \frac{c_1}{\sqrt{b_1^2+c_1^2}}$, $\sin\psi = \frac{b_1}{\sqrt{b_1^2+c_1^2}}$. Below we set $d = |\vec{\mathbf{n}}|$, $d_1 = |\vec{\mathbf{n}}_1|$.

```
> n:=[subs({x=0,y=0}, -diff(f1,x)),
  subs({x=0, y=0}, -diff(f1,y)), 1];
```

$n := [-2, 0, 1]$

```
> d1:=sqrt(n[2]^2+n[3]^2); d:=sqrt(n[1]^2+n[2]^2+n[3]^2);
```

$d_1 := 1 \qquad d := 5^{1/2}$.

Denote by R the composition of the following two rotations R_x and R_y in space: (a) The rotation R_x through an angle ψ about the axis OX transposes M_1 and S_1 into the tangent pair M_2 and S_2 with common normal vector $\vec{\mathbf{n}}_2$ at O; not that $\vec{\mathbf{n}}_2 = [\frac{a_1}{d}, 0, \frac{d_1}{d}]$ lies in the plane XZ, but in view of our assumption it is not parallel to the axis OX. Obviously, $\cos\psi = \frac{a_1}{\sqrt{a_1^2+b_1^2+c_1^2}}$, and $\sin\psi = \frac{d_1}{\sqrt{b_1^2+c_1^2}}$.

```
> Rx:=matrix([[1,0,0], [0,n[3]/d1,n[2]/d1], [0,-n[2]/d1,n[3]/d1]]):
```

(b) The rotation R_y through an angle $\varphi = \angle(\vec{\mathbf{n}}_2, OZ)$ around the axis OY translates M_2 and S_2 into the tangent pair M_3 and S_3 with common normal vector $\vec{\mathbf{n}}_3$ at O, directed along OZ.

```
> Ry:=matrix([[d1/d,0,n[1]/d], [0,1,0], [-n[1]/d,0,d1/d]]):
```

Find the rotation R and formulas for change of the coordinates:

```
> R:=multiply(Rx,Ry): n3:=multiply(n,R); # checking n3 || OZ
```

$n_3 := [0, 0, 5^{1/2}]$

```
> coor:=multiply([x1,y1,z1], transpose(R));
```

$$\text{coor} := [\ \tfrac{1}{5}5^{1/2}x_1 - \tfrac{2}{5}5^{1/2}z_1,\ y_1,\ \tfrac{2}{5}5^{1/2}x_1 + \tfrac{1}{5}5^{1/2}z_1\ ].$$

If the common normal vector $\vec{\mathbf{n}} = [a_1, b_1, c_1]$ to M_1 and S_1 at the point O is parallel to the plane XY, use the rotation in the plane XY about the point O through the angle $\psi' = \angle(\vec{\mathbf{n}}_1, OY)$. The points M and S are then translated into the tangent pair M_2' and S_2', and their common normal vector $\vec{\mathbf{n}}_2'$ at O is directed along the axis OY. Finally, change the names of the variables: $y \to z$ and $z \to y$.

(3) Now the surface M_3 and its osculating paraboloid S_3 at the point O are in an optimal position, and we can start the calculation of the equation of S_3. Let $g(x, y, z) = 0$ be the equation of M_3.

```
> g:=collect(coor[3]-(subs({x=coor[1], y=coor[2]}, f1)),
  [x1,y1,z1]); subs({x1=0, y1=0, z1=0}, grad(g, [x1,y1,z1]));
```

$$g := -\tfrac{1}{5}x_1^2 + \tfrac{4}{5}z_1x_1 + z_1 5^{1/2} - \tfrac{4}{5}z_1^2 - 2y_1^2 \quad [0, 0, 5^{1/2}] \quad \#\,\text{checking}\ \vec{\mathbf{n}}_3 \parallel OZ$$

Although the explicit equation $z = z_3(x, y)$ of the surface M_3 in a neighborhood of the point O is unknown, techniques of calculus allow us to find all derivatives of the implicit given function $z_3(x, y)$ and then derive coefficients a_{11}, a_{12}, a_{22}.

Obviously, $g_x'(0,0) = g_y'(0,0) = z_{3|x}'(0,0) = z_{3|y}'(0,0) = 0,$

$$a_{11} = \frac{1}{2}z''_{3|xx}(0,0) = \frac{g''_{xx}}{-2g'_z}(0,0), \qquad a_{12} = \frac{1}{2}z''_{3|xy}(0,0) = \frac{g''_{xy}}{-2g'_z}(0,0),$$

$$a_{22} = \frac{1}{2}z''_{3|yy}(0,0) = \frac{g''_{yy}}{-2g'_z}(0,0).$$

```
> readlib(coeftayl): readlib(mtaylor):
> mtaylor(g, [x1=0, y1=0, z1=0], 3);
```

$$-\tfrac{1}{5}x_1^2 + \tfrac{4}{5}z_1x_1 + 5^{1/2}z_1 - \tfrac{4}{5}z_1^2 - 2y_1^2$$

```
> gz:=coeftayl(g, [x1,y1,z1]=[0,0,0],[0,0,1]);
  gxx:=coeftayl(g, [x1,y1,z1]=[0,0,0],[2,0,0]);
  gyy:=coeftayl(g, [x1,y1,z1]=[0,0,0],[0,2,0]);
  gxy:=coeftayl(g, [x1,y1,z1]=[0,0,0],[1,1,0]);
```

$$g_z := 5^{1/2} \qquad g_{xx} := -\tfrac{1}{5} \qquad g_{yy} := -2 \qquad g_{xy} := 0$$

```
> a11:=evalf(-gxx/(2*gz)); a12:=evalf(-gxy/(2*gz));
  a22:=evalf(-gyy/(2*gz));
```

$$a_{11} := .0447213 \qquad a_{12} := 0 \qquad a_{22} := .447213$$

Write the final equation (19.7) of S_3.

```
> S3:=a11*x1^2+2*a12*x1*y1+a22*y1^2;
```

$$S_3 := 0.0447213\, x_1^2 + 0.447213\, y_1^2$$

Solve the characteristic (square) equation $(a_{11} - k)(a_{22} - k) - a_{12}^2 = 0$ and find coefficients k_1 and k_2.

```
> sol:=solve((a11-t)*(a22-t)-a12*a12,t):
  k1:=sol[1]; k2:=sol[2];
```

$$k_1 := 0.0447213 \qquad k_2 := 0.447213$$

```
> Scan:=k1*x^2+k2*y^2; # canonical equation of S
```

$$S_{can} := 0.0447213\, x^2 + 0.447213\, y^2$$

We plot M_3 and S_3 (in new coordinates); Fig. 19.35.

```
> plots[implicitplot3d]({g=0, S3=z1},
  x1=-2..1.4, y1=-1..1, z1=-.2..1, grid=[20,20,20]);
```

(4) The rotation R^{-1} is given by the inverse matrix, which in the case of an orthogonal matrix coincides with its transpose R^*. It transforms the pair M_3 and S_3 into the pair M_1 and S_1 and allows us to obtain the equation of S_1.

```
> co1:=multiply([x,y,z], R);
```

$$\text{co}_1 := [\, \tfrac{1}{5}5^{1/2}x + \tfrac{2}{5}5^{1/2}z,\ y,\ -\tfrac{2}{5}5^{1/2}x + \tfrac{1}{5}5^{1/2}z \,]$$

```
> S1:=simplify(subs({x1=co1[1],y1=co1[2],z1=co1[3]},z1-S3));
```

$$S_1 := -0.894427\, x + 0.447213\, z - 0.00894427\, x^2 - 0.0357770\, xz-$$
$$0.0357770\, z^2 - 0.447213\, y^2\ .$$

Plot M_1 and S_1 (in old coordinates), Fig. 19.36.

```
> plots[implicitplot3d]({z=f1,S1=0},x=-2..1,y=-1..1,z=-1..2);
```

Using formulas for parallel translation along the vector $\overline{OP}$, deduce from the equation for S_1 the equation of the osculating paraboloid of S. The figure with M and S is analogous to Fig. 19.36.

```
> S:=simplify(subs({x=x-x0, y=y-y0, z=z-z0}, S1+z0));
```

$$S := -0.840761x + 0.366715 + 0.554544z - 0.00894427x^2 -$$
$$0.0357770xz - 0.0357770z^2 - 0.447213y^2\ .$$

Exercise 19.5.1

1. Study the types of points on every surface of the second order.

2. Deduce the equation of the osculating paraboloid for the following surfaces at given points and plot them:

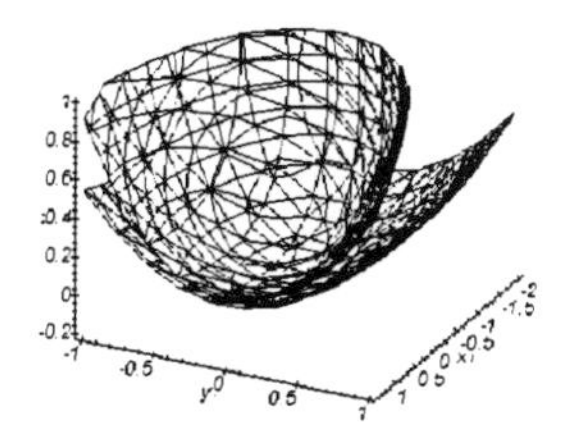
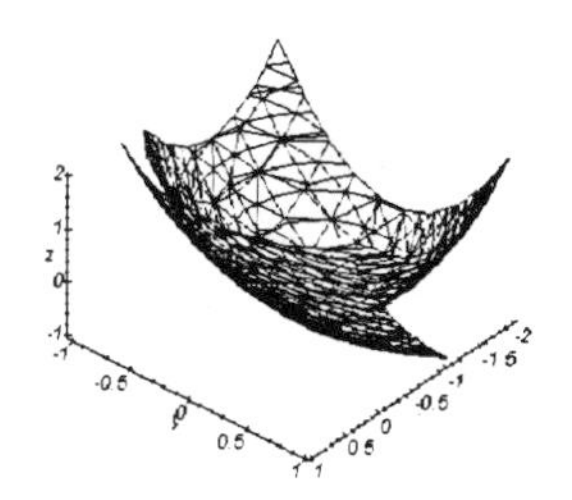

Figs. 19.35–19.36. Osculating paraboloids at the point O
(a) S_3 for M_3 (b) S_1 for M_1

(a) for the ellipsoid $\frac{x^2}{a^2}+\frac{y^2}{b^2}+\frac{z^2}{c^2}=1$ at the point $P(0,0,c)$,

Answer: $\frac{z}{c}=\left(1-\frac{1}{2}\left(\frac{x^2}{a^2}+\frac{y^2}{b^2}\right)\right)$.

(b) for the hyperboloid $x^2+y^2-z^2=1$ at the point $P(0,1,0)$,

(c) for the monkey saddle $z=3xy^2-x^3$ (Fig. 20.1) at the origin.

3. Prove that if a surface is tangent to a plane along some curve, then each point on this curve is of parabolic or planar type. Prove that elliptical and hyperbolic points on a torus of revolution form two domains with two boundary parallels of parabolic points.

4. Let α be the tangent plane of the surface M at the point P. Prove the following:

(a) If P is of elliptical (or spherical) type, then all points of the surface M close enough to P lie on the same half-space with respect to α.

(b) If the point P is hyperbolic, there exist points on M arbitrarily close to P on either side of the plane α.

(c) If the point P is parabolic or planar, both of the above cases can appear (find examples).

5. Prove that if the boundary of a surface belongs to some plane, then either this surface is a plane domain or the surface has elliptical points.

6. Prove that every compact surface has elliptical points.

7. Prove that a surface all of whose points are planar lies in a plane.

19.6 Singular Points on Surfaces

We give a test for singular points on a parametrized surface.

Lemma 19.6.1 *The point* (u_0, v_0) *on a surface* M *is singular if the vector-valued function* $\vec{\mathbf{n}}=\frac{\vec{\mathbf{r}}_u\times\vec{\mathbf{r}}_v}{|\vec{\mathbf{r}}_u\times\vec{\mathbf{r}}_v|}$ *(see Definition 19.4.3) does not tend to some limit value as* $(u,v)\to(u_0,v_0)$.

The singular curve on the surface from Exercise 2 below, see Fig. 19.38, is called the *cuspidal edge* of the surface. Each plane orthogonal to the cuspidal edge intersects the surface along the curve for which the point of the cuspidal edge is a singular point of the same type.

The structure of a parametrized surface $\vec{\mathbf{r}}(u, v)$ of class C^k, $k \geq 2$, near a singular point $P(u_0, v_0)$ is usually investigated using a Taylor expansion of second or third order. Starting with zero values of parameters u_0, v_0 from the given point P, we have

$$\vec{\mathbf{r}}(u, v) = \vec{\mathbf{r}}(p) + \vec{\mathbf{r}}_u(p)u + \vec{\mathbf{r}}_v(p)v + \frac{1}{2}(\vec{\mathbf{r}}_{uu}(p)u^2 + \vec{\mathbf{r}}_{vv}(p)v^2 + 2\vec{\mathbf{r}}_{uv}(p)uv)$$

$$+\frac{1}{6}(\vec{\mathbf{r}}_{uuu}(p)u^3 + 3\vec{\mathbf{r}}_{uuv}(p)u^2v + 3\vec{\mathbf{r}}_{uvv}(p)uv^2 + \vec{\mathbf{r}}_{vvv}(p)v^3) + \ldots + \bar{\varepsilon}_k(u^2+v^2)^{\frac{k}{2}}.$$

```
> readlib(mtaylor): r_taylor:=[mtaylor(x(u,v), [u,v], 4),
  mtaylor(y(u,v), [u,v], 4), mtaylor(z(u,v), [u,v], 4)];
```

From the definition of a singular point follows the equality $\vec{\mathbf{r}}_u \times \vec{\mathbf{r}}_v = 0$. Then the following simple cases are possible:

(1) $\vec{\mathbf{r}}_u = 0$, $\vec{\mathbf{r}}_v = 0$. If, moreover, $\vec{\mathbf{r}}_{uu}\vec{\mathbf{r}}_{uv}\vec{\mathbf{r}}_{vv} \neq 0$ holds, then the point is *edged*. Tangent rays at a singular point form a dihedral angle. The edge of this angle is directed along the vector $\vec{\mathbf{r}}_{uv}$, and its half-planes contain the vectors $\vec{\mathbf{r}}_{uu}$ and $\vec{\mathbf{r}}_{vv}$ whose starting points coincide with O.

Example. The surface $\vec{\mathbf{r}}(u, v) = [u^2, uv, v^2]$; Fig. 19.39.

```
> r:=[u^2,u*v,v^2]: ru:=diff(r,u): rv:=diff(r,v):
> with(linalg): subs({u=0,v=0}, crossprod(ru,rv));
```

$r_u := [2u, v, 0]$ $\quad r_v := [0, u, 2v]$ $\quad [0, 0, 0]$

```
> ruu:=diff(r, u$2); ruv:=diff(r,u,v);
  rvv:=diff(r, v$2); det([ruu, ruv, rvv]);
```

$r_{uu} := [2, 0, 0]$ $\quad r_{uv} := [0, 1, 0]$ $\quad r_{vv} := [0, 0, 2]$ $\quad 4$

The case $\vec{\mathbf{r}}_{uu}\vec{\mathbf{r}}_{uv}\vec{\mathbf{r}}_{vv} = 0$ requires further consideration.

(2) $\vec{\mathbf{r}}_u \neq 0$, $\vec{\mathbf{r}}_v = 0$. If, also, $\vec{\mathbf{r}}_u \times \vec{\mathbf{r}}_{vv} \neq 0$ holds, then the tangent rays at the singular point form a half-plane. The boundary line of this half-plane is directed along the vector $\vec{\mathbf{r}}_u$, and the vector $\vec{\mathbf{r}}_{vv}$ with starting point at O lies in this half-plane.

Example. The *Whitney umbrella* $\vec{\mathbf{r}}(u, v) = [uv, u, v^2]$; see Fig. 20.2.

The case $\vec{\mathbf{r}}_u \times \vec{\mathbf{r}}_{vv} = 0$ requires further consideration.

(3) $\vec{\mathbf{r}}_u \neq 0$, $\vec{\mathbf{r}}_v \neq 0$, $\vec{\mathbf{r}}_v = a\,\vec{\mathbf{r}}_u$. Changing the variables u, v to U, V by the formulas $u = U - aV$, $v = V$ reduces this case to case (2) above. If, also, $\vec{\mathbf{r}}_u \times (a^2\vec{\mathbf{r}}_{uu} - 2a\vec{\mathbf{r}}_{uv} + \vec{\mathbf{r}}_{vv}) \neq 0$, the point is called *edged*.

```
> with(linalg): r:=[u*v, u, v^2]:
  ru:=diff(r,u); rv:=diff(r,v); rvv:=diff(r,v$2);
```

$r_u := [v, 1, 0] \qquad r_v := [u, 0, 2v] \qquad r_{vv} := [0, 0, 2]$

```
> subs({u=0, v=0}, crossprod(ru, rvv));
```

$[2, 0, 0]$

The case $\vec{\mathbf{r}}_u \times (a^2\vec{\mathbf{r}}_{uu} - 2a\vec{\mathbf{r}}_{uv} + \vec{\mathbf{r}}_{vv}) = 0$ requires further consideration.

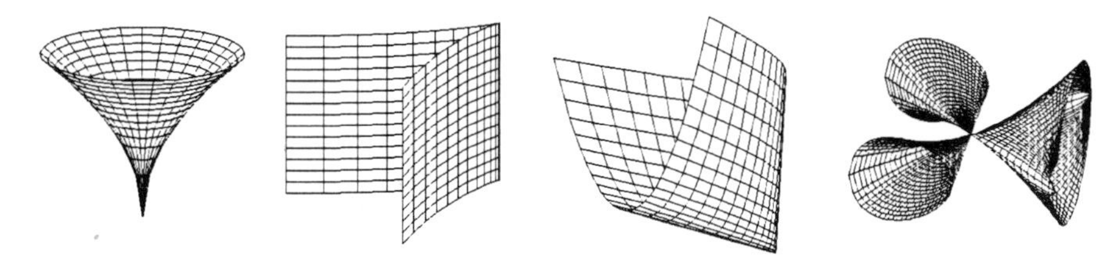

Figs. 19.37–19.40. Singular points and curves on surfaces

In studying an implicitly given surface $F(x, y, z) = 0$, where $F \in C^k,\ k \leq 2$, the notion of a singular point must be modified. In particular, self-intersections also must be kept in mind. If the necessary conditions $F'_x = F'_y = F'_z = 0$ at a given point are satisfied, then one applies Taylor's expansion to obtain information. It is convenient to assume our point is at the origin. Then

$$\begin{aligned}F(x, y, z) &= \frac{1}{2}(F_{xx}x^2 + F_{yy}y^2 + F_{zz}z^2 + F_{xy}xy + F_{yz}yz + F_{xz}xz)\\ &+ \frac{1}{6}(F_{xxx}x^3 + F_{yyy}y^3 + F_{zzz}z^3 + 3F_{xxy}x^2y + 3F_{yyx}y^2x + 3F_{yyz}y^2z\\ &+ 3F_{zzy}z^2y + 3F_{zzx}z^2x + 3F_{xxz}x^2z + 6F_{xyz}xyz) + \dots\\ &+ \varepsilon_k(x^2 + y^2 + z^2)^{k/2}.\end{aligned}$$

For a clearer picture of the shape of the surface $F = 0$ in a neighborhood of a singular point, one first calculates the second-order type of surface:

$$F_{xx}x^2 + F_{yy}y^2 + F_{zz}z^2 + F_{xy}xy + F_{yz}yz + F_{xz}xz = 0. \qquad (19.9)$$

This surface is a cone and can be one of five types. To calculate the type of cone (19.9) we need to consider the real *invariants*

$$I_1 = F_{xx} + F_{yy} + F_{zz},$$

$$I_2 = \begin{vmatrix} F_{xx} & F_{xy} \\ F_{xy} & F_{yy} \end{vmatrix} + \begin{vmatrix} F_{xx} & F_{xz} \\ F_{xz} & F_{zz} \end{vmatrix} + \begin{vmatrix} F_{yy} & F_{yz} \\ F_{yz} & F_{zz} \end{vmatrix}, \quad I_3 = \begin{vmatrix} F_{xx} & F_{xy} & F_{xz} \\ F_{xy} & F_{yy} & F_{yz} \\ F_{xz} & F_{yz} & F_{zz} \end{vmatrix}.$$

```
> with(linalg):readlib(mtaylor):F:=F1(x,y,z): # define function
> mtaylor(F,[x,y,z], 4); # Taylor formula
> Fxx:=diff(F, x$2): Fxy:=diff(F, x,y): Fyy:=diff(F, y$2):
```

```
Fxz:=diff(F,x,z): Fyz:=diff(F,y,z): Fzz:=diff(F,z$2):
> I1:=Fxx+Fyy+Fzz; I2:=det([[Fxx,Fxy],[Fxy,Fyy]])+
  det([[Fxx,Fxz],[Fxz,Fzz]])+det([[Fyy,Fyz],[Fyz,Fzz]]);
  I3:=det([[Fxx,Fxy,Fxz],[Fxy,Fyy,Fyz],[Fxz,Fyz,Fzz]]);
```

Let us list the five possible types of cones (19.9) and corresponding possible structures of the surface $F = 0$ near a singular point.

(1) $I_3 \neq 0,\ I_2 > 0,\ I_1 I_3 > 0$. The cone (19.9) consists of one singular point. This singular point of the surface $F = 0$ is *isolated.*

Example. $F(x, y, z) = (x^2 + y^2 + z^2)(z - 1)$ is the union of a plane and a point.

(2) $I_3 \neq 0,\ I_2 \leq 0$ or $I_1 I_3 < 0$. The cone (19.9) has the shape of an (oblique) circular cone. The surface near the singular point looks like a deformed neighborhood of the vertex of the above cone.

Example. The surface of revolution of the lemniscate of Bernoulli (Fig. 20.2) $(x^2 + y^2 + z^2)^2 - 2a^2(z^2 - x^2 - y^2)$, and "a figure eight" (Figs. 21.4–21.5).

(3) $I_3 = 0,\ I_2 > 0$. The cone (19.9) degenerates into a straight line. Rotating the coordinate system, we obtain the case where the axis OZ coincides with the above straight line. The equation of the surface in new coordinates takes the m $\tilde{F}(x, y, z) = 0$. Moreover,

$$\tilde{F}(x, y, z) = a^2x^2 + b^2y^2 + \tfrac{1}{6}(\tilde{F}_{xxx}x^3 + \dots) + \dots + \varepsilon_k(x^2 + y^2 + z^2)^{k/2}.$$

In this case if $\tilde{F}_{zzz} \neq 0$ holds, then a neighborhood of the singular point has the shape of a *spike*. All tangent rays of the surface at the singular point coincide.
Example. The surface $x^2 + y^2 - z^3 = 0$ from Exercise 1 below, (Fig. 19.37).
The case $\tilde{F}_{zzz} = 0$ requires further consideration.

(4) $I_3 = 0,\ I_2 < 0$. The cone (19.9) consists of two intersecting planes, and the surface $F = 0$ consists of two intersecting (along the smooth curve) sheets.
Example. The pair of intersecting planes $x^2 - y^2 = 0$.

(5) $I_3 \neq 0,\ I_2 = 0$. The cone (19.9) consists of the doubly covered plane. Rotating the coordinate system, we obtain the case where the axis OZ is orthogonal to the above plane. The equation of the surface in new coordinates takes the form $\tilde{F}(x, y, z) = 0$ for

$$\tilde{F}(x, y, z) = az^2 + \tfrac{1}{6}(\tilde{F}_{xxx}x^3 + \dots + \tilde{F}_{yyy}) + \dots + \varepsilon_k(x^2 + y^2 + z^2)^{k/2}.$$

If the following equation of third order in u,

$$\tilde{F}_{xxx} + 3\tilde{F}_{xxy}u + 3\tilde{F}_{xyy}u^2 + \tilde{F}_{yyy}u^3 = 0, \tag{19.10}$$

has a *unique real root*, then the tangent rays at a singular point form a half-plane, which is the half of the cone. If equation (19.10) has *three different real roots*, then tangent rays at the singular point form three separate sectors on the plane x, y, and the surface can be imagined as three funnels converging to the vertex, flattening to the plane (x, y) as they come near to the vertex.

Example. Consider the surface $z^2 - x^2y + y^3 = 0$, see Fig. 19.40. We plot it by gluing together six pieces (graphs):

```
> p11:=plot3d(sqrt(y^3-x^2*y), x=0..2, y=-x..0):
> p12:=plot3d(-sqrt(y^3-x^2*y), x=0..2, y=-x..0):
> p21:=plot3d(sqrt(y^3-x^2*y), x=-2..0, y=x..0):
> p22:=plot3d(-sqrt(y^3-x^2*y), x=-2..0, y=x..0):
> p31:=plot3d(sqrt(y^3-x^2*y), x=-2..2, y=abs(x)..2):
> p32:=plot3d(-sqrt(y^3-x^2*y), x=-2..2, y=abs(x)..2):
> plots[display]({p11,p12,p21,p22,p31,p32});
```

If equation (19.10) has three real roots, and one of them is of multiplicity two, then the surface near a singular point looks like the Whitney umbrella, see Fig. 20.2.

Example. The surface $z^2 - (x - y)^2(x + y) = 0$.

The case where equation (19.10) has one real root with multiplicity 3 requires further consideration.

Exercise 19.6.1

1. Prove, using Lemma 19.6.1, that the point $(0, 0)$ on the surface $\vec{\mathbf{r}}(u, v) = [u, v, (u^2 + v^2)^{1/3}]$, Fig. 19.37, is a singular point:

```
> plot3d([rho*cos(phi), rho*sin(phi), rho^(2/3)],
  rho=0..1, phi=-Pi..Pi, grid=[15,30]);
```

Solution:

```
> r:=[u, v, root(u^2+v^2, 3)]; ru:=diff(r, u); rv:=diff(r, v);
> with(linalg): crossprod(ru,rv); n1:=evalm(%/norm(%, 2));
> [limit(limit(n1[1],u=0),v=0), limit(limit(n1[2],u=0),v=0),
  limit(limit(n1[3], u=0), v=0)];
```

$$[0, \text{ undefined}, \ 0]$$

2. Prove that points on the cylindrical surface $\vec{\mathbf{r}}(u, v) = [u^2, u^3, v]$ lying on the axis OZ $(u = 0)$ form a *cuspidal edge*, see Fig. 19.38.

```
> plot3d([u^2,u^3, v], u=-1..1, v=0..2);
```

3. Prove that $O(0, 0)$ is a singular point of the surface, Fig. 19.41:

$$\vec{\mathbf{r}}(u, v) = [u^2 - v^2, 2uv, u^5].$$

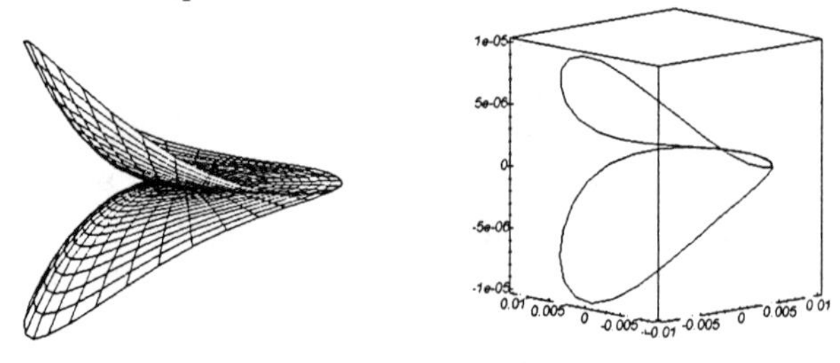

Figs. 19.41–19.42. *Wings* and the curve $\tilde{\gamma}$

```
> plot3d([rho^2*cos(2*phi),rho^2*sin(2*phi),
  (rho*sin(phi))^5],rho=0..1,phi=-Pi..Pi,grid=[15,30]);
```

Prove that while moving along the circle $\gamma\ :\ u = \rho\cos(t),\ v = \rho\sin(t)$ the projection of a point of the curve

$$\tilde{\gamma}\ :\ \vec{\mathbf{r}}(\rho\cos(t),\ \rho\sin(t)) = [\rho^2\cos(2t),\ \rho^2\sin(2t),\ (\rho\sin(t))^5]$$

(see Fig. 19.42) onto the plane XY wraps twice about the origin O.

20

Some Classes of Surfaces

In Chapter 20 we study three very important and popular classes of surfaces (algebraic, revolutionary and ruled) and the envelopes of surfaces. In Section 20.1 we use the resultant and the discriminant to deduce implicit polynomial equations of surfaces. In Section 20.2 we study surfaces of revolution, and give an example of a map on a torus that cannot be colored with six colors. In Section 20.3 we plot ruled surfaces of various types, and calculate their striction curves and distribution parameter. In Section 20.4 we use the notion of a tangent plane (see Section 19.4.1) to continue our studies from Sections 9.2–9.3. We plot some envelopes of surfaces and show that the envelope of a family of planes (as a particular case) is a ruled developable surface.

In Chapter 20 the reader will become acquainted with the commands

`resultant, discrim, vector, scalarmul, setoptions3d.`

20.1 Algebraic Surfaces

A large class of surfaces is obtained using polynomials in three variables.

Definition 20.1.1 A surface is called *algebraic* if it is given by an equation $F(x, y, z) = 0$, where F is a polynomial in variables x, y, z. The degree of the polynomial F is called the *order* of the surface. If F is the product of nonconstant polynomials, then the surface is called *decomposable*. A surface that is not algebraic is called a *transcendental surface*.

Example 20.1.1 A *plane* in space is defined by the linear equation $ax + by + cz + d = 0$, and hence it is an algebraic surface of the first order. *Surfaces of the* *second order* (the sphere and the ellipsoid, elliptic and hyperbolic paraboloids, hyperboloids of one and two sheets, the elliptic cone, cylinders over curves of second order) are studied in *analytical geometry*. The *cylinder over a sine curve* is a transcendental surface in view of the following property.

Lemma 20.1.1 *Any straight line intersects an algebraic surface of order n at no more than n distinct points, unless it is completely contained in the surface.*

Proof. Substituting parametric equations of an arbitrary straight line $x = a_1 + tb_1$, $y = a_2 + tb_2$, $z = a_3 + tb_3$ in the equation of the surface, we obtain the polynomial $F(a_1 + tb_1,\ a_2 + tb_2,\ a_3 + tb_3) = \sum_{0\le i\le n} d_i t^i$ in t of order not more than n. If all coefficients of this polynomial are zero, then the straight line lies on the surface. Otherwise, the polynomial has at most n real roots, which correspond to the points of intersection of the straight line with the surface (multiple roots correspond to points of tangency). □

Exercise 20.1.1

1. Prove that the following surfaces are algebraic:
(a) *The monkey saddle* $\vec{\mathbf{r}}(u, v) = [u, v, u^3 - 3uv^2]$ of Fig. 20.1 (third order)
(b) *The Whitney umbrella* $\vec{\mathbf{r}}(u, v) = [uv, u, v^2]$ of Fig. 20.2 (third order)
(c) *The swallow-tail* $\vec{\mathbf{r}}(u, v) = [u, -4v^3 - 2uv, 3v^4 + uv^2]$ of Fig. 20.3 (fifth order)

```
> plot3d([u,-4*v^3-2*u*v,3*v^4+u*v^2], u=-4..4,v=-1..1);
```

(d) *The wings* $\vec{\mathbf{r}}(u, v) = [u^2 - v^2, 2uv, u^5]$ of Fig. 19.41 (tenth order).

Then explain why the surface $\vec{\mathbf{r}}(u, v) = [f_1(u, v),\ f_2(u, v),\ f_3(u, v)]$, where $f_i(u, v)$ are polynomials, is an algebraic surface.

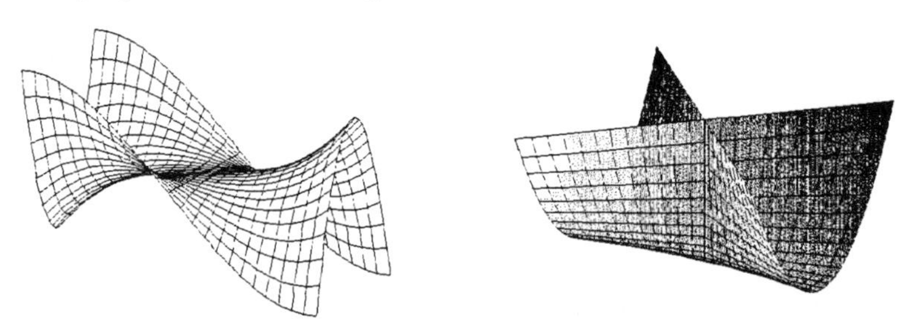

Figs. 20.1–20.2. Monkey saddle and Whitney umbrella

Solution. (a) Consider an explicit equation of this surface.

(b) Consider the function $x^2 - y^2z$.

(c) Using the `resultant` in `MAPLE` (the *resultant of two polynomials*) where $x = u$, we eliminate the variable v from the system of equations $-4v^3 - 2xv - y = 0$, $3v^4 + xv^2 - z = 0$ and obtain a polynomial in x, y, z of the fifth order:

```
> F:=resultant(-4*v^3-2*x*v-y,3*v^4+x*v^2-z,v);
```

$$F := x^3y^2 + 27y^4 - 16x^4z + 128x^2z^2 - 144xzy^2 - 256z^3.$$

(d) Using the `resultant` in `MAPLE` twice, we eliminate the variables u, v from the system of polynomial equations $u^2 - v^2 - x = 0,\ 2uv - y = 0,\ u^5 - z = 0$, and obtain a polynomial in x, y, z of the tenth order:

```
> res1:=resultant(u^2-v^2-x, 2*u*v-y, v);
```

$$\mathtt{res}_1 := 4u^4 - 4u^2x - y^2$$

```
> F:=resultant(res1, u^5-z, u);
```

$$F := -y^{10} - 320y^4z^2x - 1280z^2x^3y^2 + 1024z^4 - 1024x^5z^2.$$

2. (a) The *swallow-tail* surface has a cuspidal edge (see Fig. 20.4) and a curve of self-intersection. To better visualize this surface, consider its sections with parallel planes $x = c$, Fig. 20.5.
(b) Prove that the points (x, y, z), where the polynomial $F(t; x, y, z) = t^4 + xt^2 + yt + z$ has the multiple root t, form the set given by equations (c) of Exercise 1: $y = -4t^3 - 2xt,\ z = 3t^4 + xt^2$.
The swallow-tail divides the space of polynomials $\mathbb{R}^3(x, y, z) = \{t^4 + xt^2 + yt + z\}$ onto three domains, in one of which (looking like a pyramid) the polynomials have four real roots, in the next, two roots, and in the last domain, non-real roots.

Solution. If the polynomial $F(t; x, y, z)$ has a multiple root t, then the system $F = 0,\ F'_t = 0$ has a solution. Computing $F'_t = 4t^3 + 2xt + y$ and eliminating t from the system by the command `resultant` is equivalent to applying the command `discrim` (*discriminant*) for t to the given polynomial F.

```
> d:=discrim(t^4+x*t^2+y*t+z, t);
```

$$d := -4x^3y^2 - 27y^4 + 16x^4z - 128x^2z^2 + 144xzy^2 + 256z^3$$

```
> subs({y=-4*t^3-2*x*t, z=3*t^4+x*t^2}, d);
```

$$-4x^3(-4t^3 - 2xt)^2 - 27(-4t^3 - 2xt)^4 - 128x^2(3t^4 + xt^2)^2 + 16x^4(3t^4 + xt^2) + 144x(3t^4 + xt^2)(-4t^3 - 2xt)^2 + 256(3t^4 + xt^2)^3$$

```
> simplify(%); # checking
```

$$0$$

(c) The cuspidal edge on the swallow-tail is given by the condition that t is a triple root of the equation $F = 0$, i.e., $F''_{tt} = 0$. Prove that this singular curve can be defined by the parametric equations $\vec{\mathbf{r}}(t) = [-6t^2, 8t^3, -3t^4]$, Fig. 20.4.
Solution. Eliminate x, y, z from the system of 3 equations, then plot

```
> F:=t^4+x*t^2+y*t+z: F1:=diff(F,t): F2:=diff(F1,t):
> sol:=solve({F,F1,F2}, {x,y,z});
```

sol := $\{x = -6t^2,\ \ y = 8t^3,\ \ z = -3t^4\}$

```
> plots[spacecurve]([-6*t^2, 8*t^3, -3*t^4], t=-1..1);
```

(d) Prove that the swallow-tail is the union of all tangent lines to space curve $\vec{\mathbf{r}}(t) = [t^2, t^3, t^4]$.

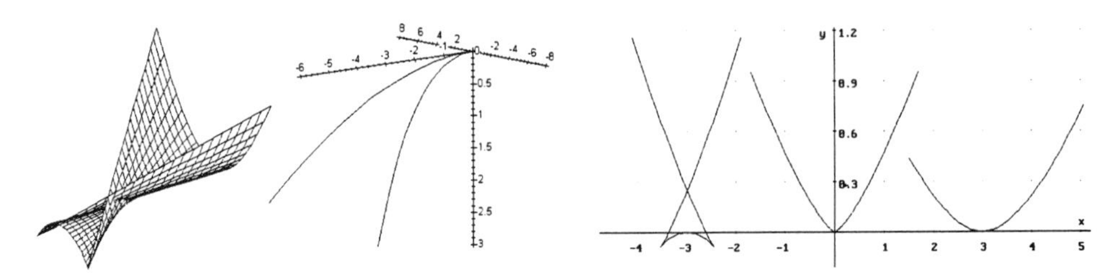

Figs. 20.3–20.5. Swallow-tail, its cuspidal edge, and sections by the planes $x = c$

We will consider below in Sections 20.2–20.4 some important classes of surfaces that can be constructed by moving one curve in space, the *ruling* along another curve, the *directrix*.

20.2 Surfaces of Revolution

The *cylinder, cone, sphere, torus, paraboloid, hyperboloids*, and *the ellipsoid with two equal axes* are simple examples of surfaces of revolution. Many examples can be seen in the real world.

Definition 20.2.1 Let α be a plane in $\mathbb{R}^3$, g a straight line in α, and γ a curve in α. If α rotates in $\mathbb{R}^3$ about g, the resulting set M is called the *surface of revolution* with *axis* g generated from the *profile* curve γ. The curve of intersection of M with any plane through the axis of revolution (in particular, the profile curve) is called a *meridian*, and the curve of intersection of M with any plane orthogonal to the axis of revolution (a circle, a point, or the empty set) is called a *parallel.*

Note that all normal vectors to a surface of revolution intersect the same straight line (axis of revolution); in fact, surfaces of revolution (and their parts) are characterized by this property.

Deleting a meridian from a surface of revolution of an elementary curve that does not intersect the axis yields an elementary surface. For convenience we choose rectangular coordinates such that α coincides with the plane XZ and such that the straight line g coincides with the axis OZ, and we denote the angle of rotation by $u \in [0, 2\pi)$. Suppose that the curve γ is given in the parametric form $x = \varphi(u) \geq 0,\ z = \psi(u)$, where $u \in I$. If the point

$P(x_0, 0, z_0)$ belongs to the profile curve, a parallel through it is given by the equations $[x_0 \cos(v), x_0 \sin(v), z_0]$. Thus the surface of revolution M can be defined by the equations

$$\vec{\mathbf{r}}(u, v) := [\varphi(u) \cos(v),\ \varphi(u) \sin(v),\ \psi(u)], \qquad 0 \le v < 2\pi, \qquad (20.1)$$

called the *standard parametrization* of the surface of revolution. The coordinate net consists of two families of curves, parallels and meridians. Substituting concrete functions $\varphi(u)$, $\psi(u)$ into (20.1) yields various examples of surfaces of revolution with the axis OZ.

The surface of revolution of the curve $\gamma(u)$: $\vec{\mathbf{r}}(u) = [u, 0, 4 + \sin(2u)]$ about the axis OZ is given in Fig. 20.6 together with its parallels and meridians.

```
> plot3d([(4+sin(2*u))*cos(v), (4+sin(2*u))*sin(v), u],
  u=0..4*Pi, v=0..2*Pi);
```

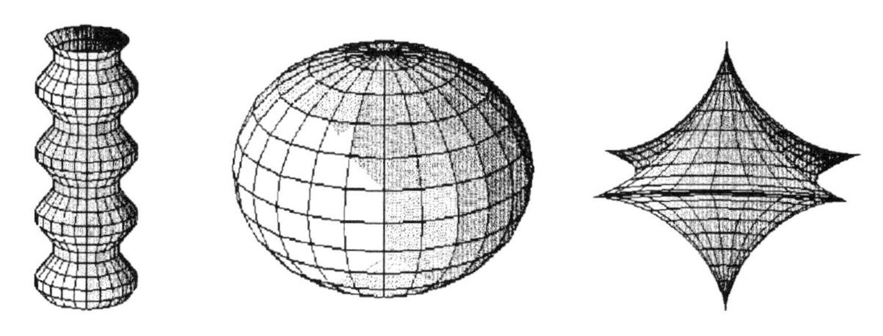

Figs. 20.6–20.8. Revolution of the curve $[t, 0, 4 + \sin(2t)]$, sphere and astroidal sphere

The reader can use the following program to plot various surfaces of revolution and their profile curves in rotation.

```
> with(plots):
> phi:=u->cos(u):psi:=u->sin(u): n:=6: u0:=-Pi/2:u1:=Pi/2:
  # enter your data
> X:=phi(u)*cos(v): Y:=phi(u)*sin(v): Z:=psi(u):
> plot([phi(u),psi(u),u=u0..u1],scaling=constrained); # curve
> a:=k->spacecurve([phi(u)*cos(k),phi(u)*sin(k),psi(u)],
  u=u0..u1, thickness=3,color=green): # profile curve
> b:=plot3d([X,Y,Z], u=u0..u1,v=-Pi..Pi, style=wireframe):
> display([seq(display([a(k*2*Pi/n), b]), k=1..n)],
  insequence=true); # surface with rotating profile curve
```

Table of some surfaces of revolution

	M	$\varphi(u)$	$\psi(u)$	Domain $G(u, v)$
1	cylinder	R	u	$v \in [0, 2\pi),\ u \in \mathbb{R}$
2	cone	u	$ku\ (k \neq 0)$	$v \in [0, 2\pi),\ u \in \mathbb{R}$
3	sphere	$R \sin u$	$R \cos u$	$v \in [0, 2\pi), u \in [0, \pi]$
4	2-hyperb.	$R \sinh u$	$R \cosh u$	$v \in [0, 2\pi),\ u \in \mathbb{R}$
5	1-hyperb.	$R \cosh u$	$R \sinh u$	$v \in [0, 2\pi),\ u \in \mathbb{R}$
6	catenoid	$a \cosh(\frac{u}{a})$	u	$v \in [0, 2\pi),\ u \in \mathbb{R}$
7	pseudosp.	$a \sin u$	$a(\cos u + \log \tan(\frac{u}{2}))$	$v \in [0, 2\pi), u \in (0, \pi)$
8	torus	$R + r \cos u$	$r \sin u\ (R > r)$	$u, v \in [0, 2\pi)$

1. Revolution of a straight line parallel to the axis yields the *circular cylinder*

$\vec{\mathbf{r}} = [R \cos v,\ R \sin v,\ u]$; Fig. 19.5.

2. Revolution of a straight line intersecting the axis yields the *cone*

$\vec{\mathbf{r}} = [u \cos v,\ u \sin v,\ au]$; Fig. 19.15.

3. Revolution of a half-circle about its diameter yields the *sphere*

$\vec{\mathbf{r}} = [R \sin u \cos v,\ R \sin u \sin v,\ R \cos u]$; Fig. 20.7.

The coordinates u and v on this sphere are the *geographical* longitude and latitude coordinates, and the coordinate curves are the geographical parallels and meridians. By deleting one meridian (the *"date line"*) we obtain an elementary surface.

4. Revolution of a branch of a hyperbola yields a *hyperboloid of two sheets* (considered a *sphere of imaginary radius* in a pseudo-Euclidean space $\mathbb{R}^3_1$) $\vec{\mathbf{r}} = [R \sinh u \cos v, R \sinh u \sin v, \pm R \cosh u]$; Fig. 19.16.

5. Revolution of another bruach of a hyperbola yields a *hyperboloid of one sheet* $\vec{\mathbf{r}} = [R \cosh u \cos v,\ R \cosh u \sin v,\ R \sinh u]$; Fig. 19.17.

6. Revolution of a catenary $\Rightarrow$ a *catenoid*

$\vec{\mathbf{r}} = [a \cosh(\frac{u}{a}) \cos v,\ a \cosh(\frac{u}{a}) \sin v,\ u]$; Fig. 20.10.

7. Revolution of a tractrix about its asymptote $\Rightarrow$ a *pseudosphere*

$\vec{\mathbf{r}} = [a \sin(u) \cos v,\ a \sin(u) \sin v,\ a(\cos(u) + \log(\tan \frac{u}{2}))]$; Fig. 20.9.

8. Revolution of a circle yields a *torus*

$\vec{\mathbf{r}} = [(R + r \sin u) \cos v,\ (R + r \sin u) \sin v,\ r \cos u]$, Fig. 19.3.

The table of surfaces of revolution (above) can be extended.

If the profile curve γ is given implicitly by $F(x, z) = 0$, then the surface of revolution with the axis OZ can be given by an equation in rectangular (or cylindrical) coordinates,

$$F(\sqrt{x^2 + y^2}, z) = 0, \qquad F(\rho, z) = 0. \tag{20.2}$$

In particular, if the profile curve is the graph of a function $z = f(x)$, then the surface of revolution is also the graph of the following function in rectangular (or cylindrical) coordinates

$$z = f(\sqrt{x^2 + y^2}), \qquad z = f(\rho). \tag{20.3}$$

In the case of parametrization, $\vec{\mathbf{r}} := [u\cos(v),\ u\sin(v),\ f(\sqrt{u^2 + v^2})]$, the coordinate net consists of parallels and meridians.

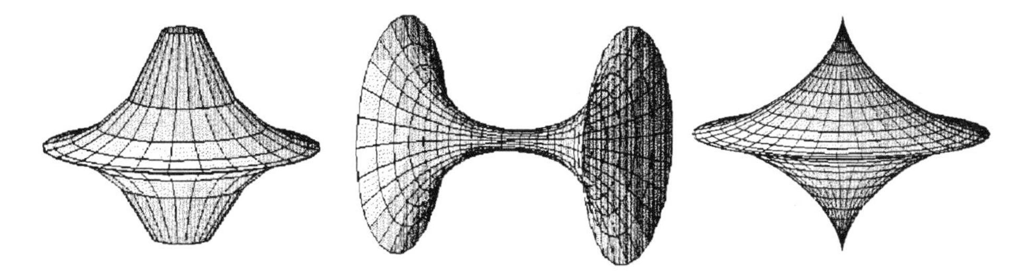

Figs. 20.9–20.11. Pseudosphere, catenoid, and revolution of the astroid

The singular (and smooth) points on a surface of revolution are unions of parallels, respectively, and can be identified by the equations of its profile curve.

Lemma 20.2.1 *If the profile curve γ has no singular points and does not intersect the axis, then the surface of revolution i also has no singular points.*

Proof. If the vectors

$$\vec{\mathbf{r}}_v = [-\varphi\sin v, \varphi\cos v, 0], \qquad \vec{\mathbf{r}}_u = [\varphi'\cos v, \varphi'\sin v, \psi']$$

are linearly independent, then all (2×2)-minors of the 3×2 matrix are nonzero:

$$\begin{vmatrix} \varphi'(u)\sin v & \psi'(u) \\ \varphi(u)\cos v & 0 \end{vmatrix} = 0 \iff \varphi(u)\cos v\psi'(u) = 0, \implies \psi'(u) = 0,$$

$$\begin{vmatrix} \varphi'(u)\cos v & \varphi(u)\sin v \\ -\varphi(u)\sin v & \varphi(u)\cos v \end{vmatrix} = 0 \iff \varphi'(u)\varphi(u) = 0 \implies \varphi'(u) = 0,$$

$$\begin{vmatrix} \varphi'(u)\cos v & \psi'(u) \\ -\varphi(u)\sin v & 0 \end{vmatrix} = 0 \iff \varphi(u)\sin v\psi'(u) = 0 \implies \psi'(u) = 0.$$

But the equalities $\varphi'(u) = \psi'(u) = 0$ are possible simultaneously only at a singular point of γ (the whole parallel through any such point consists of singular points; see Fig. 20.9 and Fig. 20.11). $\square$

Recall that a *screw motion* of space is the composition of a revolution about some axis and a parallel displacement along the same axis. Using this map we obtain the following generalization of a surface of revolution.

Definition 20.2.2 If a plane curve $\gamma(u)$ rotates about some axis in its plane and at the same time moves uniformly in the direction of this axis, then we obtain the *screw surface* with generatrix $\gamma(u)$. If the generatrix is a straight line, the screw surface is called a *helicoid*, which is also a particular case of a ruled surface (see Section 20.3).

A screw surface with generatrix $\gamma : [\varphi(u), \psi(u)]$ is defined by the equation (where $h = \text{const}$)

$$\vec{\mathbf{r}}(u, v) = [\varphi(u)\cos v,\ \varphi(u)\sin v,\ \psi(u) + h\,v] \qquad (0 \le v < 2\pi,\ u \in I).$$

Plot the *twisted sphere* (Fig. 20.14)

$$\vec{\mathbf{r}}(u, v) = [R\sin u\cos v,\ R\sin u\sin v,\ R\cos u + hv] \quad (|v| \le \pi,\ |u| \le \pi).$$

Exercise 20.2.1
1. Prove that the surface of revolution of the curve $\gamma : x = \varphi(u),\ z = \psi(u) \ge 0$, where $u \in I$, about the axis OX is given by the equations

$$\vec{\mathbf{r}}(u, v) := [\varphi(u),\ \psi(u)\cos(v),\ \psi(u)\sin(v)] \qquad (0 \le v < 2\pi). \qquad (20.1\,a)$$

2. Prove that if a profile curve is algebraic, then the surface of revolution is also algebraic.

Solution. Let $f_n(x, z) = 0$, where f_n is a polynomial of order n, be the equation of the profile algebraic curve γ. Then the surface of revolution M with the axis OZ is given by the equation $f_n(r, z) = 0$, where $r^2 - x^2 - y^2 = 0$. If the function $f_n(r, z)$ is even in the variable r, then it in fact depends on r^2. We substitute $x^2 + y^2$ for r^2 and obtain that the surface is given by a polynomial equation of degree n.

For example, in the revolution of the astroid (which is an algebraic curve of the sixth order), we eliminate r from the system of equations $(r^2 + z^2 - a^2)^3 + 27a^2z^2r^2 = 0,\ r^2 - x^2 - y^2 = 0$ and obtain $((x^2 + y^2) + z^2 - a^2)^3 + 27a^2z^2(x^2 + y^2) = 0$, i.e., a surface of the sixth order.

```
> resultant(r^2-x^2-y^2,r^2+z^2-a^2)^3+27*a^2*z^2*r^2,r);
```

$$((x^2 + y^2) + z^2 - a^2)^3 + 27a^2z^2(x^2 + y^2) = 0$$

If the function $f_n(r, z)$ is not even in r, then we eliminate the variable r from the system of equations $f_n(r, z) = 0,\ r^2 - x^2 - y^2 = 0$ using `resultant` and again the surface is given by the polynomial equation $F_m(x, y, z) = 0$, where F_m is a polynomial of degree $m \le n + 2$.

For example, in the case of revolution of a circle, we eliminate r from the system $(r-a)^2+z^2-b^2=0,\ r^2-z^2-y^2=0$ (with parameters $a>b>0$) and obtain a polynomial that defines a torus of revolution:

```
> resultant((r-a)^2+z^2-b^2, r^2-x^2-y^2, r);
```

$$-2a^2x^2-2a^2y^2+2a^2z^2-2a^2b^2-2z^2b^2+2z^2x^2+2z^2y^2-2b^2x^2-2b^2y^2+2x^2y^2+a^4+z^4+b^4+x^4+y^4,$$

which (check using MAPLE) can be simplified to the short form $4a^2(x^2+y^2)-(a^2+z^2-b^2+x^2+y^2)^2$.

3. Write down the equation and plot the *surface of revolution of the astroid* about one of its axes of symmetry (Fig. 20.11, looks like a toy top), the revolution of the lemniscate about each of its axes of symmetry, and the revolution of the four-leafed rose about an axis of symmetry.

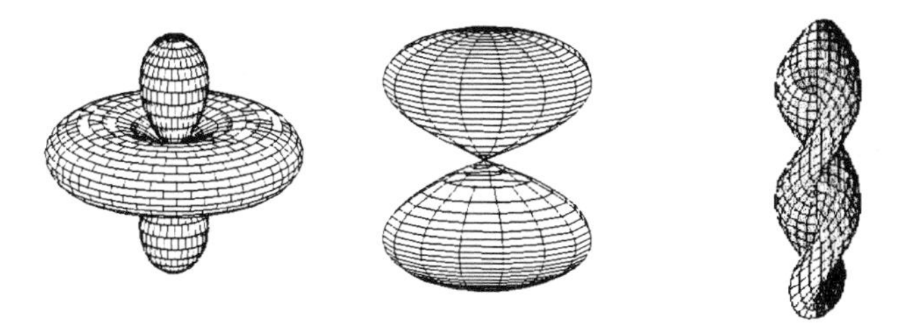

Figs. 20.12–20.14. Revolution of a four-leafed rose and lemniscate. Twisted sphere

4. The construction of a surface of revolution can be generalized if the points of a curve $\gamma(u)$ are moving not along circles (parallels), but along plane curves that are homothetic to the given curve $\gamma_1(v)$. Write down the equations of generalized surfaces of revolution in the case where $\gamma_1(v)$ is an ellipse, a parabola, a four-leafed rose, etc.

5. Seven colors are sufficient to color a map on a torus in Fig. 20.15–20.16 (see the problem of four colors [Gar 2]). Plot it using animation as in the following program.

```
> with(plots): setoptions3d(style=patchnogrid,
  projection=0.5, orientation=[40,50], scaling=constrained):
> r1:=3: r2:=1: G:=array(0..14):
> xk:=(r1+r2*sin(u))*cos(v+2*Pi*k/7):
  yk:=(r1+r2*sin(u))*sin(v+2*Pi*k/7): zk:=r2*cos(u):
> for i from 1 to 7 do G[i]:=k->plot3d([xk,yk,zk],
  u=0..2*Pi, v=(2*Pi*(2*i-2)+5*u)/14..(2*Pi*(2*i)+5*u)/14,
  color=(i-1)/7, grid=[30,10]):
> G[7+i]:=k->spacecurve([subs(v=(2*Pi*(2*i-2)+5*u)/14, xk),
```

```
  subs(v=(2*Pi*(2*i-2)+5*u)/14, yk),
  subs(v=(2*Pi*(2*i-2)+5*u)/14, zk), u=0..2*Pi],
  thickness=3, color=black, numpoints=30) od:
> G[0]:=k-> spacecurve([subs(u=2*Pi, xk), subs(u=2*Pi, yk),
  subs(u=2*Pi, zk)], thickness=3, v=0..2*Pi, color=black,
  numpoints=30):
> display([seq(display([seq(G[i](k), i=0..14)]), k=1..7)],
  insequence=true);
```

The 2-dimensional map (scheme) is the following:

```
> G2:=array(0..9):  for i from 1 to 9 do
  G2[i]:=plot([u+2*i-6,u,u=0..5], view=[0..14,0..5]) od:
> G2[0]:=polygonplot([[0,0], [14,0], [14,5], [0,5]]):
> plots[display]([seq(G2[i],i=0..9)],scaling=constrained);
```

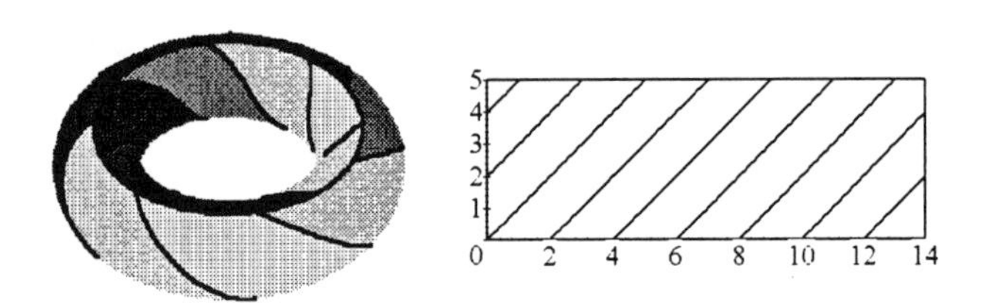

Figs. 20.15–20.16. Seven countries on the torus

6. Slowly changing the equations of a torus (using periodic flattening of the circle to a line segment) leads to a *pinched torus*; see Fig. 19.4:

```
> plot3d([(1+0.2*sin(u))*cos(v), (1+0.2*sin(u))*sin(v),
  0.2*sin(v)*cos(u)], v=0..2*Pi, u=0..2*Pi,style=patch);
```

7. Find the singular points on the surface of revolution of the astroid about its axis of symmetry; see Fig. 20.11.

8. The cubes of the coordinate functions in the equations of the sphere define a parametrization of the *astroidal sphere*. It looks like an octahedron with concave faces and edges; see Fig. 20.8:

$$\vec{\mathbf{r}} = [(R\sin u\cos v)^3,\ (R\sin u\sin v)^3,\ (R\cos u)^3].$$

Check that the equation $x^{2/3} + y^{2/3} + z^{2/3} = R^{2/3}$ also defines an astroidal sphere.

9. For plotting the *Möbius band* turn the line segment $I_0 = [R, 0, u]$, ($|u| \le 1$) with $R > 1$ about its center $(R, 0, 0)$ through the angle $\frac{v}{2} \in [0, \pi]$ in the plane XZ: $I_{v/2} = [R + u\sin(\frac{v}{2}),\ 0,\ u\cos(\frac{v}{2})]$, and then rotate the line segment $I_{v/2}$ about the axis OZ through the angle $v \in [0, 2\pi]$. We obtain the equation of a wonderful surface (see also Section 21.3), Fig. 19.6:

$$\vec{\mathbf{r}}(u,v) = [(R + u\sin(\frac{v}{2}))\cos(v),\ (R + u\sin(\frac{v}{2}))\sin(v),\ u\cos(\frac{v}{2})]$$
$$= [R\cos(v), R\sin(v), 0] + u[\sin(\frac{v}{2})\cos(v),\ \sin(\frac{v}{2})\sin(v),\ \cos(\frac{v}{2})]$$
$$(v \in [0, 2\pi),\ |u| \le 1])$$

```
> plot3d([(5+sin(v/2)*u)*cos(v), (5+sin(v/2)*u)*sin(v),
  cos(v/2)*u], v=0..2*Pi, u=-1..1, grid=[60,10],
  scaling=CONSTRAINED, orientation=[-106,70]);
```

Plot the n-times turned (through 180°) twisted strip by the analogous equations

$$\vec{\mathbf{r}}(u,v) = \Big[(R + u\sin(\frac{nv}{2}))\cos(v),\ (R + u\sin(\frac{nv}{2}))\sin(v),\ u\cos(\frac{nv}{2})\Big]$$
$$(v \in [0, 2\pi),\ |u| \le 1]).$$

The surface is homeomorphic to the cylinder for even n, and homeomorphic to the Möbius band, and it has one side for odd n.

20.3 Ruled Surfaces

20.3.1 Some Types of Ruled Surfaces

Definition 20.3.1 A surface M is called *ruled* if it is generated by a one-parameter family of straight lines (*rulings*). A space curve on a ruled surface that intersects each ruling at one point is called a *directrix curve*. For each point of some directrix $\vec{\rho}(u)$, $u \in I$, we define the unit vector $\vec{\delta}(u)$ that is parallel to the ruling g_u through this point. As a result we obtain the *standard parametrization* of the ruled surface

$$\vec{\mathbf{r}}(u,v) = \vec{\rho}(u) + v\,\vec{\delta}(u) \qquad (u \in I,\ v \in \mathbb{R}), \tag{20.4}$$

where the v-curves coincide with the rulings, and the parameter v is equal (up to sign) to the segment of the ruling between the point $\vec{\mathbf{r}}(u,v)$ and the directrix $\vec{\rho}(u)$; see Fig. 20.17.

Particular cases of ruled surfaces are the following:

• the *cylindrical surface* $\vec{\mathbf{r}}(u,v) = \vec{\rho}(u) + v\vec{\delta}_0$: the vector-valued function $\vec{\delta}(u) = \vec{\delta}_0$ is constant and represents the direction of the *axis*;

• the *conic surface* $\vec{\mathbf{r}}(u,v) = \vec{\rho}_0 + v\vec{\delta}(u)$: the vector-valued function $\vec{\rho}(u) = \vec{\rho}_0$ is constant and represents the *vertex* (the singular point);

• the *tangent developable surface* $\vec{\mathbf{r}}(u,v) = \vec{\rho}(u) + v\vec{\rho}\,'(u)$: all rulings are tangent to the directrix curve, i.e., $\vec{\delta}(u) = \vec{\rho}\,'(u)$.

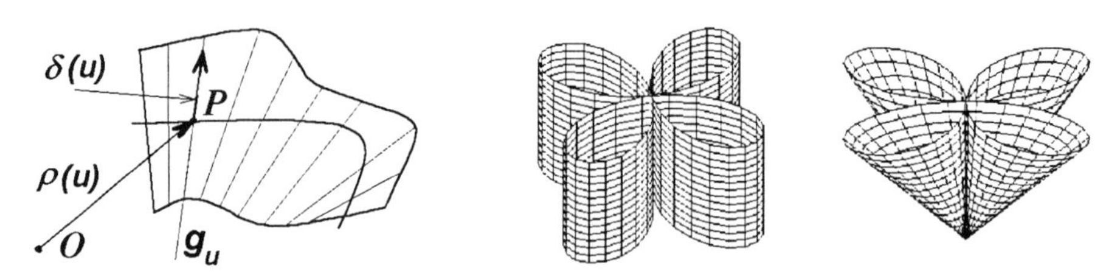

Figs. 20.17–20.19. Ruled surface.
Cylinder and cone over a four-leafed rose

Example 20.3.1 The *hyperboloid of one sheet* $\frac{x^2}{a^2}+\frac{y^2}{b^2}-\frac{x^2}{c^2}=1$:

$$\begin{aligned}\vec{\mathbf{r}}(u,v) = &\vec{\rho}(u) \pm v(\vec{\rho}\,'(u)+[0,0,c]) = \\ &[a(\cos(u) \mp v\sin(u)),\ b(\sin(u) \pm v\cos(u)),\ \pm cv],\end{aligned}$$

where $\vec{\rho}(u)=[a\cos(u), b\sin(u), 0]$, is an ellipse on the plane XY, and $\pm$ are two ways to fix the rulings.

The *hyperbolic paraboloid* $z=\frac{x^2}{a^2}-\frac{y^2}{b^2}$:

$$\vec{\mathbf{r}}(u,v)=[au,0,u^2]+v[a,\pm b,2u]=[a(u+v),\ \pm bv,\ u^2+2uv],$$

where $\pm$ are two ways to fix the rulings. There is another convenient parametrization when this surface (after some rotation) is given by the equation $z=axy$,

$$\vec{\mathbf{r}}(u,v)=[u,0,0]+v[0,1,au]=[u,v,avu].$$

The *elliptical cone* $\frac{x^2}{a^2}+\frac{y^2}{b^2}-\frac{x^2}{c^2}=0$ can be parametrized with the rulings playing the role of the v-curves,

$$\vec{\mathbf{r}}(u,v)=v[a\cos(u),\ b\sin(u),\ c] \quad (v\in\mathbb{R},\ u\in[0,2\pi)).$$

The *conical surface* with vertex $S(a,b,0)$ and a four-leafed rose $\rho=\cos(2\varphi)$ as its section by the plane $\alpha: z=1$; see Fig. 20.19:

$$\begin{aligned}\vec{\mathbf{r}}(u,v)&=[a,b,0]+v[\cos(2u)\cos(u),\ \cos(2u)\sin(u),\ 1]\\ &=[v\cos(2u)\cos(u)+a,\ v\cos(2u)\sin(u)+b,\ v]\\ &\qquad (0\le v\le 2,\ 0\le u\le 2\pi).\end{aligned}$$

The *cylinder* with axis OZ and the directrix a four-leafed rose; Fig. 20.18:

$$\begin{aligned}\vec{\mathbf{r}}(u,v)&=c[\cos(2u)\cos(u),\cos(2u)\sin(u),0]+v[0,0,1]\\ &=[c\cos(2u)\cos(u),c\cos(2u)\sin(u),v].\end{aligned}$$

The *tangent developable surface* with the helix as the directrix curve:

$$\begin{aligned}\vec{\mathbf{r}}(u,v)&=[\cos(u),\sin(u),au]+v[-\sin(u),\cos(u),a]\\ &=[\cos(u)-v\sin(u),\ \sin(u)+v\cos(u),\ a(u+v)];\ \text{seeFig. 20.26.}\end{aligned}$$

Definition 20.3.2 A ruled surface satisfying the condition $\vec{\delta}\,'(u) \neq 0$ is said to be *non-cylindrical*. (If the vector $\vec{\delta}(u)$ is not assumed to be a unit vector, then the above condition takes the form $\vec{\delta}(u) \times \vec{\delta}\,'(u) \neq 0$.)

Definition 20.3.3 A noncylindrical ruled surface whose rulings are parallel to some fixed *directrix plane* is called a *Catalan surface*.

Definition 20.3.4 A Catalan surface is called a *conoid* if all its rulings intersect a constant straight line, the so-called *axis of the conoid*. A *conoid* is called *right* if its axis is orthogonal to the directix plane.

From the definition of a conoid it follows that it is generated by a straight line that moves, guided on a fixed straight line orthogonal to it (the axis of conoid), and at the same time rotates about this straight line. If the velocity of rotation is proportional to the lifting velocity of the ruling, then the conoid is called a *right helicoid*. One can distinguish a *right-side* and *left-side* right helicoid.

Example 20.3.2 The simplest conoid is the hyperbolic paraboloid. It is defined by moving a straight line that is parallel to a fixed plane and is guided by two fixed helices (two axes!). Conversely, every conoid that differs from a plane and has two axes is a hyperbolic paraboloid.

One can construct a right helicoid as a particular case of the following surface. A *helicoid* is a surface obtained from a straight line that rotates about a fixed axis with constant angular velocity, intersects the axis at the constant angle θ, and at the same time has translational motion (with constant velocity) along the axis. If $\theta = 90°$, then the helicoid is called *right*, see Fig. 20.25, (above), and if $\theta \neq 90°$, then it is called *skew*:

$$\vec{\mathbf{r}}(u, v) = [0, 0, cu] + av[\cos(u), \sin(u), 0] = [av\cos(u), av\sin(u), cu].$$

A visual demonstration of separate rays on this surface gives the steps of spiral staircase. A generalization of the right helicoid is the *elliptical helicoid* $\vec{\mathbf{r}}(u, v) = [av\cos(u), bv\sin(u), cu]$. The following surface (*the conoid*), of Fig. 20.35, looks like the helicoid when $a = 1,\ c = 0.1$, but the velocity of rotation of the ruling is exponential:

$$\begin{aligned}\vec{\mathbf{r}}(u, v) &= [0,\ 0,\ \exp(cu)] + av[\cos(u),\ \sin(u),\ 0]\\ &= [av\cos(u),\ av\sin(u),\ \exp(cu)].\end{aligned}$$

The *Wallis conoid* (for example, with $a = 1, b = \sqrt{3},\ u \in [0, 1],\ \varphi \in [0, \pi]$) (Fig. 20.20) is given by

$$\vec{\mathbf{r}}(u, v) = [v\cos(u),\ v\sin(u),\ c\sqrt{a^2 - b^2\cos^2(u)}].$$

The *Plücker conoid* has the simple equation $z = \frac{2xy}{x^2+y^2} \iff \vec{\mathbf{r}}(u, v) = [u, v, \frac{2uv}{u^2+v^2}]$, but in this case we do not see the rulings in the picture. Using cylindrical coordinates in space allows us to write other parametric equations of this surface, where the rulings are the v-curves; see Fig. 20.21:

$$\vec{\mathbf{r}}(u, v) = [0, 0, \sin(2u)] + v[\cos(u), \sin(u), 0] = [v\cos(u), v\sin(u), \sin(2u)].$$

From these equations we see that the surface is given by the rotation of the ray about the axis OZ with simultaneous oscillatory motion (with period 2π) along the segment $[-1, 1]$ of the axis OZ.

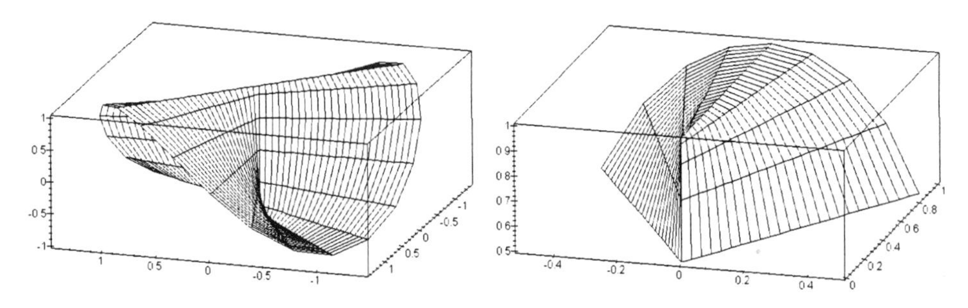

Figs. 20.20–20.21. Wallis conoid and Plücker conoid

The *generalized Plücker conoids* (having n folds instead of two) are obtained by the rotation of a ray about the axis OZ, simultaneously with the oscillatory motion (with period $2\pi n$) along the segment $[-1, 1]$ of the axis; see Figs. 20.22–20.23 with $n = 5, 8$:

$$\vec{\mathbf{r}}(u, v) = [0, 0, \sin(nu)] + v[\cos(u), \sin(u), 0] = [v\cos(u), v\sin(u), \sin(nu)].$$

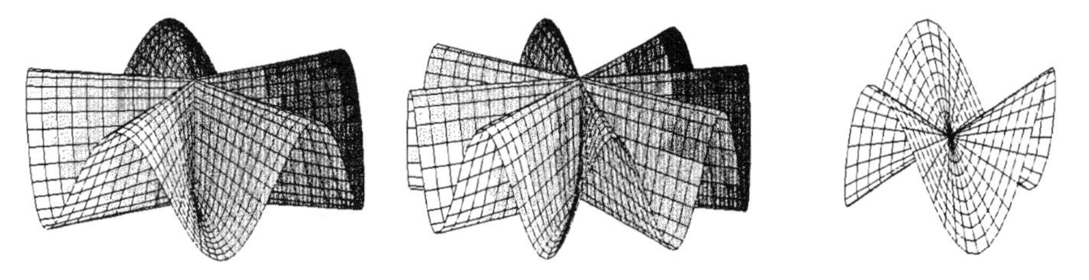

Figs. 20.22–20.24. Generalized Plücker conoids for $n = 5, 8$ and saddle-3

Definition 20.3.5 A ruled surface is called *developable* if its tangent plane is the same at all points of any ruling, or equivalently, if the normal vector $\vec{\mathbf{n}}$ preserves its direction (is stationary) along each ruling.

Examples of developable ruled surfaces are cylinders and cones.

Lemma 20.3.1 *A tangent developable surface is a developable ruled surface.*

Proof. Since $\vec{\mathbf{r}}(u, v) = \vec{\rho}(u) + v\vec{\rho}\,'(u)$, then

$$\vec{\mathbf{r}}_u = \vec{\rho}\,'(u) + v\vec{\rho}\,''(u), \qquad \vec{\mathbf{r}}_v = \vec{\rho}\,'(u),$$

and hence the normal vector $\vec{\mathbf{n}}_1 \,||\, \vec{\mathbf{r}}_u \times \vec{\mathbf{r}}_v = v\vec{\rho}\,''(u) \times \vec{\rho}\,'(u) \,||\, \vec{\rho}\,''(u) \times \vec{\rho}\,'(u)$ preserves its direction as v varies. □

The inverse statement also holds; we state it without proof.

Theorem 20.3.1 *Every developable ruled surface consists of cylindrical, conical, and tangent developable surfaces.*

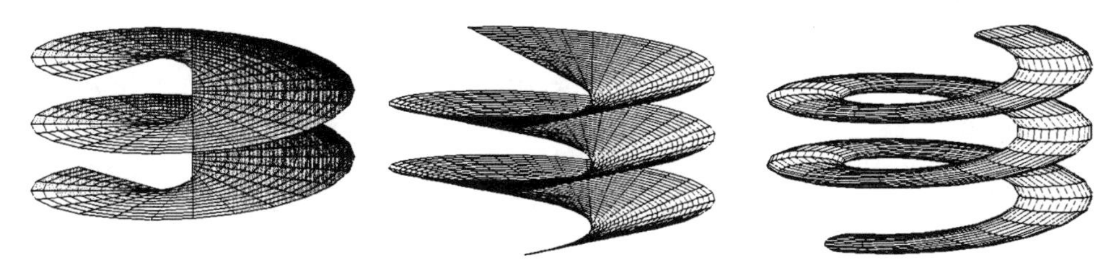

Figs. 20.25–20.27. Some surfaces generated by a helicoid: helicoid, tangent developable surface, and a tube

Exercise 20.3.1

1. Prove that:

(a) A cylinder over a smooth plane curve is also a smooth surface.

(b) The unique singular point on a cone over a smooth curve that is not contained in a plane is its vertex.

(c) The singular points of a tangent developable surface form its directrix curve, which is called the *cuspidal edge*.

2. Write equations and using MAPLE, plot the tangent developable surface over the Viviani curve, Fig. 20.29:

$$\vec{\mathbf{r}}(u) = [R\cos(u)^2,\ R\cos(u)\sin(u),\ R\sin(u)] \qquad (u \in [-\pi, \pi]).$$

3. Prove that the Catalan surface is characterized in the class of all ruled surfaces $\vec{\mathbf{r}}(u, v) = \vec{\rho}(u) + v\vec{\delta}(u)$ by the conditions

$$(\vec{\delta}, \vec{\delta}\,', \vec{\delta}\,'') = 0, \qquad \vec{\delta}\,'' \neq 0.$$

Hint: Observe that the three vectors $\vec{\delta}, \vec{\delta}\,', \vec{\delta}\,''$ are parallel to the directrix plane.

4. Prove that the Catalan surface cannot be a tangent developable surface over s any curve. *Hint*: Observe that a tangent developable surface over some curve has the property that $\vec{\delta}(u) = \vec{\rho}\,'(u)$, which leads to the plane, which is a contradiction to the noncylindrical property of the Catalan surface.

5. Find conditions for the case when the surface $z = f(x, y)$ is a ruled developable surface. Do calculations using MAPLE and give examples.

6. Show that the surface saddle-3, Fig. 20.24, generalizes both the Plücker conoid and the monkey saddle:

$$\vec{\mathbf{r}}(u, v) = [v\cos(u),\ v\sin(u),\ v^m\sin(vu)].$$

7. Write a program and plot the *normal surface* and the *binormal surface* $\vec{\mathbf{r}}_1(u, v) = \vec{\rho}(u) + v\,\vec{\nu}(u)$, $\quad \vec{\mathbf{r}}_2(u, v) = \vec{\rho}(u) + v\,\vec{\beta}(u)$, consisting of the main normals $\vec{\nu}(u)$ and binormals $\vec{\beta}(u)$, respectively, to the given space curve $\vec{\rho}(u)$. Plot the surfaces $\vec{\mathbf{r}}_1(u, v)$ and $\vec{\mathbf{r}}_2(u, v)$ for the helicoid (see the right conoid and Fig. 20.33) and for the Viviani curve, Figs. 20.30–20.31. Plot the intermediate surfaces for some values of θ (see Fig. 20.34):

$$\vec{\mathbf{r}}_\theta(u, v) = \vec{\rho}(u) + v\cos(\theta)\vec{\nu}(u) + v\sin(\theta)\vec{\beta}(u).$$

Solution.

```
> with(plots): with(linalg):
> r:=array([cos(u)^2, cos(u)*sin(u), sin(u)]);
> spacecurve(r, u=-Pi..Pi);              # Viviani, Fig. 20.28
> p1:=plot3d([cos(u)*cos(v), cos(u)*sin(v), sin(u)],
  u=-Pi/2..Pi/2, v=0..2*Pi, grid=[30,60]):
> p2:=plot3d([0.5+cos(u)/2, sin(u)/2,v],
  v=-1.2..1.2, u=0..2*Pi, grid=[20,45]):
> display3d({p1, p2});         # Viviani as intersection, Fig. 20.29

> ru:=map(diff,r,u): ruu:=map(diff,ru,u):
> tt:=scalarmul(ru, 1/norm(ru,2)):
> b:=crossprod(ru,ruu): bb:=scalarmul(b, 1/norm(b,2)):
> nn:=crossprod(bb,tt): rn:=evalm(r+scalarmul(nn,v)):
> rt:=evalm(r+scalarmul(tt,v)):rb:=evalm(r+scalarmul(bb,v)):
> rtheta:=evalm(r+scalarmul(nn,v*cos(Pi/6))+
  scalarmul(bb,v*cos(Pi/6))):
> plot3d(rt, u=-Pi..Pi, v=-1.8..1.8, grid=[70,20]); # Fig. 20.32
> plot3d(rn, u=-Pi..Pi, v=-1..1, grid=[70,20]);    # Fig. 20.30
> plot3d(rb, u=-Pi..Pi, v=-0.5..0.5, grid=[70,15]); # Fig. 20.31
> plot3d(rtheta, u=-Pi..Pi, v=-2..2, grid=[70,20]); # Fig. 20.35
```

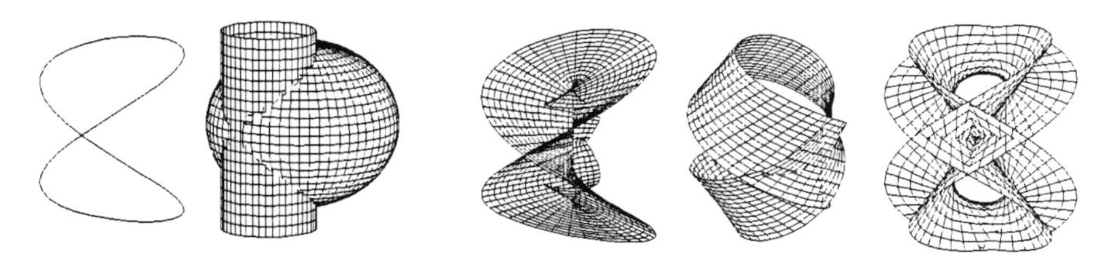

20.28–20.32. Viviani curve and its normal, binormal, and tangent developable ruled surfaces

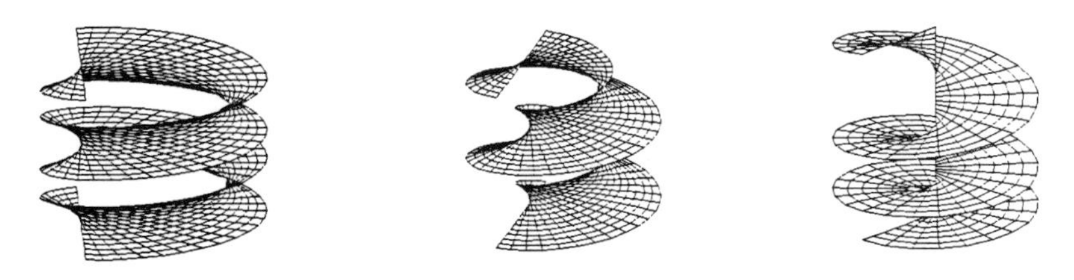

Figs. 20.33–20.35. Binormal ruled surface and $\vec{\mathbf{r}}_\theta$ $(\theta = \pi/6)$ for a helicoid and exponential helicoid

20.3.2 Striction Curve and Distribution Parameter

Consider the relative positions of two rulings g_u and $g_{u+\Delta u}$ on the ruled surface $\vec{\mathbf{r}}(u, v) = \vec{\rho}(u) + v\vec{\delta}(u)$ where $|\vec{\delta}(u)| = 1$. If the surface is non-cylindrical, then these rulings are skew lines in space. Let PP' be the *shortest segment* between rulings, where $P(u, v)$, $P'(u + \Delta u, v + \Delta v)$, and u, v are unknown, see Fig. 20.36. The segment

$$\begin{aligned} PP' &= OP' - OP = (\vec{\rho} + v\vec{\delta}) - ((\vec{\rho} + \Delta\vec{\rho}) + (v + \Delta v)(\vec{\delta} + \Delta\vec{\delta})) \\ &= \Delta\vec{\rho} + v\Delta\vec{\delta} + \Delta v(\vec{\delta} + \Delta\vec{\delta}) \end{aligned}$$

is orthogonal to the vectors $\vec{\delta}$ and $\vec{\delta} + \Delta\vec{\delta}$ and hence is orthogonal to their difference $\Delta\vec{\delta}$:

$$PP' \cdot \vec{\delta} = 0, \quad PP' \cdot \Delta\vec{\delta} = 0. \tag{20.5}$$

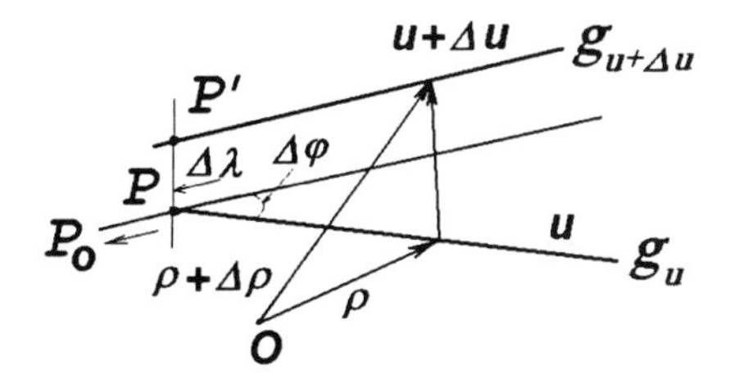

Fig. 20.36. Striction points on a ruling

Substituting these equalities into the term for PP', we obtain

$$\begin{cases} \Delta\vec{\rho}\cdot\vec{\delta}+v\Delta\vec{\delta}\cdot\vec{\delta}+\Delta v(1+\vec{\delta}\cdot\Delta\vec{\delta}) &= 0,\\ \Delta\vec{\rho}\cdot\Delta\vec{\delta}+v(\Delta\vec{\delta})^2+\Delta v(\vec{\delta}\cdot\Delta\vec{\delta}+(\Delta\vec{\delta})^2) &= 0. \end{cases} \tag{20.6}$$

Dividing the equations (20.6), respectively, by Δu and $(\Delta u)^2$, we obtain two equations in variables v and $\frac{\Delta v}{\Delta u}$:

$$\begin{cases} \frac{\Delta\vec{\rho}}{\Delta u}\cdot\vec{\delta}+v\frac{\Delta\vec{\delta}}{\Delta u}\cdot\vec{\delta}+\frac{\Delta v}{\Delta u}(1+\vec{\delta}\cdot\Delta\vec{\delta}) &= 0\\ \frac{\Delta\vec{\rho}}{\Delta u}\cdot\frac{\Delta\vec{\delta}}{\Delta u}+v(\frac{\Delta\vec{\delta}}{\Delta u})^2+\frac{\Delta v}{\Delta u}(\vec{\delta}\cdot\frac{\Delta\vec{\delta}}{\Delta u}+\frac{\Delta\vec{\delta}}{\Delta u}\cdot\Delta\vec{\delta}) &= 0 \end{cases} \tag{20.7}$$

Let us calculate the limit position of the perpendicular PP' as $\Delta u\to 0$. The limit form of equations (20.7) will be (in view of the equalities $\vec{\delta}\cdot\vec{\delta}\,'=0$, $\vec{\delta}\cdot\Delta\vec{\delta}=0$)

$$\begin{cases} \vec{\rho}\,'\cdot\vec{\delta}+w &= 0\\ \vec{\rho}\,'\cdot\vec{\delta}\,'+v\,(\vec{\delta}\,')^2 &= 0 \end{cases} \tag{20.8}$$

The determinant $\det\begin{vmatrix} 0 & 1\\ (\vec{\delta}\,')^2 & 0\end{vmatrix} = -(\vec{\delta}\,')^2$ of this system is nonzero. The functions $v_0=-\frac{\vec{\rho}\,'\cdot\vec{\delta}\,'}{(\vec{\delta}\,')^2}$, $w_0=-\vec{\rho}\,'\cdot\vec{\delta}$ form a solution to the system (20.8). Thus the bottom P of the perpendicular PP' tends along the ruling g_u to the limit value P_0 as $\Delta u\to 0$; the point P_0 is called the *point of striction* on the ruling g_u. It shows the narrowest place on the ruled surface in a neighborhood of this ruling. It is easy to write down the position vector of the striction point:

$$\vec{\mathbf{r}}(u)=\vec{\rho}(u)-\frac{\vec{\rho}\,'(u)\cdot\vec{\delta}\,'(u)}{(\vec{\delta}\,'(u))^2}\vec{\delta}(u). \tag{20.9}$$

Definition 20.3.6 The locus of striction points on a ruled surface is called the *striction curve.*

If u is considered as a variable in equation (20.9), then we have a striction curve. The geometric sense of a striction curve is the following: *it surrounds the ruled surface along the narrowest place and is independent of the choice of the directrix curve $\vec{\rho}(u)$.*

Example 20.3.3 For the hyperboloid of revolution of two sheets $\frac{x^2+y^2}{a^2}-\frac{z^2}{c^2}=1$, which is a doubly ruled (i.e., has two families of rulings) noncylindrical surface, the striction curve in both cases is the circle of its intersection with the plane XY.

From the last example we see that the striction curve, in general, does not intersect the rulings at right angle.

If the given value of the parameter u changes to a nearby value $u + \Delta u$, then the ruling g_u comes into the new position $g_{u+\Delta u}$, with rotation through some angle $\Delta\varphi$ and with a deviation from the initial position at the distance $\Delta\lambda = |PP'|$. Although both these values are small, their quotied $\frac{\Delta\lambda}{\Delta\varphi}$ has a limit value as $\Delta u \to 0$ (see below), and it is an important geometric characteristic of the ruled surface.

Theorem 20.3.2 *For a non-cylindrical ruled surface we have*

$$\lim_{\Delta u \to 0} \frac{\Delta\lambda}{\Delta\varphi} = \frac{(\vec{\delta}, \vec{\delta}\,', \vec{\rho}\,')}{|\vec{\delta}\,'|^2}. \tag{20.10}$$

Proof. Since $\Delta\varphi$ is the rotation angle of the unit vector, we have

$$\lim_{\Delta u \to 0} \frac{\Delta\varphi}{|\Delta u|} = |\vec{\delta}\,'(u)|. \tag{20.11}$$

Using the formula for the distance between two skew lines (the ratio of the volume of the cuboid $(\vec{\delta}, \vec{\delta} + \Delta\vec{\delta}, \Delta\vec{\rho})$ to the area $|\vec{\delta}| \cdot |\vec{\delta} + \Delta\vec{\delta}| \cdot \sin(\Delta\varphi)$ of its bottom face) $\Delta\lambda \approx \frac{(\vec{\delta}, \vec{\delta}+\Delta\vec{\delta}, \Delta\vec{\rho})}{|\vec{\delta}|\cdot|\vec{\delta}+\Delta\vec{\delta}|\cdot\sin(\Delta\varphi)} \approx \frac{(\vec{\delta}, \Delta\vec{\delta}, \Delta\vec{\rho})}{\sin(\Delta\varphi)}$, and in view of (20.11), we obtain

$$\lim_{\Delta u \to 0} \frac{\Delta\lambda}{\Delta u} = \lim_{\Delta u \to 0} \frac{(\vec{\delta}, \frac{\Delta\vec{\delta}}{\Delta u}, \frac{\Delta\vec{\rho}}{\Delta u})}{\frac{\sin(\Delta\varphi)}{\Delta\varphi}\frac{\Delta\varphi}{\Delta u}} = \frac{(\vec{\delta}, \vec{\delta}\,', \vec{\rho}\,')}{|\vec{\delta}\,'(u)|}. \tag{20.12}$$

The ratio of (20.12) to (20.11) gives equation (20.10) that we need. □

Definition 20.3.7 The function (20.10) is called the *distribution parameter* $p(u)$ of the non-cylindrical ruled surface (20.4).

Exercise 20.3.2

1. Prove that:

(a) The striction curve of a tangent developable surface coincides with its cuspidal edge.

(b) For the hyperboloid of one sheet of the general form $\frac{x^2}{a^2} + \frac{y^2}{b^2} - \frac{z^2}{c^2} = 1$, the striction curve does not belong to the XY plane, but it is a space curve of the fourth order, singular for each system of rulings.

(c) The axis of a right conoid is its striction curve.

2. Prove that the striction curve of the helicoid coincides with its axis and that its distribution parameter is a constant function.

3. Calculate using `MAPLE` the striction curve and the distribution parameter of the hyperbolic paraboloid.

Solution. For the parametrization $\vec{\mathbf{r}}_u = [u, 0, 0] + v[0, 1, u]$ of the hyperbolic paraboloid $z = axy$ let us calculate the striction curve $\vec{\rho}(u) = [u, 0, 0]$ and the function $\vec{\delta}(u) = \frac{[0,1,au]}{\sqrt{a^2+u^2}}$.

```
> with(linalg): d:=evalm([0,1,a*u]/sqrt(1+(a*u)^2)):
> r:=vector([u,0,0]): ru:=map(diff,r,u): du:=map(diff,d,u):
> p(u):=simplify(det(matrix(3,3,[d,du,ru])/dotprod(du,du,orthogonal));
```

$$p(u) := (1 + a^2u^2)/a$$

4. Prove that a distribution parameter of a developable (non-cylindrical) ruled surface is identically zero.

20.4 Envelope of a One-Parameter Family of Surfaces

20.4.1 Theorem on an Envelope of Surfaces

In the previous sections surfaces have been defined as figures satisfying some geometrical conditions or equations. In this section surfaces appear in another way, as envelopes of certain families of spheres, planes, and other surfaces that depend on one real parameter.

Definition 20.4.1 A set of surfaces $\{M_t\}$ that depend on the parameter t is called a one-parameter family of surfaces.

Example 20.4.1 Spheres $M_t\ :\ (x-t)^2 + y^2 + z^2 - R^2 = 0$, Fig. 20.37.

Definition 20.4.2 Let $\{M_t\}$ a the family of smooth surfaces depending on the parameter t. A smooth surface M is called an *e*nvelope of the family $\{M_t\}$ if it is tangent at each of its points to at least one surface from the family, and in each of its domains is tangent to an infinite number of surfaces from the family.

Example 20.4.2 The envelope in the previous example is the circular cylinder with axis OX; Fig. 20.38. It is easy to check that the envelope of the family of spheres of radius R with centers at the circles $\vec{\mathbf{r}}(t) = [a\cos t, a\sin t, 0]$, where $a > R$, is a torus of revolution; Figs. 20.39–20.82. This family M_t can be defined by the following equation: $(x-a\cos t)^2+(y-a\sin t)^2+z^2-R^2 = 0$.

Next we give a theorem on envelope that is analogous to the result on envelope of a one-parameter family of plane curves. Below we consider smooth functions $F(x, y, z;\ t)$.

Theorem 20.4.1 *The envelope of a one-parameter family of surfaces* $\{M_t\ :\ F(x, y, z; t) = 0\}$ *can be found from the system of equations*

$$\begin{cases} F(x, y, z;\ t) &= 0 \\ F'_t(x, y, z;\ t) &= 0, \end{cases} \tag{20.13}$$

for instance, by eliminating the parameter t.

Proof. Suppose that the envelope M of the family $\{M_t\}$ exists and try to find its equation in implicit form $\varphi(x, y, z) = 0$.

An arbitrary point $P(x, y, z)$ on an envelope also lies on some surface M_t of the given family. Moreover, the tangent planes for both surfaces at the point P coincides. In this case the real numbers x, y, z, t satisfy the system of equations $\varphi(x, y, z) = 0$, $F(x, y, z; t) = 0$. For a small displacement of $P(x, y, z)$ along the envelope we can differentiate these equalities term by term:

$$\begin{aligned} \varphi_x\, dx + \varphi_y\, dy + \varphi_z\, dz &= 0, \\ F_x\, dx + F_y\, dy + F_z\, dz + F_t\, dt &= 0, \end{aligned} \tag{20.14}$$

where dx, dy, dz, dt are differentials corresponding to the displacement of P. Since the tangent planes to both surfaces at the point $P(x, y, z)$ coincide, the normal vectors and gradients

$$\varphi_x\,\mathbf{i} + \varphi_y\,\mathbf{j} + \varphi_z\,\mathbf{k}, \qquad F_x\,\mathbf{i} + F_y\,\mathbf{j} + F_z\,\mathbf{k}$$

must be parallel. In this situation from the first equation of the system (20.14) follows $F_x\, dx + F_y\, dy + F_z\, dz = 0$, and hence the second equation of the system (20.14) takes the short form $F_t\, dt = 0$. The case $dt = 0$ means that t preserves its value under motion of $P(x, y, z)$ along the envelope, i.e., that all points $P(x, y, z)$ on the envelope belong to the same surface of the family. We exclude this case (although the system (20.13) is satisfied) because by Definition 20.4.2, the envelope must be tangent to an infinite number of surfaces of the given family. Supposing $dt \neq 0$, we obtain $F'_t = 0$. □

Example 20.4.3 As a test we apply Theorem 20.4.1 to the family of spheres $\varphi(x, y, t) = (x - t)^2 + y^2 + z^2 - R^2 = 0$ from Example 20.4.1. First calculate $F'_t = 2(x - t) = 0$. From this we obtain $t = x$, and then we substitute into the equation $F = 0$ and obtain $y^2 + z^2 = R^2$. Hence the envelope is the cylinder of radius R with axis OX; Fig. 20.38.

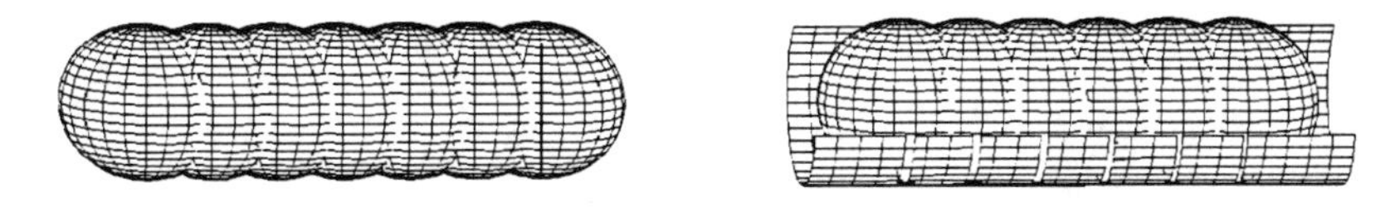

Figs. 20.37–20.38. The cylinder is the envelope of a family of spheres

```
> for t from 1 to 6 do p[t]:=plot3d([.8*t+cos(u)*cos(v),
  cos(u)*sin(v), sin(u)], u=-Pi/2..Pi/2, v=0..2*Pi) od:
> p[7]:=plot3d([v, cos(u), sin(u)], u=-Pi-1..0, v=-0.5..6):
> plots[display]([seq(p[t], t=1..7)], scaling=constrained);
```

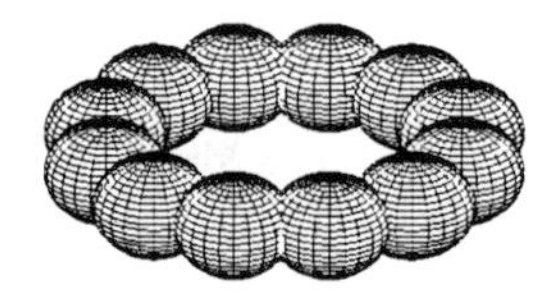
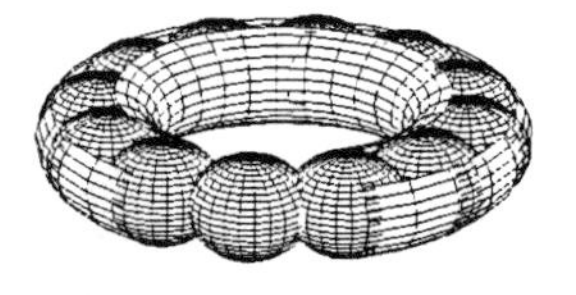

Figs. 20.39–20.40. Torus is the envelope of a family of spheres

Next we plot the envelope from Example 20.4.2; Fig. 20.40:

```
> for t from 1 to 12 do p[t]:=plot3d(
  [3*cos(t*Pi/6)+cos(u)*cos(v), 3*sin(t*Pi/6)+cos(u)*
  sin(v), sin(u)], u=-Pi/2..Pi/2, v=0..2*Pi) od:
> p[13]:=plot3d([(3+cos(u))*cos(v), (3+cos(u))*sin(v),
  sin(u)], u=-Pi-1.3..0.7, v=Pi/3..2*Pi):
> plots[display]([seq(p[t], t=1..13)], scaling=constrained);
```

In fact, the system of equations (20.13) defines the *discriminant set*, which includes extraneous solutions (analogous to the inflection points in a solution of the problem of extrema of a function $f(x)$ using the sufficient condition $f'(x) = 0$). For example, the discriminant set of the family $F(x, y, z; t) = y^3 - (x - t)^2$ of cylindrical surfaces with the axis OZ coincides with a plane $\{y = 0\}$ that is not tangent to the given surfaces but is the union of their singular points; see Fig. 20.42 (and Fig. 20.41 with the analogous two-dimensional situation).

```
> p:=t->plot3d([u^3+2*t,u^2/10,v], u=-1.5..1.5,v=0..0.1,
  grid=[60,10]): plots[display]([seq(p(t), t=1..7)]);
> p:=t->plot([u^3+2*t, u^2/10, u=-1.1..1.1]):
  plots[display]([seq(p(t), t=1..7)]);
```

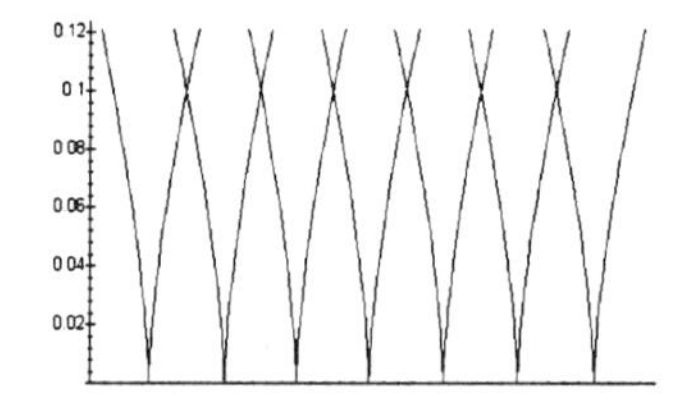

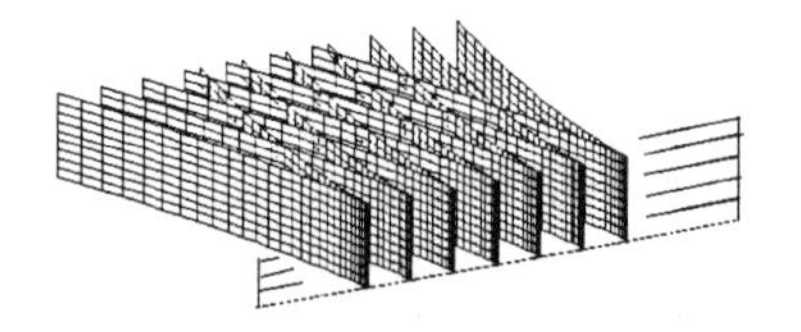

Figs. 20.41–20.42. Discriminant set (envelope is empty)

Definition 20.4.3 A curve along which the envelope is tangent to some surface from a given family is called a *characteristic*.

Equations (20.13) are at the same time the equations of a characteristic for a fixed t. Obviously, the envelope can be considered as a locus of characteristics: each surface from the family is tangent to the envelope along the characteristic, which together fill the whole envelope. In the case of an envelope of planes, the characteristics are straight lines; in the case of spheres, they are circles.

Theorem 20.4.2 *The envelope of the family of spheres*

$$M_t: \ (x - a(t))^2 + (y - b(t))^2 + (z - c(t))^2 - R(t)^2 = 0,$$

of radii $R(t)$ with a curve of centers $\vec{\mathbf{r}}(t) = [a(t), b(t), c(t)]$, having a nonzero curvature, is a canal surface (see Section 21.1)

$$\vec{\mathbf{r}}(u, v) = \vec{\mathbf{r}}(t) + R(t)(-R'(t)\,\vec{\tau}(t) \pm \sqrt{1 - (R'(t))^2}(\cos(\theta)\,\vec{\nu}(t) + \sin(\theta)\,\vec{\beta}(t))),$$

where $\vec{\tau}$, $\vec{\nu}$, $\vec{\beta}$ is a Frenet frame field of the curve $\vec{\mathbf{r}}(t)$. In case of a constant function of radius $R(t) = R > 0$ this surface is a tube (see Section 21.1). Moreover, the characteristics will be the generators (circles).

Exercise 20.4.1 Find an example of a family of surfaces whose discriminant is (a) a straight line, (b) a point.
Answer: (a) $(x - t)^2 + y^2 = t^2$, $t \neq 0$, (b) $(x - t)^2 + y^2 + z^2 = t^2$, $t \neq 0$.

20.4.2 Envelope of a One-Parameter Family of Planes

Let us study the structure of the envelope of a *one-parameter family of planes*,

$$A(t)x + B(t)y + C(t)z + D(t) = 0. \tag{20.15}$$

Note that the tangent planes of an arbitrary surface $\vec{\mathbf{r}}(u, v)$ in a neighborhood of an elliptic or hyperbolic point form a family depending on two parameters u, v. On the other hand, if all points on the surface are of parabolic type, i.e., in the case of a developable ruled surface, then the family of all tangent planes to the given surface depends, in fact, on one real parameter: the envelope coincides with the union of tangent planes of the surface at the points of any directrix curve $\vec{\rho}(u)$. The last statement is obvious for cylindrical and conic ruled surfaces, and also for a tangent developable from Lemma 20.3.1. In view of Theorem 20.4.1, we can assume that the *envelope of a one-parameter family of planes is a developable ruled surface.*

To calculate the characteristic on some plane of the given family, one must add the following to equation (20.15):

$$A'(t)x + B'(t)y + C'(t)z + D'(t) = 0. \tag{20.16}$$

This is obtained by differentiation of (20.15) with respect to the parameter t. We can suppose that equations (20.15), (20.16) are compatible and independent (in the opposite case $\frac{A'}{A} \equiv \frac{B'}{B} \equiv \frac{C'}{C}$ the planes of the family are parallel to one another, and hence the envelope does not exist).

Therefore, the characteristics are straight lines, and the envelope as their locus is a ruled surface. Note that the envelope is tangent to the plane of the given family along the characteristic (the ruling !), and hence *the envelope is a developable ruled surface.* Next let us calculate the type of this surface (cylinder, cone, or tangent developable), and in the last case find its cuspidal edge. Consider 3 planes: (20.15), (20.16), and

$$A''(t)\,x + B''(t)\,y + C''(t)\,z + D''(t) = 0, \tag{20.17}$$

the first two of which define the envelope. One can state three main hypotheses about the relative positions of these three planes:

(a) *The planes (20.15)–(20.17) have no common points for every t.*

(b) *The planes (20.15)–(20.17) intersect at a unique point P, and the same holds for all t.*

(c) *The planes (20.15)–(20.17) intersect at a unique point $P(t)$, whose position vector $\vec{\mathbf{r}}(t)$ defines a smooth curve.*

Condition **(a)** means the following inequality for the rank of the matrix of the system (20.15)–(20.17):

$$r\begin{pmatrix} A(t) & B(t) & C(t) \\ A'(t) & B'(t) & C'(t) \\ A''(t) & B''(t) & C''(t) \end{pmatrix} < r\begin{pmatrix} A(t) & B(t) & C(t) & D(t) \\ A'(t) & B'(t) & C'(t) & D'(t) \\ A''(t) & B''(t) & C''(t) & D''(t) \end{pmatrix}.$$

In case **(b)** the determinant of the system (20.15)–(20.17) is nonzero,

$$\det\begin{pmatrix} A(t) & B(t) & C(t) \\ A'(t) & B'(t) & C'(t) \\ A''(t) & B''(t) & C''(t) \end{pmatrix} \neq 0,$$

and $\vec{\mathbf{r}}(t) = [x(t),\ y(t),\ z(t)]$ is the solution of the system (20.15)–(20.17).

Theorem 20.4.3 *The envelope of a one-parameter family of planes for the main cases* **(a)**–**(c)** *is a domain either on a cylindrical surface or on a conical surface or on a tangent developable surface.*

Proof. In case **(a)** the envelope (ruled surface) is *cylindrical.* Indeed denoting by $\vec{\mathbf{n}}(t) = [A(t), B(t), C(t)]$ the normal vector to the given plane, which can be assumed of unit length, $\vec{\mathbf{b}}(t)$ is the unit vector along the characteristic g_t on the plane M_t. From equations (20.15)–(20.17) we obtain the equalities

$$\vec{\mathbf{b}}\cdot\vec{\mathbf{n}} = 0, \quad \vec{\mathbf{b}}\cdot\vec{\mathbf{n}}' = 0, \quad \vec{\mathbf{b}}\cdot\vec{\mathbf{n}}'' = 0. \tag{20.18}$$

Let us differentiate the first two equalities in (20.18) with respect to t, and then obtain

$$\vec{\mathbf{b}}'\cdot\vec{\mathbf{n}} + \vec{\mathbf{b}}\cdot\vec{\mathbf{n}}' = 0, \quad \vec{\mathbf{b}}'\cdot\vec{\mathbf{n}}' + \vec{\mathbf{b}}\cdot\vec{\mathbf{n}}'' = 0.$$

From this follows $\vec{\mathbf{b}}' \cdot \vec{\mathbf{n}} = 0$, $\vec{\mathbf{b}}' \cdot \vec{\mathbf{n}}' = 0$. Since, moreover, $\vec{\mathbf{n}}' \cdot \vec{\mathbf{n}} = 0$ holds, we have $\vec{\mathbf{b}}' = 0$. Thus all straight lines g_t are parallel to one another, which is what we needed to prove.

In case **(b)** the characteristics g_t pass through the fixed point P in space, and hence the envelope (ruled surface) is *conical.*

Finally, consider case **(c)**. We will show that the straight lines g_t that form the surface M are tangent to the directrix curve $\vec{\mathbf{r}}_p$ on the tangent developable surface. For the points $\vec{\mathbf{r}}_p(t)$ in the intersection of the planes (20.15)–(20.16) we have

$$\vec{\mathbf{r}}_p \cdot \vec{\mathbf{n}} + D = 0, \quad \vec{\mathbf{r}}_p \cdot \vec{\mathbf{n}}' + D' = 0, \quad \vec{\mathbf{r}}_p \cdot \vec{\mathbf{n}}'' + D'' = 0. \tag{20.19}$$

After differentiating the first equality of (20.19) and subtracting the second equality of (20.19) from it, we have $\vec{\mathbf{r}}'_p \cdot \vec{\mathbf{n}} = 0$. Analogously, from the second and the third equalities of (20.19) follows $\vec{\mathbf{r}}'_p \cdot \vec{\mathbf{n}}' = 0$. From this we obtain $\vec{\mathbf{r}}'_p \,||\, \vec{\mathbf{n}} \times \vec{\mathbf{n}}' = 0$. Since the straight line g_t passes through the point $P(t)$ and is orthogonal to the vectors $\vec{\mathbf{n}}(t)$ and $\vec{\mathbf{n}}'(t)$, it is parallel to the vector $\vec{\mathbf{r}}'_p$, and thus is the tangent line to the curve $\vec{\mathbf{r}}_p(t)$. □

Exercise 20.4.2

1. Find the envelope of the family of planes that cut from the coordinate angle $x, y, z > 0$ tetrahedra with constant volume V.
Answer: $xyz = \frac{2}{9}V$; see Exercise 6, Section 19.4.1.

2. Write down the equation of the family of spheres for which the envelope is the cone $x^2 + y^2 = a^2z^2$, $(z \neq 0)$ without its vertex.
Answer: $x^2 + y^2 + (z - t)^2 = \frac{a^2t^2}{a^2+1}$.

3. Let $\vec{\mathbf{r}}(s)$ be a space curve with the natural parameter s and with nonzero curvature, and let $\{\vec{\tau}, \vec{\nu}, \vec{\beta}\}$ be its Frenet frame field, $\tilde{\mathbf{r}}$ a point in space. Prove that:

(a) The envelope of the *family of osculating planes* $(\tilde{\mathbf{r}} - \vec{\mathbf{r}}(s)) \cdot \vec{\beta}(s) = 0$ is a tangent developable surface for which the given curve is the cuspidal edge; for each t the characteristic coincides with the tangent line,

(b) the envelope of the *family of normal planes* $(\tilde{\mathbf{r}} - \vec{\mathbf{r}}(s)) \cdot \vec{\tau}(s) = 0$ is a tangent developable surface that is the locus of centers of osculating spheres (for the given curve),

(c) The envelope of the *family of rectifying planes* $(\tilde{\mathbf{r}} - \vec{\mathbf{r}}(s)) \cdot \vec{\nu}(s) = 0$ contains the given curve (as its *geodesic*); the characteristics coincide with the *momentary axis of revolution* of the Frenet frame field $\{\vec{\tau}, \vec{\nu}, \vec{\beta}\}$. (This developable ruled surface is called *rectifying*, because after its unrolling onto the plane, the given curve becomes a *straight line*).

4. Prove that the family of surfaces given by the equations $F(x, y, z) = t$, where F is an arbitrary regular function in three variables x, y, z, has no envelope.

21
Some Other Classes of Surfaces

Any surface gives rise to other surfaces through a variety of general constructions. In addition to the surfaces of Chapter 20, we now study some other interesting classes of surfaces and constructions.

In Chapter 21 the reader will become acquainted with the commands

```
geom3d[inversion], draw.
```

21.1 Canal Surfaces and Tubes

Definition 21.1.1 A surface M is called a *canal surface* if it can be represented as the union of a one-parameter family of circles whose supporting planes are orthogonal to the curve of centers of these circles. By definition, a canal surface with a C^3-regular curve of centers $\vec{\alpha}(u)$ and a function of radius $R(u)$ admits a regular parametrization of the form

$$\vec{\mathbf{r}}(u, v) = \vec{\alpha}(u) + R(u)(\cos(v)\,\vec{\nu}(u) + \sin(v)\,\vec{\beta}(u)),$$

where the unit vectors of the main normal vector $\vec{\nu}(t)$ and the binormal vector $\vec{\beta}(t)$ are orthogonal to the tangent vector $\boldsymbol{\tau}(u)$ of the curve $\vec{\alpha}(u)$. For every u the vector-valued function $R(u)(\cos(v)\,\vec{\nu}(u)+\sin(v)\,\vec{\beta}(u))$, where $v \in [0, 2\pi)$, defines a circle (the *generator*) of radius $R(u)$ that belongs to M. In the case of constant radius $R(u) = R$, such a surface is called a *tube over the curve* $\vec{\alpha}(u)$ with radius R.

Example 21.1.1 A surface of revolution, see Section 20.2, is a canal surface with the curve of centers on the axis of revolution. The circular cylinder is a tube over a straight line; the torus is a tube over a circle. A tube generated by a circular helix can be seen in Fig. 20.27.

The command tubeplot in MAPLE allows us to plot a tube with an arbitrary space curve $\vec{\alpha}(t)$ of centers (see Section 8.1.1).

A program for a canal surface over an arbitrary curve:

```
> with(linalg): r:=array([x(u),y(u),z(u)]);
> R:=R(u); u0:=0; u1:=2;     # define functions x, y, z, R in u
> dr:=map(diff,r,u); ddr:=map(diff,r,u$2);
> tau:=scalarmul(dr,1/norm(dr,2)); b:=crossprod(dr, ddr);
> beta:=scalarmul(b,1/simplify(norm(b,2)));
> nu:=crossprod(beta, tau);
> rr:=evalm(r+scalarmul(nu, R*cos(v))+
  scalarmul(beta,R*sin(v)));
> plot3d(rr, u=u0..u1, v=0..2*Pi);
```

Exercise 21.1.1

1. Write down equations of the canal surface with $R(u) = ku$ and the tube over the circular and conic helix, plot surfaces (*sea shell*, Fig. 21.2).

2. Plot the tube over the torus knot, Fig. 21.3, using the following program:

```
> N:=10: t_tub:=plots[tubeplot]({[10*cos(t), 10*sin(t), 0,
  t=0..2*Pi, radius=2, numpoints=10*N, tubepoints=2*N],
  [cos(t)*(10+4*sin(9*t)), sin(t)*(10+4*sin(9*t)),
  4*cos(9*t),t=0..2*Pi,radius=1,numpoints=trunc(37.5*N),
  tubepoints=N]}): t_tub;
```

3. Plot the canal surface defined by moving the circle of radius $4 + \sin(2t)$ along the circle $\vec{\mathbf{r}}(t) = [10\cos(t),\ 10\sin(t),\ 0]$ of radius 10; Fig. 21.1

```
> plot3d([(12+(4+sin(4*v))*sin(u))*cos(v),
  (12+(4+sin(4*v))*sin(u))*sin(v), (4+sin(4*v))*cos(u)],
  u=0..2*Pi, v=0..2*Pi, grid=[20,60]);
```

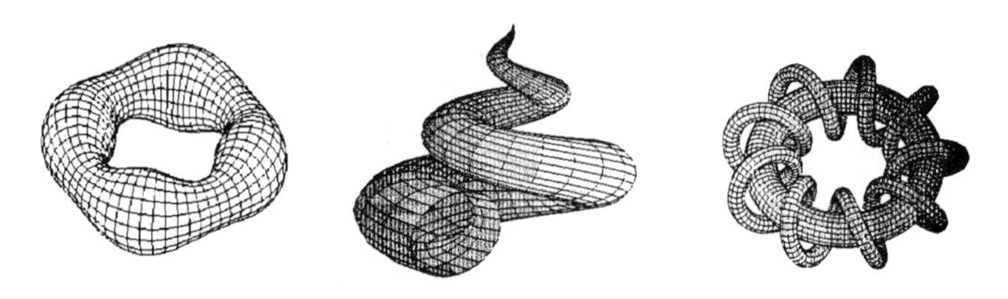

Figs. 21.1–21.3. Canal surfaces and tubes

4. Plot eight different *figure eights*.

Solution. The *Lissajous curve* $\vec{\mathbf{r}}(u) = [\sin(nu), \sin(u)]$ for $n = 2$ has the shape of a figure eight; Fig. 21.4. We plot the surface of revolution of the figure eight; Fig. 21.5.

```
> plot3d([sin(u)*sin(v),cos(u)*sin(u)*sin(v),cos(v)*sin(v)],
  v=0..2*Pi, u=0..2*Pi, grid=[40,50]);
```

We plot the surface defined by moving one figure eight along another; see Fig. 21.6.

```
> plot3d([cos(u)*sin(2*v), sin(u)*sin(2*v), sin(v)],
  v=0..2*Pi, u=0..2*Pi, grid=[40,50]);
```

Let us also plot the tube over a figure eight; see Fig. 21.7. Four more figure eights can be constructed using a lemniscate (see Section 6.2).

One can continue this experiment with figure eights using the curve $y^2 = x^2 - x^4$, similar to the lemniscate, which is related to the motion of a material point in an energy field with two symmetric potential holes, or using the *spiral curve*, i.e., the intersection of the torus with the plane parallel to its axis. (The points P on spiril curve satisfy the equality $|PA|^2 \cdot |PB|^2 = c \cdot |PO|^2 + c_1$, where A, B are fixed points, O is the midpoint of AB, and c, c_1 are real numbers. Hence these curves generalize the ovals of Cassini, see Section 6.2.)

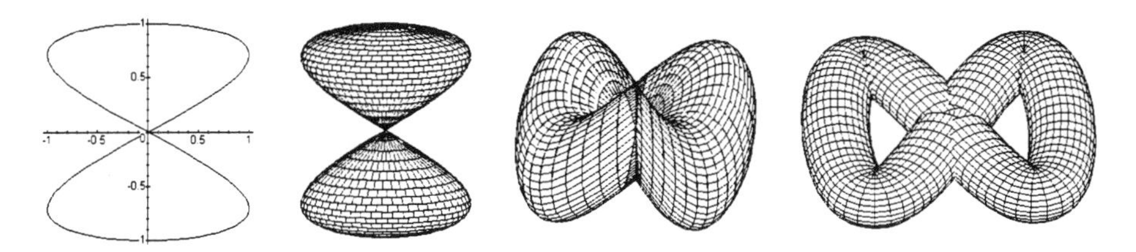

Figs. 21.4–21.7. Figure eights

21.2 Translation Surfaces

Definition 21.2.1 A surface M that admits a parametrization of the form $\vec{\mathbf{r}}(u, v) = \vec{\mathbf{r}}_1(u) + \vec{\mathbf{r}}_2(v)$ is called a *translation surface.* In other words, a translation surface is obtained by moving some (*generator*) curve that is always parallel to its initial position and is guided with one of its points on another curve, the *directrix*.

The simplest translation surface is a cylindrical surface. It can be obtained by parallel displacement of the directrix, which intersects every generator.

Exercise 21.2.1

1. Prove that the following surfaces are translation surfaces:

(a) The elliptical and hyperbolic paraboloids.

(b) The *Bohemian dome* (translation of the ellipse along the circle) $\vec{\mathbf{r}}(u, v) = [a\cos(u), a\sin(u) + b\cos(v), c\sin(v)]$; Fig. 21.8.

(c) Part of the helicoid $\vec{\mathbf{r}}(u, v) = [u\cos(v), u\sin(v), av]$ for $u \le c$. *Hint*: Replace the coordinate system by $u = c\cos\frac{\varphi-\psi}{2}$, $v = \frac{\varphi+\psi}{2}$, where $0 \le \varphi - \psi < \frac{\pi}{2}$; assume $\vec{\mathbf{r}}_1(t) = 2[c\cos(t), c\sin(t), at]$; and deduce the equation of the helicoid $\vec{\mathbf{r}}(\varphi, \psi) = \vec{\mathbf{r}}_1(\varphi) + \vec{\mathbf{r}}_1(\psi)$.

(d) The surface that is the locus of centers of segments whose endpoints belong to two given space curves.

2. Which surfaces are obtained in Exercise 1 (d) for two given screw lines, for the given circle $[R\cos(u),\ R\sin(u), 0]$ and its axis of symmetry OZ?

21.3 Twisted Surfaces

Consider a generalization of the Möbius band and the Klein bottle.

Definition 21.3.1 Let $\gamma : \vec{\mathbf{r}}(t) = [\varphi(t), \psi(t)]$ be a plane curve with the condition $\vec{\mathbf{r}}(-t) = -\vec{\mathbf{r}}(t)$. Then the *twisted surface* with profile curve γ is defined by the equations

$$\begin{aligned}\vec{\mathbf{r}}_1(u, v) &= (a + \cos(bu)\,\varphi(v) - \sin(bu)\,\psi(v))\,[\cos u, \sin u, 0]\\ &\quad + (\sin(bu)\,\varphi(v) + \cos(bu)\,\psi(v))\,[0, 0, 1].\end{aligned}$$

Example 21.3.1 The *Möbius band* (see Exercise 9 in Section 20.2) is twisted for $b = \frac{1}{2}$ using the line segment $\gamma : \vec{\mathbf{r}}(t) = [t, 0]$ ($|t| \le 1$); the *Klein bottle*, Fig. 19.7, is twisted for $b = \frac{1}{2}$ using the Lissajous curve $\gamma : \vec{\mathbf{r}}(t) = [\sin(t),\ \sin(2t)]$ ($|t| \le \pi$), and is given by the equations

$$\begin{aligned}\vec{\mathbf{r}}_1(u, v) &= (a + \cos(\frac{u}{2})\sin(v) - \sin(\frac{u}{2})\sin(2v))[\cos u, \sin u, 0]\\ &\quad + (\sin(\frac{u}{2})\sin(v) + \cos(\frac{u}{2})\sin(2v))[0, 0, 1].\end{aligned}$$

Exercise 21.3.1 Plot the Lissajous curve $\vec{\mathbf{r}}(u) = [\sin(nu), \sin(u)]$, $|u| \le \pi$, with $n = 4$, Fig. 21.9, and its twisted surface, Fig. 21.10.

```
> plot([sin(4*t), sin(t), t=0..2*Pi]);      # the curve
> plot3d([(2+cos(u/2)*sin(4*v)-sin(u/2)*sin(v))*cos(u),
  (2+cos(u/2)*sin(4*v)-sin(u/2)*sin(v))*sin(u),
  sin(u/2)*sin(4*v)+cos(u/2)*sin(v)],
```

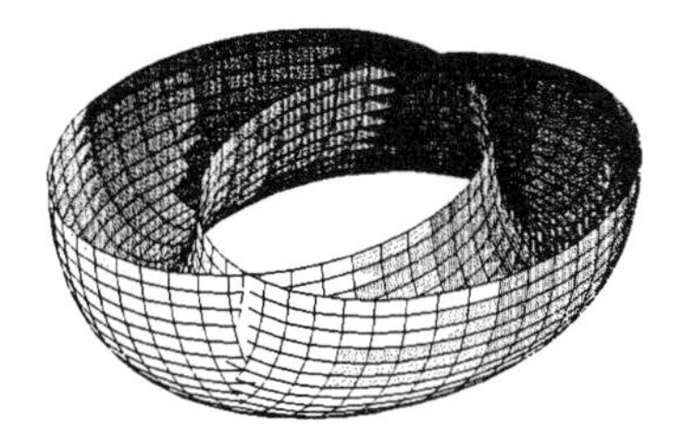

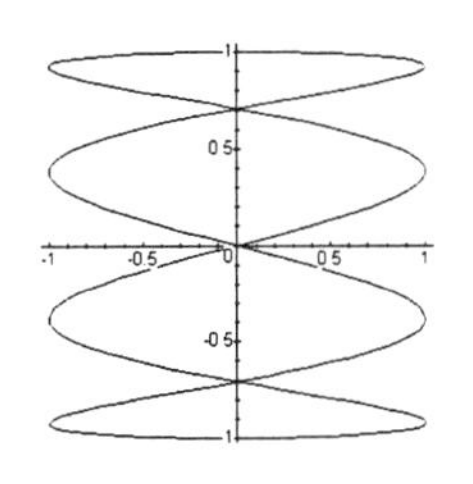

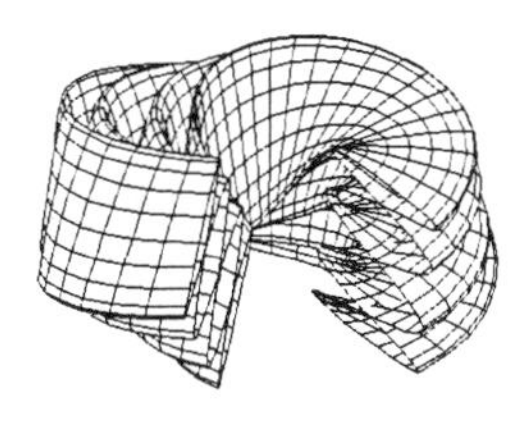

Figs. 21.8–21.10. Bohemian dome. The Lissajous curve and its twisted surface

```
u=-Pi/8..3*Pi/2, v=0..2*Pi, grid=[30,80]); # surface
```

21.4 Parallel Surfaces (Equidistants)

Definition 21.4.1 Let $M : \vec{\mathbf{r}}(u, v)$ be a smooth surface of class C^1 and $\vec{\mathbf{n}}(p)$ its normal vector field. For an arbitrary real number h plot the surface $M(h)$, placing from each point $P \in M$ the line segment of length $|h|$ in the direction of the normal vector $\vec{\mathbf{n}}(P)$, if $h > 0$, and the line segment of length $|a|$ in the direction $-\vec{\mathbf{n}}(P)$ when $h < 0$. The surface $M(h)$ is called a *parallel (or equidistant) surface* for M and is given by the equations $\vec{\mathbf{r}}(u, v; a) = \vec{\mathbf{r}}(u, v) + h\,\vec{\mathbf{n}}(u, v)$.

Example 21.4.1 A parallel surface for the plane is the plane again. The parallel surfaces for the sphere and the cylinder of radius R for $|h| < R$ are spheres and cylinders of radius $R \pm h$.

Obviously, the properties of the parallel surface $M(h)$ are defined by the properties of M and the value of h. For small h the surface $M(h)$ is regular; its normal vector $\vec{\mathbf{n}}(P, h)$ coincides with $\vec{\mathbf{n}}(P)$, and its tangent planes at corresponding points are parallel to one another. The parallelness of two surfaces is a mutual property: M is also a parallel surface for $M(h)$.

Exercise 21.4.1

1. Prove that a parallel surface for a tube of radius R over a space curve with small h is again a tube of radius $R \pm h$ over the same curve.

2. Prove that a parallel surface for a developable ruled surface is again a developable ruled surface.

3. Plot parallel surfaces for an ellipsoid, Fig. 21.11; the hyperboloid of one sheet, Fig. 21.12; the catenoid; the Möbius band, Fig. 21.13.

Solution. We plot some parallel surfaces for (part of) a ellipsoid:

```
> with(linalg): r:=[cos(u)*cos(v), cos(u)*sin(v), 2*sin(u)]:
> ru:=map(diff,r,u): rv:=map(diff,r,v):
> nn:=crossprod(ru,rv): n:=scalarmul(nn,1/norm(nn,2)):
> p1:=plot3d(evalm(r+scalarmul(n,-1.5)), u=-Pi/3..Pi/3,
  v=Pi/4..2*Pi):
```

```
  p2:=plot3d(evalm(r+scalarmul(n,-1)), u=-Pi/3..Pi/3,
  v=Pi/4..2*Pi):
  p3:=plot3d(evalm(r+scalarmul(n,-0.6)), u=-Pi/3..Pi/3,
  v=Pi/4..2*Pi):
  p4:=plot3d(evalm(r+scalarmul(n,0.2)), u=-Pi/3..Pi/3,
  v=Pi/4..2*Pi):
> plots[display3d]({p1, p2, p3, p4});
```

To plot parallel surfaces for the hyperboloid of one sheet we set in the above program

```
> r:=[cosh(u)*cos(v), cosh(u)*sin(v), sinh(u)]:
```

Since the Möbius band is a one-sided (non-orientable) surface, then one cannot select a continuous unit normal vector field along this whole surface. Thus we move twice (in the parameter u) along the Möbius band; see the formulas in the program. Moreover, the unit normal vector takes both possible values at each point of the surface. Note that a parallel surface for the Möbius band has two sides (is orientable).

```
> r:=[(5+cos(v/2)*u)*cos(v), (5+cos(v/2)*u)*sin(v),sin(v/2)*u];
  ...
> p1:=plot3d(r, u=-1..1, v=0..2*Pi, grid=[10,50]):
  p2:=plot3d(evalm(r+scalarmul(n,1)),u=-1..1, v=0..4*Pi,
  grid=[10, 80]):
> plots[display3d]({p1, p2});
```

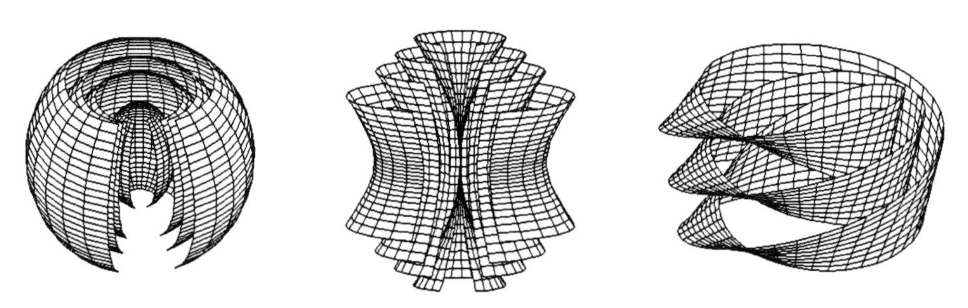

Figs. 21.11–21.13. Parallel surfaces (with a cut) for surfaces ellipsoid, hyperboloid of one sheet, Möbius band

21.5 Pedal and Podoid Surfaces

Consider the construction of a surface analogous to pedal curves.

Definition 21.5.1 A *pedal surface (French padaire, from Greek πoδoς) M(P)* consists of the bottoms of perpendiculars from the fixed point $P \in \mathbb{R}^3$ onto various tangent planes of the given surface M. If $\vec{\mathbf{r}}(u, v)$ are the parametric equations of the surface M, then the pedal surface with respect to the point P is given by the equations

$$\vec{\mathbf{r}}_1(u, v) = \vec{\mathbf{r}}_p + \frac{((\vec{\mathbf{r}}(u,v)-\vec{\mathbf{r}}_p), \vec{\mathbf{r}_u}, \vec{\mathbf{r}_v})}{|\vec{\mathbf{r}_u} \times \vec{\mathbf{r}_v}|^2} (\vec{\mathbf{r}_u} \times \vec{\mathbf{r}_v}).$$

The pedal surface for a plane, obviously, coincides with a point.

The following construction of surfaces is in a simple relation to the pedal surface.

Definition 21.5.2 The locus of points $M(P)$ that are symmetric to a given point P with respect to all possible tangent planes of the surface M is called a *podoid surface*.

Exercise 21.5.1

1. Prove that a pedal surface of the sphere is the surface of revolution of a cardioid if the point P belongs to the sphere, and plot this surface.

2. Plot the pedal surfaces of the ellipsoid with respect to its center, one of its vertices, and any external point.

```
> with(linalg): r:=([cos(u)*cos(v),2*cos(u)*sin(v),3*sin(u)]);
> ru:=map(diff,r,u): rv:=map(diff,r,v):
> nn:=crossprod(ru,rv): p:=array([3,0,0]):
> coef:=dotprod(evalm(r-p), nn)/dotprod(nn,nn):
> rr:=evalm(p+scalarmul(nn, coef)):
> plot3d(rr, u=-Pi/2..Pi/2, v=Pi/2..2*Pi);
```

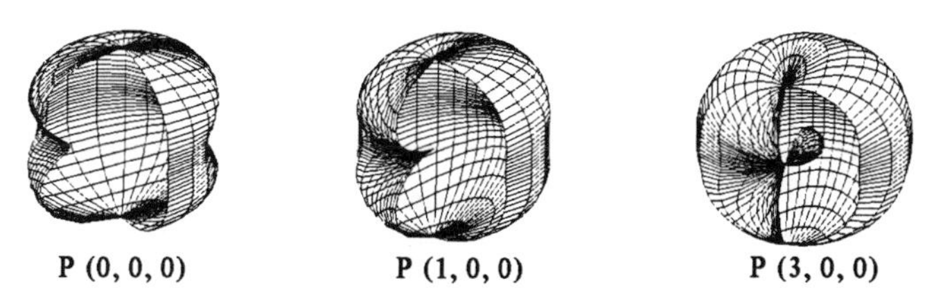

Figs. 21.14–21.16. Pedal surfaces (with a cut) for an ellipsoid

3. Plot the pedal surface of a torus with respect to its center.

4. Plot the pedal curve of a parabola with respect to its vertex (a cissoid). Then plot the analogous pedal surface of a paraboloid.

5. Prove that the podoid surface of a developable ruled surface (see Theorem 20.3.1) degenerates to a curve.

6. Prove that the podoid surface of a given surface can be obtained from a pedal surface using similarity with center at P and with coefficient equal to 2. (For example, podoid surfaces of a sphere are spheres). Write down equations of a podoid surface of a given surface.

21.6 Cissoidal and Conchoidal Maps

The following construction of surfaces is analogous to *the difference of graphs* $f(x, y) = f_1(x, y) - f_2(x, y)$ of two functions in the rectangular system of coordinates.

Definition 21.6.1 The *cissoidal map* is applied to two arbitrary surfaces (for curves see Section 6.2), whose equations in the spherical coordinate system are $\rho = \rho_1(\varphi, \theta)$ and $\rho = \rho_2(\varphi, \theta)$, with the aim of constructing the surface $\rho = \rho_1(\varphi, \theta) - \rho_2(\varphi, \theta)$. In other words, on each ray from the point O we set the line segment OM equal to the segment on the ray between the two given surfaces.

The following construction of surfaces is analogous to the known conchoidal transformation on plane curves.

Definition 21.6.2 A *conchoidal surface* for a given surface (for a curve see Section 6.2) is obtained by increasing (or decreasing) the position vector of each point of the surface by a fixed line segment l. If the given surface has the equation $\rho = \rho_1(\varphi, \theta)$ in the spherical coordinate system, then the equations of its conchoidal surface are $\rho = \rho_1(\varphi, \theta) \pm l$.

Exercise 21.6.1
1. Write down the equations and plot the cissoidal surface of a sphere with diameter $|OA| = 2a$ and tangent plane α at the point $A(2a, 0, 0)$.

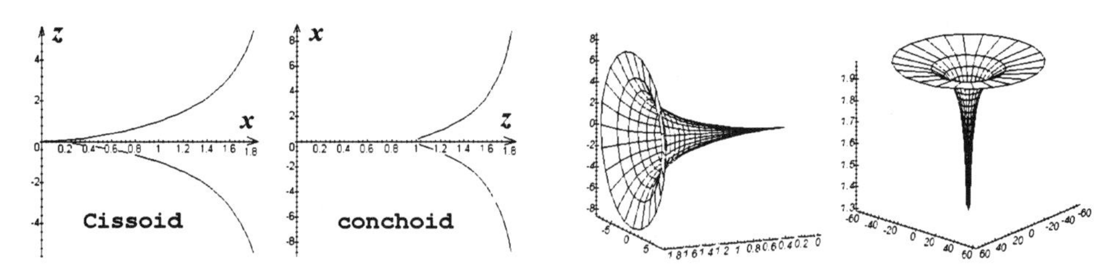

Figs. 21.17–21.20. Cissoidal surface of the sphere, conchoidal surface of the plane, and their profile curves

Solution. The *cissoid* of the straight line $x = 2a$ in the plane XZ, Fig. 21.17, has the equation $\rho = 2a\frac{\sin^2\varphi}{\cos\varphi} \iff z^2 = \frac{x^3}{2a-x}$.

The surface of revolution with axis OX of part of the cissoid with $z \geq 0$ is needed and has the equation $z^2 + y^2 = \frac{x^3}{2a-x}$. Since `implicitplot3d` gives a coarse image, we use the parametric equations of part of the cissoid

$x = t,\ z = \sqrt{\frac{t^3}{2a-t}}$ $(0 \le t < 2a)$ and plot the surface of revolution, Fig. 21.19, by formulas (20.1 a)

$$\vec{\mathbf{r}}(u, v) = [u,\ \sqrt{\tfrac{u^3}{2a-u}} \cos(v),\ \sqrt{\tfrac{u^3}{2a-u}} \sin(v)] \quad (0 \le u < 2a,\ 0 \le v < 2\pi).$$

```
> plot(sqrt(x^3/(2-x)), x=-1.8..1.8); # a=1
> plot3d([u, sqrt(u^3/(2-u))*cos(v), sqrt(u^3/(2-u))*sin(v)],
  u=0..1.9, v=0..2*Pi);
```

2. Write down the equation and plot the *conchoidal surface of the plane* $\{z = a\}$ with $l = 1$.

Solution. The conchoid of the line $z = a$ in XZ,

$$\rho = \tfrac{a}{\sin(\varphi)} \pm l \iff (z^2 + x^2)(z - a)^2 - l^2 z^2 = 0, \qquad \text{Fig. 21.18,}$$

for $l > a$ has a loop and for $0 < l < a$ has a cuspidal point. Using rotation about the axis OZ we obtain the conchoid of the plane $z = a$: $(z^2 + x^2 + y^2)(z - a)^2 - l^2 z^2 = 0$. Since `implicitplot3d` gives a coarse image, we use the parametric equations of part of the conchoid $x = \frac{a}{\sin(t)} \pm l \cos(t)$, $z = \frac{a}{\sin(t)} \pm l \sin(t)$ $(0 \le t < \frac{\pi}{2})$ and plot, using the formulas (20.1), the surface of revolution that we need with $a = 2$, $l = 1$; Fig. 21.20,

$$\vec{\mathbf{r}}(r, t) = [\left(\frac{2}{\sin(u)} \pm 1\right) \cos(u) \cos(v),$$
$$\left(\frac{2}{\sin(u)} \pm 1\right) \cos(u) \sin(v),\ \left(\frac{2}{\sin(u)} \pm 1\right) \sin(u)].$$

```
> plots[polarplot]([2/cos(t)-1, t, t=-1.5..1.5]); # profile curve
> plot3d([(2/sin(u)-1)*cos(u)*cos(v), (2/sin(u)-1)*cos(u)*
  sin(v), (2/sin(u)-1)*sin(u)], u=0..0.8, v=0..2*Pi); # surface
```

3. Prove that the conchoidal surface of the sphere is the surface of revolution of Pascal's limaçon, then plot this surface.

4. Describe the analogous *cissoidal and conchoidal mappings of surfaces* using the straight line (the axis OZ) instead of the point O and the idea of cylindrical coordinates. Plot examples of such surfaces.

21.7 Inversion of a Surface

Inversion is the simplest (nonlinear) transformation of space after rigid motions and affine maps. Inversion with respect to a sphere can be considered as a generalization of symmetry with respect to a plane, but in our case the points near the center (lying inside the sphere) are mapped far from the center (outside the sphere) and conversely. We give a more exact definition of the inversion of the plane in Section 6.4.

Definition 21.7.1 *Inversion* with respect to the sphere $S(O, R)$ with *center* O and *radius* R is defined as the map $i_{O,R} : \mathbb{R}^3 \setminus \{O\} \to \mathbb{R}^3 \setminus \{O\}$, for which the image of an arbitrary point m is the point m' on the ray Om such that the product of the distances of both points to the origin O is equal to R^2, i.e., $\vec{Om'} = \frac{R^2}{|Om|^2}\vec{Om}$. If we take the origin of the coordinate system at the point O, then the position vectors of the points m and m' are related by the formula $\vec{\mathbf{r}}_{m'} = \frac{R^2}{|\vec{\mathbf{r}}_m|^2}\vec{\mathbf{r}}_m$.

Inversion is an involutive map, i.e., $(i_{O,R})^2 = Id$ (identity map). The sphere $S(O, R)$ is the set of stationary points of the inversion. We plot the inversion images of space figures using MAPLE.

Example 21.7.1

1. Inversion with center O and radius 1 of the circular helix, Fig. 21.21, $\vec{\mathbf{r}}(t) = [3 + \cos(t), \sin(t), 0.1t]$ is the curve on Fig. 21.22.

```
> with(linalg): r:=[3+cos(u), sin(u), 0.1*u]:
> ri:=(scalarmul(r, 1/norm(r,2)^2)):
> plots[spacecurve](r, u=0..8*Pi, numpoints=200);
> plots[spacecurve](ri, u=-25*Pi..25*Pi, numpoints=999);
```

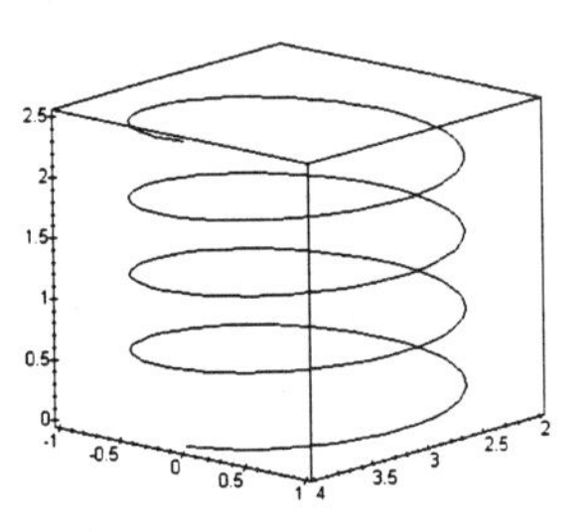

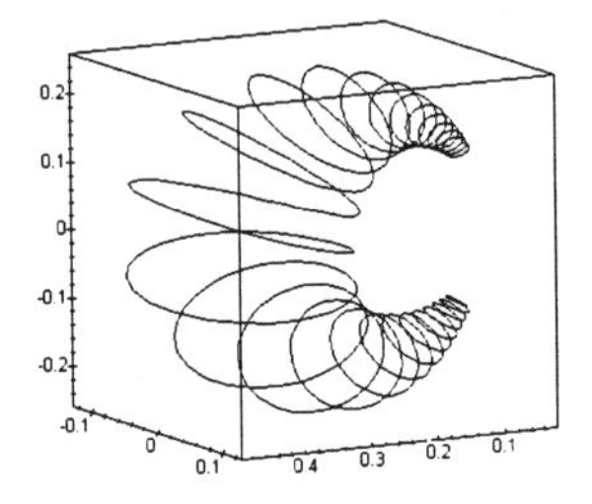

Figs. 21.21–21.22. The circular helix (4 turns) and its inversion image (30 turns)

2. The command `inversion(Q, P, S)` from the library `geom3d` plots the inversion (the object `Q`) of a point, plane, or sphere `P` with respect to the sphere `S`.

```
> restart: with(geom3d):
> point(A, 1,2,-1): point(B, 0,0,-1): point(C, 3,0,0):
> plane(P, [A,B,C]): sphere(S, [point(O,0,0,0),1]):
> inversion(F,P,S):
> draw([S, F, P(style=patchnogrid,color=maroon)],
  style=wireframe,view=[-1..1,-1..1,-2..1]);
```

We will formulate some elementary properties of inversion.

• *Inversion maps a plane (or a straight line) that does not contain the center of the inversion onto a sphere (respectively, circle) passing through the center of the inversion. Conversely, a sphere (or a circle) that contains the center of the inversion is transformed by inversion into a plane (respectively, straight line) that does not pass through the center.*

• *Inversion maps a sphere (or circle) that does not contain the center into a sphere (respectively, circle) that also does not contain the center of the inversion.*

• *Inversion preserves the angles between vectors, and hence preserves the angles between intersecting curves.*

Using inversion one can compare the behavior of very similar non-compact surfaces (and curves) at infinity.

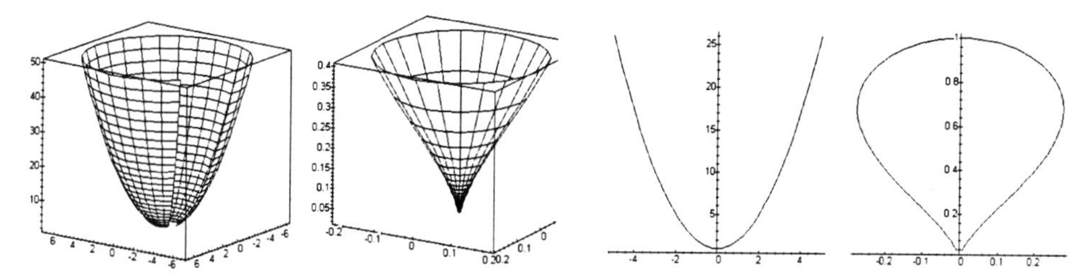

Figs. 21.23–21.26. Paraboloid of revolution
and its inversion (outside the circle $x^2 + y^2 \leq 1$)

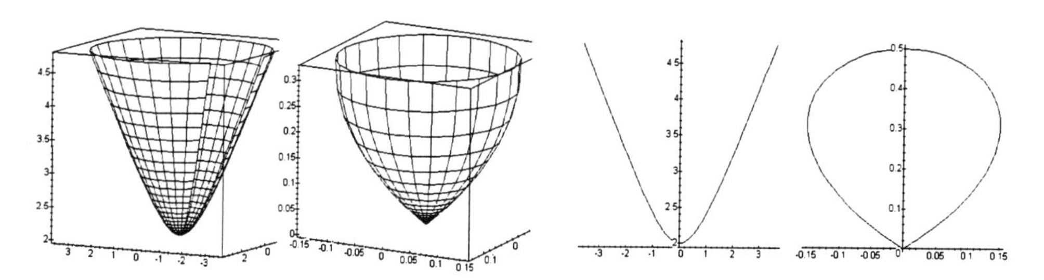

Figs. 21.27–21.30. Hyperboloid of one sheet
and its inversion (outside the circle $x^2 + y^2 \leq 1$)

Let us plot an inversion image of a paraboloid of revolution; Figs. 21.23–21.24; the hyperboloid of one sheet, Figs. 21.27–21.28, and the inversion images of their profiles, Figs. 21.25–21.26, and 21.29–21.30:

```
> with(linalg): r:=[u*cos(v),u*sin(v),u^2+1]:
> ri:=(scalarmul(r,1/norm(r,2)^2)):
> plot3d(r, u=1..7, v=-Pi..(3/4)*Pi); # paraboloid
> plot3d(ri, u=1..7, v=-Pi..Pi);        # inversion of paraboloid
```

```
> r:=[sinh(u)*cos(v), sinh(u)*sin(v), cosh(u)+1]:
> ri:=scalarmul(r, 1/norm(r,2)^2);   # hyperboloid
> plot3d(ri, u=1..7, v=-Pi..Pi);     # inversion of hyperboloid
```

```
> r2:=[u, u^2+1]:                     # parabola
> r2:=[sinh(u), cosh(u)+1]:          # hyperbola
> r2i:=scalarmul(r2, 1/norm(r2,2)^2):
> plot([r2[1], r2[2], u=-2..2]);      # curve
> plot([r2i[1], r2i[2], u=-6..6]); # inversion of curve
```

Exercise 21.7.1 The image of a torus under inversion is called the *cyclide of Dupin*. Show that this class of surfaces includes cones and cylinders. Prove that the cyclide of Dupin is a particular case of a canal surface that can be generated by each of two families of circles (images of parallels and meridians on a torus) forming an orthogonal net. Write down the equation and plot some cyclides of Dupin.

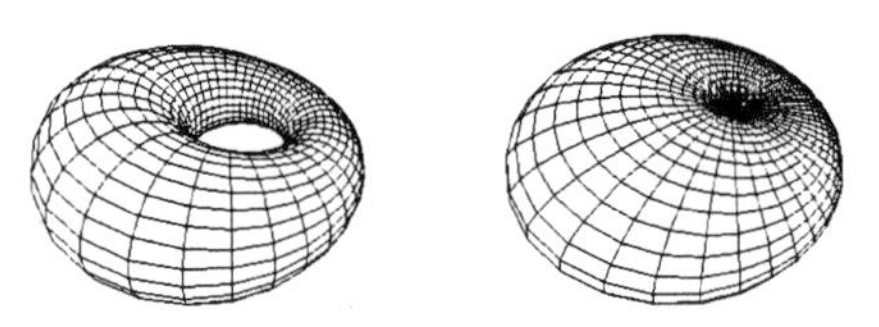

Figs. 21.31–21.32. Cyclides of Dupin

Hint. We plot two images, Figs. 21.31–21.32, of the torus

$$\vec{\mathbf{r}}(u, v) = [b + (a + \cos(u))\cos(v),\ (a + \cos(u))\sin(v),\ \sin(u)]$$

under inversion with center O and radius 1: when $a = 2,\ b = 8$, and when $a = 1,\ b = 4$.

```
> r:=[8+(2+cos(u))*cos(v), (2+cos(u))*sin(v), sin(u)]:
> r:=[4+(1+cos(u))*cos(v), (1+cos(u))*sin(v), sin(u)]:
> with(linalg): invr:=scalarmul(r, 1/(norm(r,2)^2)):
> plot3d(invr, u=-Pi..Pi, v=-Pi..Pi, grid=[40,60]);
```

References

[Amm] Ammeral L., *Graphics Programming in Turbo C*. John Wiley & Sons, 1989.

[Arg] Arganbright D., *Practical Handbook of Spreadsheet Curves and Geometric Constructions*. CRC Press, Boca Raton, FL, 199.

[AH] Anton H., Herr A., MAPLE *Supplement for Calculus with Analytic Geometry*, Fourth Edition. John Wiley and Sons, 1992.

[AI] Abbasian R., Ionescu A., *Vector Calculus with* MAPLE. McGraw-Hill, 1996.

[AS] Abramovich M., Stegun I., *A Handbook of Mathematical Functions*. Dover Publications, New York, 1965.

[Ban] Banchoff T., *Beyond the Third Dimension, Geometry, Computer Graphics, and Higher Dimension*. Sci. Amer. Library, New York, 1990.

[Bea] Beach R. C., *An Introduction to Curves and Surfaces of Computer-Aided Design*. Van Nostrand Reinhold, New York, 1991.

[Ber] Berdon A.F., *The Geometry of Discrete Groups*. Springer-Verlag, 1983.

[Bor] Borisov Yu. F., *Removing of a priori restrictions in a theorem on complete system of invariants of a curve in* $\mathbb{E}^n_l$. Siberian Math. Journal, **38**(3)(1997), 485–503.

[Bur] Bursky A., *Computer Graphics and Geometric Modeling using Beta-splines*. Springer, 1987.

[BK] Brieskorn E. and Knörrer H., *Plane Algebraic Curves*. Birkhäuser, 1986.

[CJ] Carlson J., Johnson, J., *Multivariable Mathematics with* MAPLE. *Linear Algebra, Vector Calculus and Differential Equations*. Prentice Hall, 1997.

[Cro] Cromwell P.R., *Polyhedra*, Cambridge University Press, 1997.

[EVGL] Efremovič V., Vainstein A., Gorelik E., Loginov E., *Metric fibrations of Riemannian manifolds*. Colloq. Math. Soc. J. Bolyai, 1979, 195–198.

[Gar 1] Gardner M., *Mathematical Games from Scientific American*. Moscow, 1984.

[Gar 2] Gardner M., *Mathematical Puzzles and Diversions*. London, Bell and Sons, 1980

[Gra] Gray A., *Modern Differential Geometry of Curves and Surfaces with* Mathematica. *Studies in Advanced Mathematics*. Springer, Second Ed., 1998.

[GH] Gander W., Hřebíček, J. *Solving Problems in Scientific Computing Using* MAPLE *and* MATLAB, Third Edition. Springer-Verlag, 1997.

[GCV] Gilbert D., Cohn-Vossen S., *Geometry and the Imagination*. Chelsea Publ. Co., New York, 1952.

[HP] Hege H.C., Polthier K., *Mathematical Visualizing. Algorithms, Applications, and Numerics*. Springer-Verlag, 1998.

[KK] Klimek G., Klimek M., *Discovering Curves and Surfaces with* MAPLE. Springer-Verlag, 1997.

[Kos] Kosniowski C., *Fun Mathematics on your Computer*. Cambridge Univ. Press, 1984.

[Law] Lawrence J.D., *A Catalog of Special Plane Curves*. New York, Dover Publ., 1972.

[Las] Laszlo M., *Computational Geometry and Computer Graphics in C++*. Prentice Hall, 1996.

[Loc] Lockwood E.H., *A Book of Curves*. Cambridge Univ. Press, 1961.

[Man] Manzon, B.M., `MAPLE V` *Power Edition*. Filin Publisher, 1998.

[MK] Marlin J.A., Kim H., *Calculus I–III with* `MAPLE`, 1994–1995.

[O'n] O'Neill B., *Elementary Differential Geometry*. Academic Press, London, New York, 1966.

[Ped] Pedow D., *Geometry and the Liberal Arts*. Penguin Books, 1976.

[Pon] Pontryagin L., *Method of Coordinates. Studies of Higher Mathematics*. Moscow, 1987.

[Shi] Shikin E., Pliss A., *Handbook on Splines for the User*. Springer-Verlag, 1995.

[Spi] Spivak M., *A Comprehensive Introduction to Differential Geometry*. Publish or Perish, Wilmington, 1979.

[Sto] Stroeker R., Kaashoek J., *Discovering Mathematica with* `MAPLE`. Birkhäuser, 1999.

[Su] Su Pu-ch'ing, *Computational Geometry – Curve and Surface modeling*. Academic Press, Inc., 1989.

[Tik] Tikhomirov V., *Stories About Maxima and Minima*. Math. World, vol. 1, 1991 (Reprinted in 1996).

[Tho] Thorpe J., *Elementary Topics in Differential Geometry*. Springer-Verlag, 1979.

[vC] von Seggern D., *CRC Handbook of Mathematical Curves and Surfaces*. CRC Press, Boca Raton, 1990.

[Zal] Zalgaller B.A., *Convex Polyhedra with Regular Faces*. Moscow, 1966.

[Wen] Wenninger M.J., *Polyhedron Models*. Cambridge University Press, Cambridge, 1971.

Index